Cooperative Control of Dynamical Systems

Zhihua Qu

Cooperative Control
of Dynamical Systems

Applications to Autonomous Vehicles

 Springer

Zhihua Qu, PhD
School of Electrical Engineering
and Computer Science
University of Central Florida
Orlando, FL 32816
USA

ISBN 978-1-84882-324-2   e-ISBN 978-1-84882-325-9

DOI 10.1007/978-1-84882-325-9

A catalogue record for this book is available from the British Library

Library of Congress Control Number: 2008940267

Cover design: eStudio Calamar S.L., Girona, Spain

Printed on acid-free paper

9 8 7 6 5 4 3 2 1

springer.com

To
Xinfan Zhang
E. Vivian Qu and M. Willa Qu

Preface

Stability theory has allowed us to study both qualitative and quantitative properties of dynamical systems, and control theory has played a key role in designing numerous systems. Contemporary sensing and communication networks enable collection and subscription of geographically-distributed information and such information can be used to enhance significantly the performance of many of existing systems. Through a shared sensing/communication network, heterogeneous systems can now be controlled to operate robustly and autonomously; cooperative control is to make the systems act as one group and exhibit certain cooperative behavior, and it must be pliable to physical and environmental constraints as well as be robust to intermittency, latency and changing patterns of the information flow in the network. This book attempts to provide a detailed coverage on the tools of and the results on analyzing and synthesizing cooperative systems. Dynamical systems under consideration can be either continuous-time or discrete-time, either linear or non-linear, and either unconstrained or constrained.

Technical contents of the book are divided into three parts. The first part consists of Chapters 1, 2, and 4. Chapter 1 provides an overview of cooperative behaviors, kinematical and dynamical modeling approaches, and typical vehicle models. Chapter 2 contains a review of standard analysis and design tools in both linear control theory and non-linear control theory. Chapter 4 is a focused treatment of non-negative matrices and their properties, multiplicative sequence convergence of non-negative and row-stochastic matrices, and the presence of these matrices and sequences in linear cooperative systems.

The second part of the book deals with cooperative control designs that synthesize cooperative behaviors for dynamical systems. In Chapter 5, linear dynamical systems are considered, the matrix-theoretical approach developed in Chapter 4 is used to conclude cooperative stability in the presence of local, intermittent, and unpredictable changes in their sensing and communication network, and a class of linear cooperative controls is designed based only on relative measurements of neighbors' outputs. In Chapter 6, cooperative stability of heterogeneous non-linear systems is considered, a comparative and topology-based Lyapunov argument and the corresponding comparison theorem on cooperative stability are introduced, and cooperative controls are designed for several classes of non-linear networked systems.

As the third part, the aforementioned results are applied to a team of unmanned ground and aerial vehicles. It is revealed in Chapter 1 that these vehicles belong to the class of so-called non-holonomic systems. Accordingly, in Chapter 3, properties of non-holonomic systems are investigated, their canonical form is derived, and path planning and control designs for an individual non-holonomic system are carried out. Application of cooperative control to vehicle systems can be found in Sections 5.3, 6.5 and 6.6.

During the last 18 years at University of Central Florida, I have developed and taught several new courses, including "EEL4664 Autonomous Robotic Systems," "EEL6667 Planning and Control for Mobile Robotic Systems," and "EEL6683 Cooperative Control of Networked and Autonomous Systems." In recent years I also taught summer short courses and seminars on these topics at several universities abroad. This book is the outgrowth of my course notes, and it incorporates many research results in the most recent literature. When teaching senior undergraduate students, I have chosen to cover mainly all the matrix results in Chapters 2 and 4, to focus upon linear cooperative systems in Chapter 5, and to apply cooperative control to simple vehicle models (with the aid of dynamic feedback linearization in Chapter 3). At the graduate level, many of our students have already taken our courses on linear system theory and on non-linear systems, and hence they are able to go through most of the materials in the book. In analyzing and designing cooperative systems, autonomous vehicles are used as examples. Most students appear to find this a happy pedagogical practice, since they become familiar with both theory and application(s).

I wish to express my indebtedness to the following research collaborators for their useful comments and suggestions: Kevin L. Conrad, Mark Falash, Richard A. Hull, Clinton E. Plaisted, Eytan Pollak, and Jing Wang. My thanks go to former postdoctors and students in our Controls and Robotics Laboratories; in particular, Jing Wang generated many of the simulation results for the book and provided assistance in preparing a few sections of Chapters 1 and 3, and Jian Yang coded the real-time path planning algorithm and generated several figures in Chapter 3. I also wish to thank Thomas Ditzinger, Oliver Jackson, Sorina Moosdorf, and Aislinn Bunning at Springer and Cornelia Kresser at le-tex publishing services oHG for their assistance in getting the manuscript ready for publication.

My special appreciation goes to the following individuals who professionally inspired and supported me in different ways over the years: Tamer Basar, Theodore Djaferis, John F. Dorsey, Erol Gelenbe, Abraham H. Haddad, Mohamed Kamel, Edward W. Kamen, Miroslav Krstic, Hassan K. Khalil, Petar V. Kokotovic, Frank L. Lewis, Marwan A. Simaan, Mark W. Spong, and Yorai Wardi.

Finally, I would like to acknowledge the following agencies and companies which provided me research grants over the last eight years: Army Research Laboratory, Department of Defense, Florida High Tech Council, Florida Space Grant Consortium, Florida Space Research Initiative, L-3 Communications Link Simulation & Training, Lockheed Martin Corporation, Microtronic Inc., NASA Kennedy Space Center, National Science Foundation (CISE, CMMI, and MRI), Oak Ridge National Laboratory, and Science Applications International Corporation (SAIC).

Orlando, Florida   Zhihua Qu
October 2008

Contents

List of Figures

1

Introduction

In this chapter, the so-called cooperative and pliable systems and their characteristics are described. It is shown through simple examples that cooperative and pliable behaviors emerge naturally in many complex systems and are also desirable for the operation of engineering systems such as autonomous vehicles. Indeed, a team of heterogeneous vehicles can autonomously interact with each other and their surroundings and exhibit cooperative and pliable behaviors upon implementing a multi-level networked-based local-feedback control on the vehicles.

Heterogeneous entities in cooperative systems may be as simple as scalar agents without any dynamics, or may be those described by a second-order point-mass model, or may be as complicated as nonlinear dynamical systems with nonholonomic motion constraints. In this chapter, standard modeling methods and basic models of typical vehicles are reviewed, providing the foundation for analysis and control of autonomous vehicle systems.

1.1 Cooperative, Pliable and Robust Systems

Although there isn't a universal definition, a *cooperative, pliable and robust* system consisting of several sub-systems should have the following characteristics:

(a) Trajectories of all the sub-systems in a cooperative system move collaboratively toward achieving a common objective. In stability and control theory, the common objective correspond to *equilibrium points*. Different from a standard stability analysis problem, *equilibrium points* of a cooperative system may not be selected *a priori*. Rather, the equilibrium point reached by the cooperative system may depend upon such factors as initial conditions, changes of the system dynamics, and influence from its environment. Later in Chapter 5, *cooperative stability* will be defined to quantify mathematically the characteristic of sub-systems moving toward the common objective.

(b) Associated with a cooperative system, there is a network of either sensors or communication links or a mix of both. The *sensing/communication network* provides the means of information exchange among the sub-systems and, unless sensor control or a communication protocol is explicitly considered, its changes can be modeled over time as binary variables of either on or off. Different from a standard control problem, feedback patterns of the network over time may not be known *a priori*. As such, not only do individual sub-systems satisfy certain controllability conditions, but also the networked system must require *cooperative controllability*. In addition to intermittency, the network may also suffer from limited bandwidth, latency, noises, *etc.*

(c) The cooperative system generally evolves in a dynamically changing physical environment. As exogenous dynamics, the environment has a direct impact on the status of sensing and communication network. Another major impact of the environment is the geometrical constraints on the maneuverable sub-space. In the process of achieving the common goal, motion of all the sub-systems must be *robust* in the sense that their trajectories comply with changes of the environment.

(d) Motion by each of the sub-systems may also be subject to constraints in kinematics or dynamics or both. As a result, control design must also be *pliable* in the sense that the resulting motion trajectory complies with these constraints.

Control through an intermittent network, cooperative behaviors of multiple entities, and pliable and robust systems are of particular interest and hence are discussed subsequently.

1.1.1 Control Through an Intermittent Network

Figure 1.1 shows the standard configuration of a feedback control system, in which the reference input $r(t)$ is given and the control objective is to make output $y(t)$ converge to $r(t)$. Typically, feedback of sensor $\mathbf{S}$ is of most importance in the system, the sensor model is usually known, and its information flow is either continuous or continually available at a prescribed series of time instants. Plant $\mathbf{P}$ is mostly known, while unknowns such as disturbances and uncertainties may be considered. Depending upon plant dynamics and its unknowns, control $\mathbf{C}$ can be designed. For instance, linear and non-linear controls can be designed, an adaptive law can be introduced if $\mathbf{P}$ contains unknown parameters, and robust control can be added to compensate for uncertainties in $\mathbf{P}$, *etc.* It is well known that, under any control, the sensitivity function of the closed-loop system with respect to uncertainties in $\mathbf{S}$ always has a value close to 1. As such, most feedback control systems do not perform well if the feedback is significantly delayed or if the feedback is available on an irregular, sporadic, or unpredictable basis.

Through a high-speed and reliable data network, feedback control systems can be implemented. Figure 1.2 shows a network control system in which

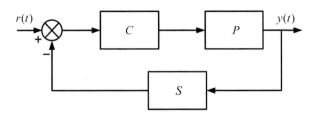

Fig. 1.1. Block diagram of a standard feedback control system

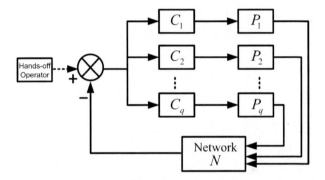

Fig. 1.2. Block diagram of a networked control system

multiple plants are admissible. Compared to Fig. 1.1, the networked control
system has the same setup. The main difference is that information generally
flows through sensing or communication network **N** because the plants as well
as control stations are often geographically distributed. A dedicated network
can be used to ensure stability and performance of networked control systems.
If feasible and deemed to be necessary, a centralized control with global information can be designed to achieve the optimal performance. For robustness,
it is better to have each of the plants stabilized by its local control, while
the network is to provide information sharing toward synthesizing or adjusting reference inputs to the plants. Nonetheless, the presence of the networked
loop could cause problems for stability and performance of the overall system.
Indeed, for large-scale interconnected systems, decentralized control (feedback
control individually installed at each of the plants) is often desired.

   Unless stated otherwise, the networked control problem studied in this
book is the one in which information flow through network **N** is intermittent, of limited bandwidth and range, and in certain directions, cannot be
predicted by any deterministic or stochastic model, and may have significant
latency. Examples of such a network include certain wireless communication

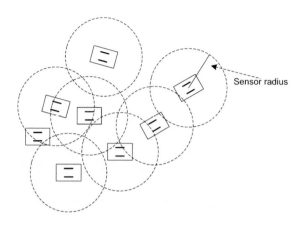

Fig. 1.3. *Ad hoc* sensor network for mobile robots

networks, *ad hoc* sensor networks, *etc.* Figure 1.3 shows an *ad hoc* sensor network among a group of mobile robots. As robots move relative to each other, feedback information available to a given robot is only about those neighboring robots in a certain region of its vicinity. Consequently, motion control of the mobile robot is restricted by the network to be local feedback (only from those neighboring robots). And, the control objective is for the robotic vehicles to exhibit certain group behavior (which is the topic of the subsequent subsection). Should an operator be involved to inject command signals through the same network, a virtual vehicle can be added into the group to represent the operator. The questions to be answered in this book include the following. (i) Controllability *via* a network and networked control design: given the intermittent nature of the network, the (necessary and sufficient) condition needs to be found under which a network-enabled local feedback control can be designed to achieve the control objective. (ii) Vehicle control design and performance: based on models of robotic vehicles, vehicle feedback controls are designed to ensure certain performance.

1.1.2 Cooperative Behaviors

Complex systems such as social systems, biological systems, science and engineering systems often consist of heterogeneous entities which have individual characteristics, interact with each other in various ways, and exhibit certain group phenomena. Since interactions existing among the entities may be quite sophisticated, it is often necessary to model these complex systems as networked systems. Cooperative behaviors refer to their group phenomena

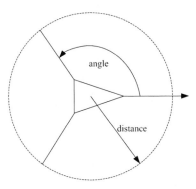

Fig. 1.4. Boid's neighborhood

and characteristics, and they are manifestation of the algorithms or controls governing each of the entities in relationship to its local environment. To illustrate the basics without undue complexity, let us consider a group of finite entities whose dynamics are overlooked in this subsection. In what follows, sample systems and their typical cooperative behaviors are outlined. Mathematically, these cooperative behaviors could all be captured by the limit of $[\mathcal{G}_i(y_i) - \mathcal{G}_j(y_j)] \to 0$, where y_i is the output of the ith entity, and $\mathcal{G}_i(\cdot)$ is a certain (set-based) mapping associated with the ith entity. It is based on this abstraction that the so-called cooperative stability will be defined in Chapter 5. Analysis and synthesis of linear and non-linear cooperative models as well as their behaviors will systematically be carried out in Chapters 5 and 6.

Boid's Model of Animal Flocking

As often observed, biological systems such as the groups of ants, fish, birds and bacteria reveal some amazing cooperative behaviors in their motion. Computer simulation and animation have been used to generate generic simulated flocking creatures called *boids* [217]. To describe interactions in a flock of birds or a school of fish, the so-called boid's neighborhood can be used. As shown in Fig. 1.4, it is parameterized in terms of a distance and an angle, and it has been used as a computer animation/simulation model. The behavior of each boid is influenced by other boids within its neighborhood. To generate or capture cooperative behaviors, three common steering rules of cohesion, separation and alignment can be applied. By cohesion, an entity moves toward the center of mass of its neighboring entities; by separation, the entity is moving away from the nearest boids in order to avoid possible collision; and by alignment, the entity adjusts its heading according to the average heading of those boids in the neighborhood. Graphically, the rules of separation, alignment and cohesion are depicted in Fig. 1.5.

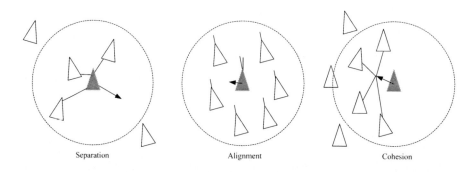

Fig. 1.5. Cooperative behaviors of animal flocking

Couzin Model of Animal Motion

The Couzin model [45] can be used to describe how the animals in a group make their movement decision with a limited pertinent information. For a group of q individuals moving at a constant velocity, heading ϕ'_i of the ith individual is determined by

$$\phi'_i(k+1) = \frac{\phi_i(k+1) + \omega_i \phi_i^d}{\|\phi_i(k+1) + \omega_i \phi_i^d\|}, \tag{1.1}$$

where ϕ_i^d is the desired direction (along certain segment of a migration route to some resource), $0 \leq \omega_i \leq 1$ is a weighting constant,

$$\begin{cases} \phi_i(k+1) = -\sum_{j \in N'_i} \frac{x_j(k) - x_i(k)}{d_{ij}(k)}, & \text{if } N'_i \text{ is not empty} \\ \phi_i(k+1) = \sum_{j \in N_i} \frac{x_j(k) - x_i(k)}{d_{ij}(k)} + \sum_{j \in N_i \cup \{i\}} \frac{\phi_j(k)}{\|\phi_j(k)\|}, & \text{if } N'_i \text{ is empty} \end{cases} \tag{1.2}$$

$x_i(t)$ is the position vector of the ith individual, $d_{ij}(k) = \|x_j(k) - x_i(k)\|$ is the norm-based distance between the ith and jth individuals, α is the minimum distance between i and j, β represents the range of interaction, $N_i(k) = \{j : 1 \leq j \leq q, d_{ij}(k) \leq \beta\}$ is the index set of individual i's neighbors, and $N'_i = \{j : 1 \leq j \leq q, d_{ij}(k) < \alpha\}$ is the index set of those neighbors that are too close. In (1.1), the individuals with $\omega_i > 0$ are informed about their preferred motion direction, and the rest of individuals with $\omega_i = 0$ are not. According to (1.2), collision avoidance is the highest priority (which is achieved by moving away from each other) and, if collision is not expected, the individuals will tend to attract toward and align with each other. Experiments show that, using the Couzin model of (1.1), only a very small proportion of individuals needs to be informed in order to guide correctly the group motion and that the informed individuals do not have to know their roles.

Aggregation and Flocking of Swarms

Aggregation is a basic group behavior of biological swarms. Simulation models have long been studied and used by biologists to animate aggregation [29, 178]. Indeed, an aggregation behavior can be generated using a simple model. For instance, an aggregation model based on an artificial potential field function is given by

$$\dot{x}_i = \sum_{j=1, j \neq i}^{q} \left(c_1 - c_2 e^{-\frac{\|x_i - x_j\|^2}{c_3}} \right) (x_j - x_i), \qquad (1.3)$$

where $x_i \in \Re^n$, $i \in \{1, \cdots, q\}$ is the index integer, and c_i are positive constants. It is shown in [76] that, if all the members move simultaneously and share position information with each other, aggregation is achieved under Model 1.3 in the sense that all the members of the swarm converge into a small region around the swarm center defined by $\sum_{i=1}^{q} x_i / q$. Model 1.3 is similar to the so-called artificial social potential functions [212, 270]. Aggregation behaviors can also be studied by using probabilistic method [240] or discrete-time models [75, 137, 138].

Flocking motion is another basic group behavior of biological swarms. Loosely speaking, flocking motion of a swarm is characterized by the common velocity to which velocities of the swarm members all converge. A simple flocking model is

$$v_i(k+1) = v_i(k) + \sum_{j=1}^{q} a_{ij}(x_i, x_j)[v_j(k) - v_i(k)], \qquad (1.4)$$

where $v_i(t)$ is the velocity of the ith member, the influence on the ith member by the jth member is quantified by the non-linear weighting term of

$$a_{ij}(x_i, x_j) = \frac{k_0}{(k_1^2 + \|x_i - x_j\|^2)^{k_2}},$$

k_l are positive constants, and x_i is the location of the ith member. Characteristics of the flocking behavior can be analyzed in terms of parameters k_i and initial conditions [46].

Synchronized Motion

Consider a group of q particles which move in a plane and have the same linear speed v. As shown in Fig. 1.6, motion of the particles can be synchronized by aligning their heading angles. A localized alignment algorithm is given by the so-called Vicsek model [263]:

$$\theta_i(k+1) = \frac{1}{1 + n_i(k)} \left(\theta_i(k) + \sum_{j \in N_i(k)} \theta_j(k) \right), \qquad (1.5)$$

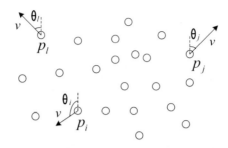

Fig. 1.6. Synchronized motion of planar motion particles

where $\aleph$ is the set of all non-negative integers, $k \in \aleph$ is the discrete-time index, $N_i(k)$ is the index set of particle i's neighbors at time k, and $n_i(k)$ is the number of entries in set $N_i(k)$. Essentially, Model 1.5 states that the heading of any given particle is adjusted regularly to the average of those headings of its neighbors. It is shown by experimentation and computer simulation [263] or by a graph theoretical analysis [98] that, if sufficient information is shared among the particles, headings of the particles all converge to the same value. Closely related to particle alignment is the stability problem of nano-particles for friction control [82, 83].

Synchronization arises naturally from such physical systems as coupled oscillators. For the set of q oscillators with angles θ_i and natural frequencies ω_i, their coupling dynamics can be described by the so-called Kuramoto model [247]:

$$\dot{\theta}_i = \omega_i + \sum_{j \in N_i} k_{ij} \sin(\theta_j - \theta_i), \qquad (1.6)$$

where N_i is the coupling set of the ith oscillator, and k_{ij} are positive constants of coupling strength. In the case that $\omega_i = \omega$ for all i and that there is no isolated oscillator, it can be shown [136] that angles $\theta_i(t)$ have the same limit. Alternatively, by considering the special case of $\theta_i \in (-\pi/2, \pi/2)$ and $w_i = 0$ for all i, Model 1.6 is converted under transformation $x_i = \tan \theta_i$ to

$$\dot{x}_i = \sum_{j \in N_i} k_{ji} \frac{\sqrt{1 + x_j^2}}{\sqrt{1 + x_i^2}} (x_j - x_i). \qquad (1.7)$$

Model 1.7 can be used as the alternative to analyze local synchronization of coupled oscillators [161].

Artificial Behaviors of Autonomous Vehicles

In [13, 150], artificial behaviors of mobile robots are achieved by instituting certain basic maneuvers and a set of rules governing robots' interactions

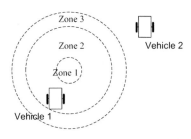

Fig. 1.7. Zones for the maintain-formation motor schema

and decision-making. Specifically, artificial formation behaviors are defined in
[13] in terms of such maneuvers as move-to-goal, avoid-static-obstacle, avoid-robot, maintain-formation, *etc.* These maneuvers are called *motor schemas*,
and each of them generates a motion vector of the direction and magnitude
to represent the corresponding artificial behavior under the current sensory
information. For instance, for a vehicle undertaking the maintain-formation
motor schema, its direction and magnitude of motion vector is determined by
geometrical relationship between its current location and the corresponding
desired formation position. As an example, Fig. 1.7 depicts one circular zone
and two ring-like zones, all concentric at the desired formation location of
vehicle 1 (in order to maintain a formation with respect to vehicle 2). Then,
the motion direction of vehicle 1 is that pointing toward the desired location
from its current location, and the motion magnitude is set to be zero if the
vehicle is in zone 1 (the circular zone called dead zone), or to be proportional to the distance between its desired and current locations if the vehicle
lies in zone 2 (the inner ring, also called the controlled zone), or to be the
maximum value if the vehicle enters zone 3 (the outer ring). Once motion
vectors are calculated under all relevant motor schemas, the resulting motion
of artificial behavior can be computed using a set of specific rules. Clearly,
rule-based artificial behaviors are intuitive and imitate certain animal group
behaviors. On-going research is to design analytical algorithms and/or explicit
protocols of neighboring-feedback cooperation for the vehicles by synthesizing
analytical algorithms of emulating animal group behaviors, by making vehicle
motion comply with a dynamically changing and uncertain environment, and
by guaranteeing their mission completion and success.

Coverage Control of Mobile Sensor Network

A group of mobile vehicles can be deployed to undertake such missions as
search and exploration, surveillance and monitoring, and rescue and recovery.
In these applications, vehicles need to coordinate their motion to form an ad-

hoc communication network or act individually as a mobile sensor. Toward the goal of motion cooperation, a coverage control problem needs to be solved to determine the optimal spatial resource allocation [44]. Given a convex polytope Ω and a distribution density function $\phi(\cdot)$, the coverage control problem of finding optimal static locations of $P = [p_1, \cdots, p_n]^T$ for n sensors/robots can be mathematically formulated as

$$P = \min_{p_i, W_i} \sum_{i=1}^{n} \int_{W_i} \|q - p_i\| d\phi(q),$$

where partition $W_i \subset \Omega$ is the designated region for the ith sensor at position $p_i \in W_i$. If distribution density function $\phi(\cdot)$ is uniform in Ω, partitions W_i are determined solely by distance as

$$W_i = \{q \in Q \mid \|q - p_i\| \leq \|q - p_j\|, \ \forall j \neq i\},$$

and so are p_i. In this case, the solutions are given by the Voronoi diagram in Fig. 1.8: a partition of space into cells, each of which consists of the points closer to one p_i than to any others p_j. If distribution density function $\phi(\cdot)$ is not uniform, solutions of W_i and p_i are weighted by the distribution density function. Dynamically, gradient descent algorithms can be used to solve this coverage control problem in an adaptive, distributed, and asynchronous manner [44].

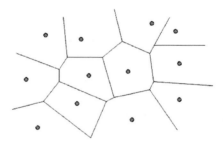

Fig. 1.8. Voronoi diagram for locating mobile sensors

1.1.3 Pliable and Robust Systems

A control system shown in Fig. 1.1 is said to be pliable if all the constraints of the plant are met. Should the plant be a mechanical system, it will be shown in Sections 1.2 and 1.3 that the constraints may be kinematic or dynamic or both. A networked control system in Fig. 1.2 is pliable if all the plants satisfy their constraints and if all the network-based controls conform with information flow in the network.

In addition to the sensing and communication network, the networked control system must also be robust by complying to the changes (if any) in its physical environment. For an autonomous multi-vehicle system, this means that the vehicles can avoid static and moving obstacles and that they can also avoid each other. The latter is of particular importance for cooperative control because many of cooperative behaviors require that vehicles stay in close vicinity of each other while moving.

Indeed, as performance measures, autonomous vehicles are required to be both cooperative, pliable, and robust. The multiple control objectives for the vehicle systems and for cooperative, robust and pliable systems in general imply that multi-level controls need to be properly designed, which is the subject of Section 1.4.

1.2 Modeling of Constrained Mechanical Systems

Most vehicles are constrained mechanical systems, and their models can be derived using fundamental principles in rigid-body mechanics and in terms of lumped parameters. In this section, standard modeling techniques and steps are summarized. Unless stated otherwise, all functions are assumed to be smooth.

1.2.1 Motion Constraints

Let $\Re^n$ be the configuration space, $q \in \Re^n$ be the vector of generalized coordinates, and $\dot{q} \in \Re^n$ be the vector of generalized velocities. A matrix expression of k motion constraints is

$$A(q,t)\dot{q} + B(q,t) = 0 \in \Re^k, \tag{1.8}$$

where $0 < k < n$ and $A(q,t) \in \Re^{k \times n}$. Constraints 1.8 are said to be *holonomic* or *integrable* if they correspond to algebraic equations only in terms of configuration variables of q as

$$\beta(q,t) = 0 \in \Re^k.$$

Therefore, given the set of holonomic constraints, the matrices in (1.8) can be calculated as

$$A(q,t) = \frac{\partial \beta(q,t)}{\partial q}, \quad B(q,t) = \frac{\partial \beta(q,t)}{\partial t};$$

conversely, given the constraints in (1.8), their integrability can be determined (using the Frobenius theorem which is to be stated in Section 2.6.2 and gives necessary and sufficient conditions), or the following equations can be verified as necessary conditions: for all meaningful indices i, j, l:

$$\frac{\partial a_{ij}(q,t)}{\partial q_l} = \frac{\partial a_{il}(q,t)}{\partial q_j} = \frac{\partial^2 \beta_i(q,t)}{\partial q_l \partial q_j}, \quad \frac{\partial a_{ij}(q,t)}{\partial t} = \frac{\partial b_i(q,t)}{\partial q_j} = \frac{\partial^2 \beta_i(q,t)}{\partial t \partial q_j},$$

where $a_{ij}(q,t)$ denote the (i,j)th element of $A(\cdot)$, and $b_j(q,t)$ denote the jth element of $B(\cdot)$. If constraints are not integrable, they are said to be *non-holonomic*.

Constraints 1.8 are called *rheonomic* if they explicitly depend upon time, and *scleronomic* otherwise. Constraints 1.8 are called *driftless* if they are linear in velocities or $B(q,t) = 0$. If $B(q,t) \neq 0$, the constraints have drift, which will be discussed in Section 1.2.6. Should Constraints 1.8 be both scleronomic and driftless, they reduce to the so-called *Pfaffian constraints* of form

$$A(q)\dot{q} = 0 \in \Re^k. \tag{1.9}$$

If Pfaffian constraints are integrable, the holonomic constraints become

$$\beta(q) = 0 \in \Re^k. \tag{1.10}$$

Obviously, holonomic constraints of (1.10) restrict the motion of q on an $(n - k)$-dimensional hypersurface (or sub-manifold). These geometric constraints can always be satisfied by reducing the number of free configuration variables from n to $(n - k)$. For instance, both a closed-chain mechanical device and the two robots holding a common object are holonomic constraints. On the other hand, a rolling motion without side slipping is the typical non-holonomic constraint in the form of (1.9). These non-holonomic constraints require that all admissible velocities belong to the null space of matrix $A(q)$, while the motion in the configuration space of $q \in \Re^n$ is not limited. To ensure non-holonomic constraints, a kinematic model and the corresponding kinematic control need to be derived, which is the subject of the next subsection.

1.2.2 Kinematic Model

Consider a mechanical system kinematically subject to non-holonomic velocity constraints in the form of (1.9). Since $A(q) \in \Re^{k \times n}$ with $k < n$, it is always possible to find a rank-$(n - k)$ matrix $G(q) \in \Re^{n \times (n-k)}$ such that columns of $G(q)$, $g_j(q)$, consist of smooth and linearly independent vector fields spanning the null space of $A(q)$, that is,

$$A(q)G(q) = 0, \quad \text{or} \quad A(q)g_j(q) = 0, \; j = 1, \cdots, n - k. \tag{1.11}$$

It then follows from (1.9) and (1.11) that $\dot{q}$ can be expressed in terms of a linear combination of vector fields $g_i(q)$ and as

$$\dot{q} = G(q)u = \sum_{i=1}^{n-k} g_i(q)u_i, \tag{1.12}$$

where auxiliary input $u = \begin{bmatrix} u_1 \cdots u_{n-k} \end{bmatrix}^T \in \Re^m$ is called *kinematic control*. Equation 1.12 gives the kinematic model of a non-holonomic system, and the non-holonomic constraints in (1.9) are met if the corresponding velocities are planned by or steered according to (1.12). Properties of Model 1.12 will be studied in Section 3.1.

1.2.3 Dynamic Model

The dynamic equation of a constrained (non-holonomic) mechanical system can be derived using variational principles. To this end, let $L(q, \dot{q})$ be the so-called *Lagrangian* defined by

$$L(q, \dot{q}) = K(q, \dot{q}) - V(q),$$

where $K(\cdot)$ is the kinetic energy of the system, and $V(q)$ is the potential energy. Given time interval $[t_0, t_f]$ and terminal conditions $q(t_0) = q_0$ and $q(t_f) = q_f$, we can define the so-called *variation* $q(t, \epsilon)$ as a smooth mapping satisfying $q(t, 0) = q(t)$, $q(t_0, \epsilon) = q_0$ and $q(t_f, \epsilon) = q_f$. Then, quantity

$$\delta q(t) \triangleq \left. \frac{\partial q(t, \epsilon)}{\partial \epsilon} \right|_{\epsilon=0}$$

is the *virtual displacement* corresponding to the variation, and its boundary conditions are

$$\delta q(t_0) = \delta q(t_f) = 0. \tag{1.13}$$

In the absence of external force and constraint, the *Hamilton principle* states that the system trajectory $q(t)$ is the stationary solution with respect to the time integral of the Lagrangian, that is,

$$\left. \frac{\partial}{\partial \epsilon} \int_{t_0}^{t_f} L(q(t, \epsilon), \dot{q}(t, \epsilon)) dt \right|_{\epsilon=0} = 0.$$

Using the chain rule, we can compute the variation of the integral corresponding to variation $q(t, \epsilon)$ and express it in terms of its virtual displacement as

$$\int_{t_0}^{t_f} \left[\left(\frac{\partial L}{\partial q} \right)^T \delta q + \left(\frac{\partial L}{\partial \dot{q}} \right)^T \delta \dot{q} \right] dt = 0.$$

Noting that $\delta \dot{q} = d(\delta q)/dt$, integrating by parts, and substituting the boundary conditions in (1.13) yield

$$\int_{t_0}^{t_f} \left[\left(-\frac{d}{dt} \frac{\partial L}{\partial \dot{q}} + \frac{\partial L}{\partial q} \right)^T \delta q \right] dt = 0. \tag{1.14}$$

In the presence of any input $\tau \in \Re^m$, the *Lagrange-d'Alembert principle* generalizes the Hamilton principle as

$$\left. \frac{\partial}{\partial \epsilon} \int_{t_0}^{t_f} L(q(t, \epsilon), \dot{q}(t, \epsilon)) dt \right|_{\epsilon=0} + \int_{t_0}^{t_f} \left\{ [B(q)\tau]^T \delta q \right\} dt = 0, \tag{1.15}$$

where $B(q) \in \Re^{n \times m}$ maps input τ into forces/torques, and $[B(q)\tau]^T \delta q$ is the virtual work done by force τ with respect to virtual displacement δq.

Repeating the derivations leading to (1.14) yields the so-called *Lagrange-Euler equation of motion*:

$$\frac{d}{dt}\left(\frac{\partial L}{\partial \dot{q}}\right) - \frac{\partial L}{\partial q} = B(q)\tau. \tag{1.16}$$

If present, constraints in (1.9) exert a constrained force vector F on the system which undergoes a virtual displacement δq. As long as the non-holonomic constraints are enforced by these forces, the system can be thought to be holonomic but subject to the constrained forces, that is, its equation of motion is given by (1.16) with $B(q)\tau$ replaced by F. The work done by these forces is $F^T \delta q$. The *d'Alembert principle* about the constrained forces is that, with respect to any virtual displacement consistent with the constraints, the constrained forces do no work as

$$F^T \delta q = 0, \tag{1.17}$$

where virtual displacement δq is assumed to satisfy the constraint equation of (1.9), that is,

$$A(q)\delta q = 0. \tag{1.18}$$

Note that δq and $\dot{q}$ are different since, while virtual displacement δq satisfies only the constraints, the generalized velocity $\dot{q}$ satisfies both the velocity constraints and the equation of motion. Comparing (1.17) and (1.18) yields

$$F = A^T(q)\lambda,$$

where $\lambda \in \Re^k$ is the Lagrange multiplier. The above equation can also be established using the same argument that proves the Lagrange multiplier theorem.

In the presence of both external input $\tau \in \Re^m$ and the non-holonomic constraints of (1.9), the *Lagrange-d'Alembert principle* also gives (1.15) except that δq satisfies (1.18). Combining the external forces and constrained forces, we obtain the following Euler-Lagrange equation, also called *Lagrange-d'Alembert equation*:

$$\frac{d}{dt}\left(\frac{\partial L}{\partial \dot{q}}\right) - \frac{\partial L}{\partial q} = A^T(q)\lambda + B(q)\tau. \tag{1.19}$$

Note that, in general, substituting a set of Pfaffian constraints into the Lagrangian and then applying the Lagrange-d'Alembert equations render equations that are *not* correct. See Section 1.4 in Chapter 6 of [168] for an illustrative example.

The Euler-Lagrange equation or Lagrange-d'Alembert equation is convenient to derive and use because it does not need to account for any forces internal to the system and it is independent of coordinate systems. These two advantages are beneficial, especially in handling a multi-body mechanical system with moving coordinate frames. The alternative Newton-Euler method will be introduced in Section 1.3.5.

1.2.4 Hamiltonian and Energy

If *inertia matrix*

$$M(q) \overset{\triangle}{=} \frac{\partial^2 L}{\partial \dot{q}^2}$$

is non-singular, Lagrangian L is said to be *regular*. In this case, the *Legendre transformation* changes the state from $[q^T \; \dot{q}^T]^T$ to $[q^T \; p^T]^T$, where

$$p = \frac{\partial L}{\partial \dot{q}}$$

is the so-called *momentum*. Then, the so-called *Hamiltonian* is defined as

$$H(q,p) \overset{\triangle}{=} p^T \dot{q} - L(q,\dot{q}). \tag{1.20}$$

It is straightforward to verify that, if $\tau = 0$, the Lagrange-Euler Equation of (1.16) becomes the following *Hamilton equations*:

$$\dot{q} = \frac{\partial H}{\partial p}, \quad \dot{p} = -\frac{\partial H}{\partial q}.$$

In terms of symmetrical inertia matrix $M(q)$, the Lagrangian and Hamiltonian can be expressed as

$$L = \frac{1}{2}\dot{q}^T M(q)\dot{q} - V(q), \quad H = \frac{1}{2}\dot{q}^T M(q)\dot{q} + V(q). \tag{1.21}$$

That is, the Lagrangian is kinetic energy minus potential energy, the Hamiltonian is kinetic energy plus potential energy, and hence the Hamiltonian is the energy function. To see that non-holonomic systems are conservative under $\tau = 0$, we use energy function H and take its time derivative along the solution of (1.19), that is,

$$\frac{dH(q,p)}{dt} = \left[\frac{d}{dt}\left(\frac{\partial L}{\partial \dot{q}}\right)^T\right]\dot{q} + \left(\frac{\partial L}{\partial \dot{q}}\right)^T \ddot{q} - \left(\frac{\partial L}{\partial q}\right)^T \dot{q} - \left(\frac{\partial L}{\partial \dot{q}}\right)^T \ddot{q}$$
$$= \lambda^T A(q)\dot{q},$$

which is zero according to (1.9).

1.2.5 Reduced-order Model

It follows from (1.21), (1.19) and (1.9) that the dynamic equations of a non-holonomic system are given by

$$M(q)\ddot{q} + N(q,\dot{q}) = A^T(q)\lambda + B(q)\tau,$$
$$A(q)\dot{q} = 0, \tag{1.22}$$

where $q \in \Re^n$, $\lambda \in \Re^k$, $\tau \in \Re^m$, $N(q,\dot{q}) = C(q,\dot{q})\dot{q} + f_g(q)$, $f_g(q) = \partial V(q)/\partial q$ is the vector containing gravity terms, and $C(q,\dot{q}) \in \Re^{n \times n}$ is the matrix containing centrifugal forces (of the terms q_i^2) and Coriolis forces (of terms $q_i q_j$). Matrix $C(q,\dot{q})$ can be expressed in tensor as

$$C(q,\dot{q}) = \dot{M}(q) - \frac{1}{2}\dot{q}^T\left(\frac{\partial}{\partial q}M(q)\right),$$

or element-by-element as

$$C_{lj}(q,\dot{q}) = \frac{1}{2}\sum_{i=1}^{n}\left[\frac{\partial m_{lj}(q)}{\partial q_i} + \frac{\partial m_{li}(q)}{\partial q_j} - \frac{\partial m_{ij}(q)}{\partial q_l}\right]\dot{q}_i, \qquad (1.23)$$

where $m_{ij}(q)$ are the elements of $M(q)$. Embedded in (1.23) is the inherent property that $\dot{M}(q) - 2C(q,\dot{q})$ is skew symmetrical, which is explored extensively in robotics texts [130, 195, 242]. There are $(2n+k)$ equations in (1.22), and there are $(2n+k)$ variables in q, $\dot{q}$, and λ.

Since the state is of dimension $2n$ and there are also k constraints, a total of $(2n-k)$ reduced-order dynamic equations are preferred for analysis and design. To this end, we need to pre-multiply $G^T(q)$ on both sides of the first equation in (1.22) and have

$$G^T(q)M(q)\ddot{q} + G^T(q)N(q,\dot{q}) = G^T(q)B(q)\tau,$$

in which λ is eliminated by invoking (1.11). On the other hand, differentiating (1.12) yields

$$\ddot{q} = \dot{G}(q)u + G(q)\dot{u}.$$

Combining the above two equations and recalling (1.12) render the following reduced dynamic model of non-holonomic systems:

$$\begin{aligned}\dot{q} &= G(q)u \\ M'(q)\dot{u} + N'(q,u) &= G^T(q)B(q)\tau,\end{aligned} \qquad (1.24)$$

where

$$M'(q) = G^T(q)M(q)G(q), \quad N'(q,u) = G^T(q)M(q)\dot{G}(q)u + G^T(q)N(q,G(q)u).$$

Equation 1.24 contains $(2n-k)$ differential equations for exactly $(2n-k)$ variables. If $m = n - k$ and if $G^T(q)B(q)$ is invertible, the second equation in (1.24) is identical to those of holonomic fully actuated robots, and it is feedback linearizable (see Section 2.6.2 for details), and various controls can easily be designed for τ to make u track any desired trajectory $u^d(t)$ [130, 195, 242]. As a result, the focus of controlling non-holonomic systems is to design kinematic control u for Kinematic Model 1.12. Once control u^d is designed for Kinematic Model 1.12, control τ can be found using the backstepping procedure (see Section 2.6.1 for details).

Upon solving for q and $\dot{q}$ from (1.24), λ can also be determined. It follows from the second equation in (1.22) that

$$A(q)\ddot{q} + \dot{A}(q)\dot{q} = 0.$$

Pre-multiplying first $M^{-1}(q)$ and then $A(q)$ on both sides of the first equation in (1.22), we know from the above equation that

$$\lambda = [A(q)M^{-1}(q)A^T(q)]^{-1}\left\{A(q)M^{-1}(q)\left[N(q,\dot{q}) - B(q)\tau\right] - \dot{A}(q)\dot{q}\right\}.$$

1.2.6 Underactuated Systems

Generally, non-holonomic constraints with drift arise naturally from dynamics of underactuated mechanical systems. For example, the reduced dynamic model of (1.24) is underactuated if the dimension of torque-level input is less than the dimension of kinematic input, or simply $m < n - k$. To illustrate the point, consider the simple case that, in (1.24),

$$n - k = 2, \quad m = 1, \quad G^T(q)B(q) = \begin{bmatrix} 0 & 1 \end{bmatrix}^T, \quad M'(q) \triangleq \begin{bmatrix} m'_{11}(q) & m'_{12}(q) \\ m'_{21}(q) & m'_{22}(q) \end{bmatrix},$$

and

$$N'(q,u) \triangleq \begin{bmatrix} n'_1(q,u) \\ n'_2(q,u) \end{bmatrix}.$$

That is, the reduced-order dynamic equations are

$$\begin{aligned} m'_{11}(q)\dot{u}_1 + m'_{12}(q)\dot{u}_2 + n'_1(q,u) &= 0, \\ m'_{21}(q)\dot{u}_1 + m'_{22}(q)\dot{u}_2 + n'_2(q,u) &= \tau. \end{aligned} \tag{1.25}$$

If the underactuated mechanical system is properly designed, controllability of Dynamic Sub-system 1.25 and System 1.24 can be ensured (and, if needed, also checked using the conditions presented in Section 2.5). Solving for $\dot{u}_i$ from (1.25) yields

$$\dot{u}_1 = \frac{-m'_{12}(q)\tau - m'_{22}(q)n'_1(q,u) + m'_{12}(q)n'_2(q,u)}{m'_{11}(q)m'_{22}(q) - m'_{21}(q)m'_{12}(q)}, \tag{1.26}$$

$$\dot{u}_2 = \frac{m'_{11}(q)\tau + m'_{21}(q)n'_1(q,u) - m'_{11}(q)n'_2(q,u)}{m'_{11}(q)m'_{22}(q) - m'_{21}(q)m'_{12}(q)}. \tag{1.27}$$

If u_1 were the primary variable to be controlled and u_1^d were the desired trajectory for u_1, we would design τ according to (1.26), for instance,

$$\begin{aligned} \tau = \frac{1}{m'_{12}(q)} \Big\{ &-m'_{22}(q)n'_1(q,u) + m'_{12}(q)n'_2(q,u) \\ &+ [u_1 - u_1^d - \dot{u}_1^d][m'_{11}(q)m'_{22}(q) - m'_{21}(q)m'_{12}(q)] \Big\}, \end{aligned} \tag{1.28}$$

under which

$$\frac{d}{dt}[u_1 - u_1^d] = -[u_1 - u_1^d].$$

Substituting (1.28) into (1.27) yields a constraint which is in the form of (1.8) and in general non-holonomic and with drift. As will be shown in Section 2.6.2, the above process is input-output feedback linearization (also called partial feedback linearization in [241]), and the resulting non-holonomic constraint is the internal dynamics. It will be shown in Chapter 2 that the internal dynamics being input-to-state stable is critical for control design and stability of the overall system.

If a mechanical system does not have Pfaffian constraints but is under-actuated (*i.e.*, $m < n$), non-holonomic constraints arises directly from its Euler-Lagrange Equation 1.16. For example, if $n = 2$ and $m = 1$ with $B(q) = \begin{bmatrix} 0 & 1 \end{bmatrix}^T$, Dynamic Equation 1.16 becomes

$$\begin{aligned} m_{11}(q)\ddot{q}_1 + m_{12}(q)\ddot{q}_2 + n_1(q, \dot{q}) = 0, \\ m_{21}(q)\ddot{q}_1 + m_{22}(q)\ddot{q}_2 + n_2(q, \dot{q}) = \tau. \end{aligned} \quad (1.29)$$

Although differential equations of underactuated systems such as those in (1.29) are of second-order (and one of them is a non-holonomic constraint referred to as non-holonomic acceleration constraint [181]), there is little difference from those of first-order. For example, (1.25) and (1.29) become essentially identical after introducing the state variable $u = \dot{q}$, and hence the aforementioned process can be applied to (1.29) and to the class of underactuated mechanical systems. More discussions on the class of underactuated mechanical systems can be found in [216].

1.3 Vehicle Models

In this section, vehicle models and other examples are reviewed. Pfaffian constraints normally arise from two types of systems: (i) in contact with a certain surface, a body rolls without side slipping; (ii) conservation of angular momentum of rotational components in a multi-body system. Ground vehicles belong to the first type, and aerial and space vehicles are of the second type.

1.3.1 Differential-drive Vehicle

Among all the vehicles, the differential-drive vehicle shown in Fig. 1.9 is the simplest. If there are more than one pair of wheels, it is assumed for simplicity that all the wheels on the same side are of the same size and have the same angular velocity. Let the center of the vehicle be the guidepoint whose generalized coordinate is $q = \begin{bmatrix} x & y & \theta \end{bmatrix}^T$, where (x, y) is the 2-D position of the

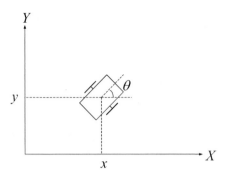

Fig. 1.9. A differential-drive vehicle

vehicle, and θ is the heading angle of its body. For 2-D motion, the Lagrangian of the vehicle is merely the kinetic energy, that is,

$$L = \frac{1}{2}m(\dot{x}^2 + \dot{y}^2) + \frac{1}{2}J\dot{\theta}^2, \qquad (1.30)$$

where m is the mass of the vehicle, and J is the inertia of the vehicle with respect to the vertical axis passing through the guidepoint.

Under the assumption that the vehicle rolls without side slipping, the vehicle does not experience any motion along the direction of the wheel axle, which is described by the following non-holonomic constraint:

$$\dot{x}\sin\theta - \dot{y}\cos\theta = 0,$$

or in matrix form,

$$0 = \begin{bmatrix} \sin\theta & -\cos\theta & 0 \end{bmatrix} \begin{bmatrix} \dot{x} \\ \dot{y} \\ \dot{\theta} \end{bmatrix} \triangleq A(q)\dot{q}.$$

As stated in (1.11), the null space of matrix $A(q)$ is the span of column vectors in matrix

$$G(q) = \begin{bmatrix} \cos\theta & 0 \\ \sin\theta & 0 \\ 0 & 1 \end{bmatrix}.$$

Therefore, by (1.12), kinematic model of the differential-drive vehicle is

$$\begin{bmatrix} \dot{x} \\ \dot{y} \\ \dot{\theta} \end{bmatrix} = \begin{bmatrix} \cos\theta \\ \sin\theta \\ 0 \end{bmatrix} u_1 + \begin{bmatrix} 0 \\ 0 \\ 1 \end{bmatrix} u_2, \qquad (1.31)$$

where u_1 is the driving velocity, and u_2 is the steering velocity. Kinematic control variables u_1 and u_2 are related to physical kinematic variables as

$$u_1 = \frac{\rho}{2}(\omega_r + \omega_l), \quad u_2 = \frac{\rho}{2}(\omega_r - \omega_l),$$

where ω_l is the angular velocity of left-side wheels, ω_r is the angular velocity of right-side wheels, and ρ is the radius of all the wheels.

It follows from (1.30) and (1.19) that the vehicle's dynamic model is

$$M\ddot{q} = A^T(q)\lambda + B(q)\tau,$$

that is,

$$\begin{bmatrix} m & 0 & 0 \\ 0 & m & 0 \\ 0 & 0 & J \end{bmatrix} \begin{bmatrix} \ddot{x} \\ \ddot{y} \\ \ddot{\theta} \end{bmatrix} = \begin{bmatrix} \sin\theta \\ -\cos\theta \\ 0 \end{bmatrix} \lambda + \begin{bmatrix} \cos\theta & 0 \\ \sin\theta & 0 \\ 0 & 1 \end{bmatrix} \begin{bmatrix} \tau_1 \\ \tau_2 \end{bmatrix}, \tag{1.32}$$

where λ is the Lagrange multiplier, τ_1 is the driving force, and τ_2 is the steering torque. It is obvious that

$$G^T(q)B(q) = I_{2\times 2}, \quad G^T(q)M\dot{G}(q) = 0, \quad G^T(q)MG(q) = \begin{bmatrix} m & 0 \\ 0 & I \end{bmatrix}.$$

Thus, the reduced-order dynamic model in the form of (1.24) becomes

$$m\dot{u}_1 = \tau_1, \quad J\dot{u}_2 = \tau_2. \tag{1.33}$$

In summary, the model of the differential-drive vehicle consists of the two cascaded equations of (1.31) and (1.33).

1.3.2 A Car-like Vehicle

A car-like vehicle is shown in Fig. 1.10; its front wheels steer the vehicle while its rear wheels have a fixed orientation with respect to the body. As shown in the figure, l is the distance between the midpoints of two wheel axles, and the center of the back axle is the guidepoint. The generalized coordinate of the guidepoint is represented by $q = [x \; y \; \theta \; \phi]^T$, where (x, y) are the 2-D Cartesian coordinates, θ is the orientation of the body with respect to the x-axis, and ϕ is the steering angle.

During its normal operation, the vehicle's wheels roll but do not slip sideways, which translates into the following motion constraints:

$$v_x^f \sin(\theta + \phi) - v_y^f \cos(\theta + \phi) = 0,$$
$$v_x^b \sin\theta - v_y^b \cos\theta = 0, \tag{1.34}$$

where (v_x^f, v_y^f) and (v_x^b, v_y^b) are the pairs of x-axis and y-axis velocities for the front and back wheels, respectively. As evident from Fig. 1.10, the coordinates of the midpoints of the front and back axles are $(x + l\cos\theta, y + l\sin\theta)$ and (x, y), respectively. Hence, we have

$$v_x^b = \dot{x}, \quad v_y^b = \dot{y}, \quad v_x^f = \dot{x} - l\dot{\theta}\sin\theta, \quad v_y^f = \dot{y} + l\dot{\theta}\cos\theta.$$

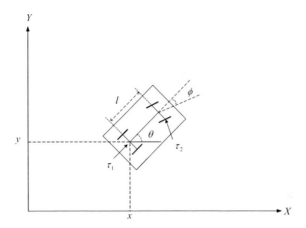

Fig. 1.10. A car-like robot

Substituting the above expressions into the equations in (1.34) yields the following non-holonomic constraints in matrix form:

$$0 = \begin{bmatrix} \sin(\theta + \phi) & -\cos(\theta + \phi) & -l\cos\phi & 0 \\ \sin\theta & -\cos\theta & 0 & 0 \end{bmatrix} \dot{q} \overset{\triangle}{=} A(q)\dot{q}.$$

Accordingly, we can find the corresponding matrix $G(q)$ (as stated for (1.11) and (1.12)) and conclude that the kinematic model of a car-like vehicle is

$$\begin{bmatrix} \dot{x} \\ \dot{y} \\ \dot{\theta} \\ \dot{\phi} \end{bmatrix} = \begin{bmatrix} \cos\theta & 0 \\ \sin\theta & 0 \\ \frac{1}{l}\tan\phi & 0 \\ 0 & 1 \end{bmatrix} \begin{bmatrix} u_1 \\ u_2 \end{bmatrix}, \tag{1.35}$$

where $u_1 \geq 0$ (or $u_1 \leq 0$) and u_2 are kinematic control inputs which need to be related back to physical kinematic inputs. It follows from (1.35) that

$$\sqrt{\dot{x}^2 + \dot{y}^2} = u_1, \quad \dot{\phi} = u_2,$$

by which the physical meanings of u_1 and u_2 are the linear velocity and steering rate of the body, respectively.

For a front-steering and back-driving vehicle, it follows from the physical meanings that

$$u_1 = \rho\omega_1, \quad u_2 = \omega_2,$$

where ρ is the radius of the back wheels, ω_1 is the angular velocity of the back wheels, and ω_2 is the steering rate of the front wheels. Substituting the above equation into (1.35) yields the kinematic control model:

$$\begin{bmatrix} \dot{x} \\ \dot{y} \\ \dot{\theta} \\ \dot{\phi} \end{bmatrix} = \begin{bmatrix} \rho\cos\theta & 0 \\ \rho\sin\theta & 0 \\ \dfrac{\rho}{l}\tan\phi & 0 \\ 0 & 1 \end{bmatrix} \begin{bmatrix} \omega_1 \\ \omega_2 \end{bmatrix}, \qquad (1.36)$$

Kinematic Model 1.36 has singularity at the hyperplane of $\phi = \pm\pi/2$, which can be avoided (mathematically and in practice) by limiting the range of ϕ to the interval $(-\pi/2, \pi/2)$.

For a front-steering and front-driving vehicle, we know from their physical meanings that

$$u_1 = \rho\omega_1\cos\phi, \quad u_2 = \omega_2,$$

where ρ is the radius of the front wheels, ω_1 is the angular velocity of the front wheels, and ω_2 is the steering rate of the front wheels. Thus, the corresponding kinematic control model is

$$\begin{bmatrix} \dot{x} \\ \dot{y} \\ \dot{\theta} \\ \dot{\phi} \end{bmatrix} = \begin{bmatrix} \rho\cos\theta\cos\phi \\ \rho\sin\theta\cos\phi \\ \dfrac{\rho}{l}\sin\phi \\ 0 \end{bmatrix} \omega_1 + \begin{bmatrix} 0 \\ 0 \\ 0 \\ 1 \end{bmatrix} \omega_2,$$

which is free of any singularity.

To derive the dynamic model for car-like vehicles, we know from the Lagrangian being the kinetic energy that

$$L = \frac{1}{2}m\dot{x}^2 + \frac{1}{2}m\dot{y}^2 + \frac{1}{2}J_b\dot{\theta}^2 + \frac{1}{2}J_s\dot{\phi}^2,$$

where m is the mass of the vehicle, J_b is the body's total rotational inertia around the vertical axis, and J_s is the inertial of the front-steering mechanism. It follows from (1.19) that the dynamic equation is

$$M\ddot{q} = A^T(q)\lambda + B(q)\tau,$$

where

$$M = \begin{bmatrix} m & 0 & 0 & 0 \\ 0 & m & 0 & 0 \\ 0 & 0 & J_b & 0 \\ 0 & 0 & 0 & J_s \end{bmatrix}, \quad B(q) = \begin{bmatrix} \cos\theta & 0 \\ \sin\theta & 0 \\ 0 & 0 \\ 0 & 1 \end{bmatrix},$$

$\lambda \in \Re^2$ is the Lagrange multiplier, $\tau = [\tau_1, \tau_2]^T$, τ_1 is the torque acting on the driving wheels, and τ_2 is the steering torque. Following the procedure from (1.22) to (1.24), we obtain the following reduced-order dynamic model:

$$\left(m + \frac{J_b}{l^2}\tan^2\phi\right)\dot{u}_1 = \frac{J_b}{l^2}u_1u_2\tan\phi\sec^2\phi + \tau_1, \quad J_s\dot{u}_2 = \tau_2. \qquad (1.37)$$

Note that reduced-order dynamic equations of (1.37) and (1.33) are similar as both have the same form as the second equation in (1.24). In comparison, kinematic models of different types of vehicles are usually distinct. Accordingly, we will focus upon kinematic models in the subsequent discussions of vehicle modeling.

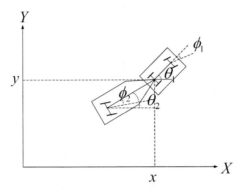

Fig. 1.11. A fire truck

1.3.3 Tractor-trailer Systems

Consider first the fire truck shown in Fig. 1.11; it consists of a tractor-trailer pair that is connected at the middle axle while the first and third axles are allowed to be steered [34]. Let us choose the guidepoint to be the midpoint of the second axle (the rear axle of the tractor) and define the generalized coordinate as $q = \begin{bmatrix} x & y & \phi_1 & \theta_1 & \phi_2 & \theta_2 \end{bmatrix}^T$, where (x, y) is the position of the guidepoint, ϕ_1 is the steering angle of the front wheels, θ_1 is the orientation of the tractor, ϕ_2 is the steering angle of the third axle, and θ_2 is the orientation of the trailer. It follows that

$$\begin{cases} x_f = x + l_f \cos\theta_1, \quad y_f = y + l_f \sin\theta_1, \\ x_m = x, \quad y_m = y, \\ x_b = x - l_b \cos\theta_2, \quad y_b = y - l_b \sin\theta_2, \end{cases} \tag{1.38}$$

where l_f is the distance between the midpoints of the front and middle axles, l_b is the distance between the midpoints of the middle and rear axles, and (x_f, y_f), (x_m, y_m) (x_b, y_b) are the midpoints of the front, middle and back axles, respectively. If the vehicle operates without experiencing side slippage of the wheels, the corresponding non-slip constraints are:

$$\begin{cases} \dot{x}_f \sin(\theta_1 + \phi_1) - \dot{y}_f \cos(\theta_1 + \phi_1) = 0, \\ \dot{x}_m \sin\theta_1 - \dot{y}_m \cos\theta_1 = 0, \\ \dot{x}_b \sin(\theta_2 + \phi_2) - \dot{y}_b \cos(\theta_2 + \phi_2) = 0. \end{cases} \tag{1.39}$$

Combining (1.38) and (1.39) yields the non-holonomic constraints in matrix form as $A(q)\dot{q} = 0$, where

$$A(q) = \begin{bmatrix} \sin(\theta_1 + \phi_1) & -\cos(\theta_1 + \phi_1) & 0 & -l_f \cos\phi_1 & 0 & 0 \\ \sin\theta_1 & -\cos\theta_1 & 0 & 0 & 0 & 0 \\ \sin(\theta_2 + \phi_2) & -\cos(\theta_2 + \phi_2) & 0 & 0 & 0 & l_b \cos\phi_2 \end{bmatrix}.$$

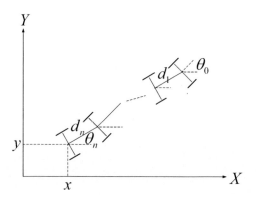

Fig. 1.12. An n-trailer system

Following the step from (1.11) to (1.12), we obtain the following kinematic model for the fire truck:

$$
\dot{q} =
\begin{bmatrix}
\cos\theta_1 \\
\sin\theta_1 \\
0 \\
\frac{1}{l_f}\tan\phi_1 \\
0 \\
-\frac{1}{l_b}\sec\phi_2\sin(\phi_2 - \theta_1 + \theta_2)
\end{bmatrix}
u_1 +
\begin{bmatrix}
0 \\ 0 \\ 1 \\ 0 \\ 0 \\ 0
\end{bmatrix}
u_2 +
\begin{bmatrix}
0 \\ 0 \\ 0 \\ 0 \\ 1 \\ 0
\end{bmatrix}
u_3,
\qquad (1.40)
$$

where $u_1 \geq 0$ is the linear body velocity of the tractor, u_2 is the steering rate of the tractor's front axle, and u_3 is the steering rate of the trailer's axle.

The process of deriving Kinematic Equation 1.40 can be applied to other tractor-trailer systems. For instance, consider the n-trailer system studied in [237] and shown in Fig. 1.12. The system consists of a differential-drive tractor (also referred to as "trailer 0") pulling a chain of n trailers each of which is hinged to the midpoint of the preceding trailer's wheel axle. It can be shown analogously that kinematic model of the n-trailer system is given by

$$
\begin{aligned}
\dot{x} &= v_n \cos\theta_n \\
\dot{y} &= v_n \sin\theta_n \\
\dot{\theta}_n &= \frac{1}{d_n}v_{n-1}\sin(\theta_{n-1} - \theta_n) \\
&\ \vdots \\
\dot{\theta}_i &= \frac{1}{d_i}v_{i-1}\sin(\theta_{i-1} - \theta_i), \quad i = n-1,\cdots,2 \\
&\ \vdots \\
\dot{\theta}_1 &= \frac{1}{d_1}u_1\sin(\theta_0 - \theta_1) \\
\dot{\theta}_0 &= u_2,
\end{aligned}
$$

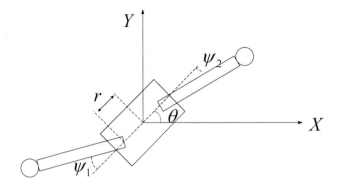

Fig. 1.13. A planar space robot

where (x, y) is the guidepoint located at the axle midpoint of the last trailer, θ_j is the orientation angle of the jth trailer (for $j = 0, \cdots, n$), d_j is the distance between the midpoints of the jth and $(j-1)$th trailers' axles, v_i is the tangential velocity of trailer i as defined by

$$v_i = \prod_{k=1}^{i} \cos(\theta_{k-1} - \theta_k) u_1, \quad i = 1, \cdots, n,$$

u_1 is the tangential velocity of the tractor, and u_2 is the steering angular velocity of the tractor.

1.3.4 A Planar Space Robot

Consider planar and frictionless motion of a space robot which, as shown in Fig. 1.13, consists of a main body and two small point-mass objects at the ends of the rigid and weightless revolute arms. If the center location of the main body is denoted by (x, y), then the positions of the small objects are at (x_1, y_1) and (x_2, y_2) respectively, where

$$x_1 = x - r\cos\theta - l\cos(\theta - \psi_1), \quad y_1 = y - r\sin\theta - l\sin(\theta - \psi_1),$$
$$x_2 = x + r\cos\theta + l\cos(\theta - \psi_2), \quad y_2 = y + r\sin\theta + l\sin(\theta - \psi_2),$$

θ is the orientation of the main body, ψ_1 and ψ_2 are the angles of the arms with respect to the main body, and l is the length of the arms, and r is the distance from the revolute joints to the center of the main body.

It follows that the Lagrangian (*i.e.*, kinematic energy) of the system is given by

$$L = \frac{1}{2}m_0(\dot{x}^2 + \dot{y}^2) + \frac{1}{2}J\dot{\theta}^2 + \frac{1}{2}m(\dot{x}_1^2 + \dot{y}_1^2) + \frac{1}{2}m(\dot{x}_2^2 + \dot{y}_2^2)$$
$$= \frac{1}{2}[\dot{x}\ \dot{y}\ \dot{\psi}_1\ \dot{\psi}_2\ \dot{\theta}]M[\dot{x}\ \dot{y}\ \dot{\psi}_1\ \dot{\psi}_2\ \dot{\theta}]^T,$$

where m_0 is the mass of the main body, J is the inertia of the main body, m is the mass of the two small objects, and inertia matrix $M = [M_{ij}] \in \Re^{5 \times 5}$ is symmetric and has the following entries:

$$M_{11} = M_{22} = m_0 + 2m, \quad M_{12} = 0,$$
$$M_{13} = -ml \sin(\theta - \psi_1), \quad M_{14} = ml \sin(\theta - \psi_2), \quad M_{15} = 0,$$
$$M_{23} = ml \cos(\theta - \psi_1), \quad M_{24} = -ml \cos(\theta - \psi_2), \quad M_{25} = 0,$$
$$M_{33} = ml^2, \quad M_{34} = 0, \quad M_{35} = -ml^2 - mlr \cos \psi_1,$$
$$M_{44} = ml^2, \quad M_{45} = -ml^2 - mlr \cos \psi_2,$$
$$M_{55} = J + 2mr^2 + 2ml^2 + 2mlr \cos \psi_1 + 2mlr \cos \psi_2.$$

Then, the dynamic equation of the space robot is given by (1.16).

If the robot is free floating in the sense that it is not subject to any external force/torque and that $x(t_0) = y(t_0) = 0$, $x(t) = y(t) = 0$ for all $t > t_0$, and hence the Lagrangian becomes independent of θ. It follows from (1.16) that

$$\frac{\partial L}{\partial \dot{\theta}} = M_{35} \dot{\psi}_1 + M_{45} \dot{\psi}_2 + M_{55} \dot{\theta}$$

must be a constant representing conservation of the angular momentum. If the robot has zero initial angular momentum, its conservation is in essence described by the non-holonomic constraint that

$$\dot{q} = \begin{bmatrix} 1 \\ 0 \\ -\dfrac{M_{35}}{M_{55}} \end{bmatrix} u_1 + \begin{bmatrix} 0 \\ 1 \\ -\dfrac{M_{45}}{M_{55}} \end{bmatrix} u_2,$$

where $q = [\psi_1, \psi_2, \theta]^T$, $u_1 = \dot{\psi}_1$, and $u_2 = \dot{\psi}_2$.

For orbital and attitude maneuvers of satellites and space vehicles, momentum precession and adjustment must be taken into account. Details on underlying principles of spacecraft dynamics and control can be found in [107, 272].

1.3.5 Newton's Model of Rigid-body Motion

In general, equations of 3-D motion for a rigid-body can be derived using Newton's principle as follows. First, establish the fixed frame $\{X_f, Y_f, Z_f\}$ (or earth inertia frame), where Z_f is the vertical axis, and axes $\{X_f, Y_f, Z_f\}$ are perpendicular and satisfy the right-hand rule. The coordinates in the fixed frame are position components of $\{x_f, y_f, z_f\}$ and Euler angles of $\{\theta, \phi, \psi\}$, and their rates are defined by $\{\dot{x}_f, \dot{y}_f, \dot{z}_f\}$ and $\{\dot{\theta}, \dot{\phi}, \dot{\psi}\}$, respectively. Second, establish an appropriate body frame $\{X_b, Y_b, Z_b\}$ which satisfies the right-hand rule. The coordinates in the body frame are position components of $\{x_b, y_b, z_b\}$, while the linear velocities are denoted by $\{u, v, w\}$ (or equivalently $\{\dot{x}_b, \dot{y}_b, \dot{z}_b\}$) and angular velocities are denoted by $\{p, q, r\}$ (or $\{\omega_x, \omega_y, \omega_z\}$). There is no need to determine orientation angles in the body frame as they are specified by Euler angles (with respect to the fixed frame, by convention).

Euler angles of roll ϕ, pitch θ, and yaw ψ are defined by the roll-pitch-yaw convention, that is, orientation of the fixed frame is brought to that of the body frame through the following right-handed rotations in the given sequence: (i) rotate $\{X_f, Y_f, Z_f\}$ about the Z_f axis through yaw angle ψ to generate frame $\{X_1, Y_1, Z_1\}$; (ii) rotate $\{X_1, Y_1, Z_1\}$ about the Y_1 axis through pitch angle θ to generate frame $\{X_2, Y_2, Z_2\}$; (iii) rotate $\{X_2, Y_2, Z_2\}$ about the X_2 axis through roll angle ψ to body frame $\{X_b, Y_b, Z_b\}$. That is, the 3-D rotation matrix from the fixed frame to the body frame is

$$
R_b^f = \begin{bmatrix} 1 & 0 & 0 \\ 0 & \cos\phi & \sin\phi \\ 0 & -\sin\phi & \cos\phi \end{bmatrix} \times \begin{bmatrix} \cos\theta & 0 & -\sin\theta \\ 0 & 1 & 0 \\ \sin\theta & 0 & \cos\theta \end{bmatrix} \times \begin{bmatrix} \cos\psi & \sin\psi & 0 \\ -\sin\psi & \cos\psi & 0 \\ 0 & 0 & 1 \end{bmatrix}.
$$

Conversely, the 3-D rotation matrix from the body frame back to the body frame is defined by

$$
R = R_z(\psi)R_y(\theta)R_x(\phi),
$$

that is,

$$
R = \begin{bmatrix} \cos\theta\cos\psi & \sin\theta\sin\phi\cos\psi - \cos\phi\sin\psi & \sin\theta\cos\phi\cos\psi + \sin\phi\sin\psi \\ \cos\theta\sin\psi & \sin\theta\sin\phi\sin\psi + \cos\phi\cos\psi & \sin\theta\cos\phi\sin\psi - \sin\phi\cos\psi \\ -\sin\theta & \sin\phi\cos\theta & \cos\phi\cos\theta \end{bmatrix},
$$

$$(1.41)$$

where $R_z(\psi), R_y(\theta), R_x(\phi)$ are the elementary rotation matrices about z, y, x axes, respectively, and they are given by

$$
R_z(\psi) = \begin{bmatrix} \cos\psi & -\sin\psi & 0 \\ \sin\psi & \cos\psi & 0 \\ 0 & 0 & 1 \end{bmatrix}, \quad R_y(\theta) = \begin{bmatrix} \cos\theta & 0 & \sin\theta \\ 0 & 1 & 0 \\ -\sin\theta & 0 & \cos\theta \end{bmatrix},
$$

and

$$
R_x(\phi) = \begin{bmatrix} 1 & 0 & 0 \\ 0 & \cos\phi & -\sin\phi \\ 0 & \sin\phi & \cos\phi \end{bmatrix}.
$$

It follows that linear velocities in the body frame are transformed to those in the inertia frame according to

$$
\begin{bmatrix} \dot{x}_f & \dot{y}_f & \dot{z}_f \end{bmatrix}^T = R \begin{bmatrix} u & v & w \end{bmatrix}^T.
$$

$$(1.42)$$

On the other hand, it follows that $R^T R = I$ and hence

$$
\dot{R}^T R + R^T \dot{R} = 0.
$$

Defining matrix

$$
\hat{\omega}(t) \triangleq R^T \dot{R},
$$

$$(1.43)$$

we know that $\hat{\omega}(t)$ is skew-symmetric and that, by direction computation

$$\hat{\omega}(t) = \begin{bmatrix} 0 & -r & q \\ r & 0 & -p \\ -q & p & 0 \end{bmatrix}. \tag{1.44}$$

Pre-multiplying both sides of (1.43) by R yields

$$\dot{R} = R\hat{\omega}(t) \tag{1.45}$$

It is straightforward to verify that (1.45) is equivalent to the following transformation from angular velocities in the body frame to those in the inertia frame:

$$\begin{bmatrix} \dot{\phi} \\ \dot{\theta} \\ \dot{\psi} \end{bmatrix} = \begin{bmatrix} 1 & \sin\phi\tan\theta & \cos\phi\tan\theta \\ 0 & \cos\phi & -\sin\phi \\ 0 & \sin\phi\sec\theta & \cos\phi\sec\theta \end{bmatrix} \begin{bmatrix} p \\ q \\ r \end{bmatrix}. \tag{1.46}$$

Equations 1.42 and 1.46 constitute the kinematic model of rigid-body motion.

Note that, given any rotation matrix R in (1.41), Euler angles can always be solved, but not globally. For instance, at $\theta = -\pi/2$, matrix R reduces to

$$R = \begin{bmatrix} 0 & -\sin(\phi+\psi) & -\cos(\phi+\psi) \\ 0 & \cos(\phi+\psi) & -\sin(\phi+\psi) \\ 1 & 0 & 0 \end{bmatrix},$$

and hence there are an infinite number of solutions ϕ and ψ for such a matrix R. The lack of existence of global and smooth solutions to the inverse problem of determining the Euler angles from the rotation is the singularity issue of the rotation space. The singularity can be overcome by using four-parameter quaternions which generalize complex numbers and provide a global parameterization of the rotation space.

In the derivations of Lagrange-Euler equation in Section 1.2.3, it is assumed implicitly that the configuration space (*i.e.*, both position and rotation spaces) can be parameterized by generalized coordinate $q \in \Re^n$. It is due to the singularity issue of Euler angles that Lagrange-Euler Equation 1.16 cannot be used to determine global dynamics of 3-D rigid-body motion. In what follows, a global characterization of the dynamics of one rigid-body subject to external forces and torques is derived using the *Newton-Euler* method. The Newton-Euler method can also be applied to a multi-body mechanical system, but it needs to be executed recursively by accounting for all interactive forces/torques within the system [80].

Given angular velocity vector $\omega = \begin{bmatrix} p & q & r \end{bmatrix}^T$ in the body frame, the angular velocity vector ω_f in the fixed frame is given by $\omega_f = R\omega$, and the angular momentum in the fixed frame is given by $J_f(t)\omega_f$, where J is the inertia of the rigid-body, and $J_f(t) = RJR^T$ is the instantaneous inertia with respect to the fixed frame. It follows from Newton's law that

$$\frac{d}{dt}[J_f(t)\omega_f] = \tau_f,$$

where $\tau \in \Re^3$ is the external torque vector with respect to the body frame, and $\tau_f = R\tau$ is external torque vector with respect to the fixed frame. Invoking the property of $RR^T = R^T R = I$, differentiating the left hand side of the above equation, and utilizing (1.45) yield

$$\tau_f = \frac{d}{dt}[RJ\omega]$$
$$= RJ\dot{\omega} + \dot{R}J\omega$$
$$= RJ\dot{\omega} + R\hat{\omega}J\omega,$$

or equivalently

$$J\dot{\omega} + \hat{\omega}J\omega = \tau. \tag{1.47}$$

On the other hand, given linear velocity vector $V_b = \begin{bmatrix} u & v & w \end{bmatrix}^T$ in the body frame, the linear velocity V_f in the fixed frame is given by $V_f = RV_b$, and the linear momentum in the fixed frame is given by mV_f. It follows from Newton's law that

$$\frac{d}{dt}[mV_f] = F_f,$$

where $F \in \Re^3$ is the external force vector with respect to the body frame, and $F_f = RF$ is external force vector with respect to the fixed frame. Differentiating the left hand side of the above equation and substituting (1.45) into the expression yield

$$m\dot{V}_b + \hat{\omega}mV_b = F. \tag{1.48}$$

Equations 1.47 and 1.48 constitute the dynamic model of 3-D rigid-body motion and are called *Newton-Euler equations*.

In the subsequent subsections, Kinematic Equations 1.42 and 1.46 and Dynamic Equations 1.47 and 1.48 are used to model several types of 3-D or 2-D vehicles.

1.3.6 Underwater Vehicle and Surface Vessel

In addition to rigid-body motion equations of (1.47) and (1.48), modeling of underwater vehicles and surface vessels may involve their operational modes and wave climate modeling. In what follows, two simplified modes of operation are considered, and the corresponding models are presented.

Cruising of an Underwater Vehicle

Consider an underwater vehicle cruising at a constant linear speed. Without loss of any generality, choose the body frame $\{x_b, \ y_b, \ z_b\}$ such that $V_b = [v_x \ 0 \ 0]^T \in \Re^3$ is the velocity in the body frame, and let $[x \ y \ z]^T \in \Re^3$ be the

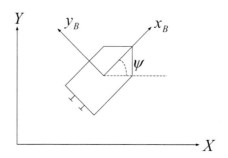

Fig. 1.14. A surface vessel

position of the center of mass of the vehicle in the inertia frame. It follows from (1.42) and (1.46) that kinematic equations of the vehicle reduce to [57, 171]

$$
\begin{bmatrix} \dot{x} \\ \dot{y} \\ \dot{z} \\ \dot{\phi} \\ \dot{\theta} \\ \dot{\psi} \end{bmatrix} =
\begin{bmatrix}
\cos\theta\cos\psi & 0 & 0 & 0 \\
\sin\psi\cos\theta & 0 & 0 & 0 \\
-\sin\theta & 0 & 0 & 0 \\
0 & 1 & \sin\phi\tan\theta & \cos\phi\tan\theta \\
0 & 0 & \cos\phi & -\sin\phi \\
0 & 0 & \sin\phi\sec\theta & \cos\phi\sec\theta
\end{bmatrix}
\begin{bmatrix} v_x \\ \omega_x \\ \omega_y \\ \omega_z \end{bmatrix}
\triangleq G(q)
\begin{bmatrix} v_x \\ \omega_x \\ \omega_y \\ \omega_z \end{bmatrix},
$$

where $\{\theta, \phi, \psi\}$ are the Euler angles, and $\{\omega_x, \omega_y, \omega_z\}$ are the angular velocities along x, y and z axes in the body frame, respectively. The above model is non-holonomic due to zero linear velocity along the y and z axes in the body frame. It follows from the expression of $G(q)$ that the two non-holonomic constraints can be expressed as $A(q)\dot{q} = 0$, where matrix $A(q)$ is given by

$$
\begin{bmatrix}
\cos\psi\sin\theta\sin\phi - \sin\psi\cos\phi & \sin\psi\sin\theta\sin\phi + \cos\psi\cos\phi & \cos\theta\sin\phi & 0 & 0 & 0 \\
\sin\psi\sin\theta\cos\phi - \sin\psi\sin\phi & \sin\psi\sin\theta\cos\phi - \cos\psi\sin\phi & \cos\theta\cos\phi & 0 & 0 & 0
\end{bmatrix}.
$$

A Surface Vessel

Consider the planar motion of an underactuated surface vessel shown in Fig. 1.14. It is controlled by two independent propellers which generate the force in surge and the torque in yaw. It follows from the relationship between the body and inertia frames that vessel's kinematic model is [215, 271]:

$$
\begin{bmatrix} \dot{x} \\ \dot{y} \\ \dot{\psi} \end{bmatrix} =
\begin{bmatrix}
\cos\psi & -\sin\psi & 0 \\
\sin\psi & \cos\psi & 0 \\
0 & 0 & 1
\end{bmatrix}
\begin{bmatrix} v_x \\ v_y \\ \omega_z \end{bmatrix},
\tag{1.49}
$$

where (x, y) denote the position of the center of mass in the inertia frame $(X - Y)$, ψ is the yaw orientation, and (v_x, v_y) and ω_z are the linear velocities and angular velocity in the body frame. By ignoring wind and wave forces, one can consolidate (1.47) and (1.48) and obtain the following simplified dynamic model [68, 215]:

$$
\begin{cases}
\dot{v}_x = \dfrac{m_{22}}{m_{11}} v_y \omega_z - \dfrac{d_{11}}{m_{11}} v_x + \dfrac{1}{m_{11}} \tau_1, \\[2mm]
\dot{v}_y = -\dfrac{m_{11}}{m_{22}} v_x \omega_z - \dfrac{d_{22}}{m_{22}} v_y, \\[2mm]
\dot{\omega}_z = \dfrac{m_{11} - m_{22}}{m_{33}} v_x v_y - \dfrac{d_{33}}{m_{33}} \omega_z + \dfrac{1}{m_{33}} \tau_2,
\end{cases}
\tag{1.50}
$$

where positive constants m_{ii} are determined by vessel's inertia and its added mass effects, positive constants d_{ii} are due to hydrodynamic damping, and τ_1 and τ_2 are the external force and torque generated by the two propellers. Model 1.50 is non-holonomic due to the non-integrable constraint of

$$
m_{22} \dot{v}_y = -m_{11} v_x \omega_z - d_{22} v_y.
$$

In the inertia frame, the above constraint can be expressed as

$$
m_{22}(\ddot{x} \sin \psi - \ddot{y} \cos \psi) + (m_{22} - m_{11})\dot{\psi}(\dot{x} \cos \psi + \dot{y} \sin \psi) + d_{22}(\dot{x} \sin \psi - \dot{y} \cos \psi) = 0.
$$

1.3.7 Aerial Vehicles

Depending on the type and operational mode of an aerial vehicle, its dynamic model can be modeled by properly consolidating or elaborating Kinematic Equations 1.42 and 1.46 and Dynamic Equations 1.47 and 1.48.

Fixed-wing Aircraft

In addition to Dynamic Equations 1.47 and 1.48, dynamic modeling of a fixed-wing aircraft typically involves establishment of the so-called flow/stability frame (whose coordinates are angle of attack α and angle of sideslip β). In the stability frame, aerodynamic coefficients can be calculated, and aerodynamic forces (lift, drag and side forces) and aerodynamic moments (in pitch, roll and yaw) can be determined with respect to the body frame. Should aerodynamic coefficients depend upon $\dot{\alpha}$ or $\dot{\beta}$ or both, the resulting equations obtained by substituting the expressions of aerodynamic and thrust forces/torques into (1.47) and (1.48) would have to be rearranged as accelerations would appear in both sides of the equations. This process of detailed modeling is quite involved, and readers are referred to [174, 245, 265].

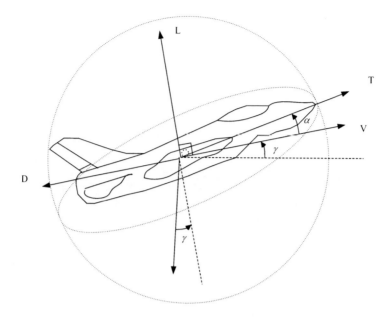

Fig. 1.15. An unmanned aerial vehicle

For the fixed-wing aerial vehicle shown in Fig. 1.15, its simplified dynamic model is given by [265]

$$\begin{cases}
\dot{x} = V \cos \gamma \cos \phi, \\
\dot{y} = V \cos \gamma \sin \phi, \\
\dot{h} = V \sin \gamma, \\
\dot{V} = \dfrac{T \cos \alpha - D}{m} - g \sin \gamma, \\
\dot{\gamma} = \dfrac{T \sin \alpha + L}{mV} \cos \delta - \dfrac{g}{V} \cos \gamma, \\
\dot{\phi} = \dfrac{T \sin \alpha + L}{mV} \dfrac{\sin \delta}{\cos \gamma},
\end{cases} \tag{1.51}$$

where x is the down-range displacement, y is the cross-range displacement, h is the altitude, V is the ground speed and is assumed to be equal to the airspeed, T is the aircraft engine thrust, α is the angle of attack, D is the aerodynamic drag, m is the aircraft mass, g is the gravity constant, γ is the flight path angle, L is the lift force, δ is the bank angle, and ϕ is the heading angle. In the case that altitude variations are not considered, Model 1.51 reduces to the following planar aerial vehicle model:

$$\begin{cases}
\dot{x} = u_1 \cos \phi \\
\dot{y} = u_1 \sin \phi \\
\dot{\phi} = u_2,
\end{cases}$$

where
$$u_1 \overset{\triangle}{=} V \cos \gamma, \quad \text{and} \quad u_2 \overset{\triangle}{=} \frac{(T \sin \alpha + L) \sin \delta}{mV \cos \gamma}.$$

Note that the above simplified planar model is identical to Model 1.31 of a differential-drive ground vehicle.

By using tilt-rotors or by utilizing directed jet thrust, fixed-wing aircraft can lift off vertically. These aircraft are called vertical take-off and landing (VTOL) aircraft. If only the planar motion is considered, the dynamic equations can be simplified to [223]

$$\begin{cases} \ddot{x} = -u_1 \sin \theta + \varepsilon u_2 \cos \theta \\ \ddot{y} = u_1 \cos \theta + \varepsilon u_2 \sin \theta - g \\ \ddot{\theta} = u_2, \end{cases}$$

where (x, y, θ) denote the position and orientation of the center of mass of the aircraft, g is the gravitational constant, control inputs u_1 and u_2 are the thrust (directed downwards) and rolling moment of the jets, respectively, and $\varepsilon > 0$ is the small coefficient representing the coupling between the rolling moment and lateral acceleration of the aircraft. In this case, the VTOL aircraft is an underactuated system with three degrees of freedom but two control inputs.

Helicopter

Parallel to the derivations of Dynamic Equations 1.47 and 1.48, one can show in a straightforward manner as done in [223] that the *helicopter* model is given by

$$\begin{cases} \dot{p}_f = v_f, \\ m\dot{v}_f = RF, \\ J\dot{\omega} = -\hat{\omega}J\omega + \tau, \end{cases}$$

where $p_f \in \Re^3$ and $v_f \in \Re^3$ denote the position and velocity of the center of mass of the helicopter in the inertia frame, respectively; $\omega \in \Re^3$, $F \in \Re^3$, and $\tau \in \Re^3$ are the angular velocity, the force and the torque in the body frame, respectively; m and J are the mass and inertia in the body frame, and R and $\hat{\omega}$ are defined in (1.41) and (1.44), respectively.

Autonomous Munitions

Smart munitions range from ballistic missiles to self-guided bullets. For these munitions, basic equations of motion remain in the form of (1.47) and (1.48). Details on missile modeling, guidance and control can be found in [230].

1.3.8 Other Models

Operation of a vehicle system may involve other entities in addition to physical vehicles themselves. Those entities, real or virtual, can be described by other models such as the following l-integrator model:

$$\dot{z}_j = z_{j+1}, \ j = 1, \cdots, l-1; \quad \dot{z}_l = v, \quad \psi = z_1, \tag{1.52}$$

where $z_j \in \Re^m$ are the state sub-vectors, $\psi \in \Re^m$ is the output, and $v \in \Re^m$ is the input. For example, if $l = 1$, Model 1.52 reduces to the single integrator model of $\dot{\psi} = v$ which has been used to characterize dynamics of generic agents. For $l = 2$, Model 1.52 becomes the double integrator model which is often used as the model of life forms.

1.4 Control of Heterogeneous Vehicles

Consider a team of q heterogeneous vehicles. Without loss of any generality, the model of the ith vehicle can be expressed as

$$\dot{\phi}_i = \Phi_i(\phi_i, v_i), \quad \psi_i = h_i(\phi_i), \tag{1.53}$$

where $i \in \{1, \cdots, q\}$, $\phi_i(t) \in \Re^{n_i}$ is the state, and $\psi_i(t) \in \Re^m$ is the output in the configuration space, and $v_i(t) \in \Re^m$ is the control input. As shown in Section 1.3, Model 1.53 often contains non-holonomic constraints. The team could interact with remote and hands-off operator(s) represented by virtual vehicle(s). In order to make the vehicles be cooperative, pliable and robust, control must be designed to achieve cooperative behavior(s) while meeting a set of constraints. Figure 1.16 shows a hierarchical control structure for an autonomous vehicle system.

While the hierarchy in Fig. 1.16 can have many variations, the key is its multi-level control and autonomy:

Vehicle-level autonomy: navigation and control algorithms are designed and implemented (as the bottom level in Fig. 1.16) such that each of the vehicles is pliable to its constraints, is robust to environmental changes, and is also capable of best following any given command signal.

Team-level autonomy: cooperative control algorithm is designed and implemented (as the second level in Fig. 1.16) to account for an intermittent sensing and communication network and to achieve cooperative behavior(s). Through the network and by the means of a virtual vehicle, an operator can adjust the status of cooperative behavior of the vehicles.

Mission-level autonomy and intelligence: high-level tactical decisions are enabled through human-machine interaction and automated by a multi-objective decision-making model with online learning capabilities.

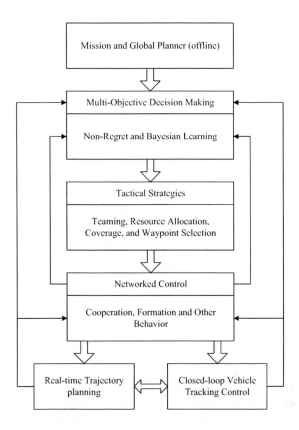

Fig. 1.16. Hierarchical control structure for autonomous vehicle system

To achieve motion autonomy for each vehicle, the navigation and control problems need to be solved. The navigation problem is to find a desired trajectory ψ_i^d and its open-loop steering control v_i^d. Typically, the trajectory ψ_i^d should satisfy the boundary conditions and motion constraints (such as non-holonomic constraints in the model as well as geometrical constraints imposed by the environment), and it needs to be updated. To find ψ_i^d and v_i^d in real-time, an online trajectory planning algorithm (such as the one in Section 3.2) can be used. Then, the control problem is to find a closed-loop feedback control of form

$$v_i = v_i(\psi_i^d, \phi_i, v_i^d) \tag{1.54}$$

such that the desired trajectory is followed asymptotically as $q_i \rightarrow q_i^d$. This will be pursued in Section 3.3.1. Alternatively, the navigation and control problems can be combined, in which case q_i^d is merely a desired (moving) target location and the control needs to be improved to be

$$v_i = v_i'(\psi_i^d, \phi_i, r_i), \tag{1.55}$$

where r_i is the reactive control capable of making the vehicle conform to the changes of the environment (whenever possible). Comparison of Controls 1.54 and 1.55 will be made in Section 3.4.2.

To achieve team autonomy, a cooperative control should be designed in terms of the available feedback information as

$$u_i = u_i(t, s_{i1}(t)\psi_1, \cdots, s_{iq}(t)\psi_q), \tag{1.56}$$

where $s_{ij}(t)$ are binary time functions, $s_{ii} \equiv 1$; $s_{ij}(t) = 1$ if $\psi_j(t)$ (or its equivalence) is known to the ith vehicle at time t, and $s_{ij} = 0$ otherwise. Since information flow over the network changes over time, Cooperative Control 1.56 must react accordingly but ensure group cooperative behavior(s), which is the major issue in analysis and synthesis. Besides observing the group behavior ψ_g, each vehicle can also exhibit an individual behavior ψ_i^d within the group. Then, cooperative control can be implemented on top of vehicle-level control by embedding u_i into v_i and by modifying (1.55) as

$$v_i = v_i''(\psi_i^d, \phi_i, r_i + u_i). \tag{1.57}$$

Control 1.57 is to ensure $\psi_i \rightarrow \psi_i^d + \psi_g$, which is the subject of Chapters 5 and 6. In contrast, it is often too difficult to integrate cooperative control into Feedback Control 1.54, since changing the desired trajectory induces large transient responses and since Control 1.54 is a purely tracking control and is not pliable or robust by itself.

Although both cooperative control and navigation control are implemented on the same vehicle, the hierarchy is physically necessary: the navigation control is a self-feedback control required to make any specific vehicle follow any motion command it received, and the cooperative control is needed to synthesize a motion command based on the information available from all the sensors/receivers on a vehicle. From the point view of analysis and design, the hierarchy is also necessary since, by separating the two, canonical forms can be developed so that heterogeneous vehicles can be handled systematically.

Finally, successful operation of autonomous vehicles requires that certain tactical decisions be made real-time and automatically by each of the autonomous vehicles. Often, a decision must be made according to a number of factors which may or may not agree with each other. In [92], a decision-making model based on no-regret is mathematically formulated to describe qualitatively the possible courses of action, to measure the outcomes, to quantify the success likelihood, and to enable the optimal multi-objective decision-making. Using the Bayesian formula, the decision-making model is capable of online learning through which the weightings are updated to improve decision-making over time. In addition to deriving tactical decisions, the decision-making model can also be used to choose adequate way-points for the trajectory planning algorithm, to determine fidelity of obstacle/velocity detection from noisy sensor feedbacks, to adjust the formation of a multi-vehicle team, *etc*. In these applications, the decision-making model increases the level of autonomy and improves the system performance. Nonetheless, decision-making models and high-level autonomy/learning are beyond the scope of this book and hence not discussed further.

1.5 Notes and Summary

Cooperative control of autonomous vehicles is the primary application considered in the book. Most vehicles and underactuated mechanical systems are non-holonomic. Holonomic systems are the systems subject to constraints that limit their possible configurations, and non-holonomic systems are those with non-integrable constraints on their velocities. The terminology of holonomic and non-holonomic systems can be traced back to [89], and the word holonomic (or holonomous) in Greek contains the words meaning integral. In-depth analysis of non-holonomic systems as well as other interesting examples (such as a unicycle with rotor, a skate on an inclined plane, a Chaplygin sleigh, a roller racer, a car towing several trailers, a rattleback, a satellite or spacecraft with momentum wheels, Toda lattice, and hopping robot) can be found in texts [23, 79, 145, 168, 169, 238]. In this chapter, analysis of holonomic and non-holonomic systems are reviewed, and common vehicles are modeled. In Chapter 3, a canonical form is developed for kinematic models of non-holonomic systems, and the kinematic control problem of a non-holonomic system is solved. By physical nature, the kinematic and dynamic models of a vehicle are a set of cascaded equations. It will be shown in Chapter 2 that, if the vehicle can be kinematically controlled, its dynamic control can easily designed using the backstepping design, a conceptually intuitive, mathematically simple and physically rooted technique.

Cooperative control of dynamical systems is the main subject of this book. Even for the simplest entities with no dynamics, analysis and synthesis of cooperative behaviors are quite involved because the interactions among the entities are time-varying and may not be known *a priori*. This is illustrated by the sample systems in Section 1.1.2. To investigate cooperative systems, useful results on analysis and design of control systems are reviewed first in Chapter 2. Later in Chapters 4, 5 and 6, these results are extended to the systems equipped with an intermittent sensing/communication network, and tools and techniques of analyzing and designing linear and non-linear cooperative systems are developed. Combined with the design in Chapter 3, cooperative systems can be synthesized and constructed for heterogeneous entities that may have high-order dynamics and may be non-linear and non-holonomic.

2

Preliminaries on Systems Theory

In this chapter, basic concepts and analysis tools in systems theory are summarized. We begin with matrix algebra and matrix norms, the standard in-
struments for qualitatively and quantitatively analyzing linear time-invariant
systems and their properties. While analysis of time-varying and non-linear
systems often requires advanced tools and particular results, the simple mathematical concept of a contraction mapping can be used to describe such qualitative characteristics as stability and convergence. Accordingly, the contraction mapping theorem is introduced, and it, together with the Barbalat lemma
and the comparison theorem, will facilitate the development of analysis tools
for cooperative control.

Lyapunov direct method is the universal approach for analyzing general
dynamical systems and their stability. Search for a successful Lyapunov function is the key, and relevant results on linear systems, non-linear systems, and
switching systems are outlined. Standard results on controllability as well as
control design methodologies are also reviewed.

2.1 Matrix Algebra

Let $\Re$ and $\mathcal{C}$ be the set of real and complex numbers, respectively. Then, $\Re^n$
represents the set of $n-$tuples for which all components belong to $\Re$, and $\Re^{n \times m}$
is the set of n-by-m matrices over $\Re$. Let $\Re_+ \triangleq [0, +\infty)$ and $\Re_{+/0} \triangleq (0, +\infty)$,
and let $\Re_+^n$ denote the set of all n-tuples whose components belong to $\Re_+$.

Definition 2.1. *For any matrix $A \in \Re^{n \times n}$, its* eigenvalues *λ_i are the roots of
the characteristic equation of $|\lambda I - A| = 0$, its* eigenvectors *are the non-zero
vectors s_i satisfying the equation $As_i = \lambda_i s_i$. The set of all its eigenvalues
$\lambda \in \mathcal{C}$ is called* spectrum *of A and is denoted by $\sigma(A)$, and its* spectral radius
is defined by $\rho(A) \triangleq \max\{|\lambda| : \lambda \in \sigma(A)\}$.

Set $\mathcal{X}$ is said to be a *linear vector space* if $x, y, z \in \mathcal{X}$ implies $x+y = y+x \in \mathcal{X}$, $x+(y+z) = (x+y)+z \in \mathcal{X}$, $-x \in \mathcal{X}$ with $x+(-x) = 0 \in \mathcal{X}$, $\alpha x \in \mathcal{X}$, and $\alpha(x+y) = \alpha x + \alpha y$, where $\alpha \in \Re$ is arbitrary. Clearly, $\Re^{n \times m}$ is a linear vector space. Operations of addition, subtraction, scalar multiplication, and matrix multiplication are standard in matrix algebra. The so-called *Kronecker product* is defined as $D \otimes E = [d_{ij}E]$. Among the properties of Kronecker product are: if matrix dimensions are compatible,

$$A \otimes (B+C) = A \otimes B + A \otimes C, \quad (A \otimes B)^T = (A^T \otimes B^T),$$

$$(A \otimes B)^{-1} = (A^{-1} \otimes B^{-1}), \quad \text{and} \quad (A \otimes B)(C \otimes D) = ((AC) \otimes (BD)).$$

Definition 2.2. *A set of vectors $v_i \in \mathcal{X}$ for $i = 1, \cdots, l$ is said to be* linearly independent *if the equation*

$$\alpha_1 v_1 + \cdots + \alpha_l v_l = 0$$

has only the trivial solution $\alpha_1 = \cdots = \alpha_l = 0$. If there is a non-trivial solution α_i to the above equation, say $\alpha_k \neq 0$, the set of vector is said to be linear dependent, in which case one of the vectors can be expressed as a linear combination of the rest, i.e.,

$$v_k = \sum_{j=1, j \neq k}^{n} -\frac{\alpha_j}{\alpha_k} v_j.$$

Set $\mathcal{X} \subset \Re^n$ is said to be of rank p *if $\mathcal{X}$ is composed of exactly p linearly independent vectors.*

Matrix $S \in \Re^{n \times n}$ is invertible and its inverse S^{-1} exists as $SS^{-1} = S^{-1}S = I$ if and only if its rows (columns) are linearly independent. For matrix $A \in \Re^{n \times n}$, there are always exactly n eigenvalues, either complex or real, but there may not be n linearly independent eigenvectors. A square matrix whose number of linearly independent eigenvectors is less than its order is said to be *defective*. A sufficient condition under which there are n linearly independent eigenvectors is that there are n distinct eigenvalues. Matrix product $S^{-1}AS$ is called a *similarity transformation* on matrix A, and eigenvalues are invariant under such a transformation. Matrix A is said to be *diagonalizable* as $S^{-1}AS = \Lambda$ if and only if matrix A has n linearly independent eigenvectors s_i, where $S = [s_1 \ s_2 \ \cdots \ s_n]$ and $\Lambda = \text{diag}\{\lambda_1, \cdots, \lambda_n\}$. Transformation matrix S is said to be a *unitary matrix* if $S^{-1} = S^T$. It is known that any symmetric matrix A can be diagonalized under a similarity transformation of unitary matrix S. A very special unitary matrix is the *permutation matrix* which is obtained by permuting the rows of an $n \times n$ identity matrix according to some new ordering of the numbers 1 to n. Therefore, for any permutation matrix P, every row and column contains precisely a single 1 with 0s everywhere else, its determinant is always ± 1, and $P^{-1} = P^T$ as a permutation is reversed by permutating according to the new (*i.e.* transposed) order.

Diagonalization fails only if (but not necessarily if) there are repeated eigenvalues. For an nth-order matrix whose eigenvalue λ_i repeats k times, the *null space* of matrix $(A - \lambda_i I)$ determines linearly independent eigenvectors s_i associated with λ_i. Specifically, the linearly independent eigenvectors of matrix $(A - \lambda_i I)$ span its null space, and their number is the dimension of the null space and can assume any integer value between 1 and k. Obviously, defectiveness of matrix A occurs when the dimension of the null space is less than k. If matrix $A \in \Re^{n \times n}$ is defective, its set of eigenvectors is incomplete in the sense that there is no similarity transformation (or an invertible eigenvector matrix) that diagonalizes A. Hence, the best one can do is to choose a similarity transformation matrix S such that $S^{-1}AS$ is as nearly diagonal as possible. A standardized matrix defined to be the closest to diagonal is the so-called *Jordan canonical form*. In (2.1) below, order n_i is determined by $(A - \lambda_i I)^{n_i} s_i' = 0$ but $(A - \lambda_i I)^{n_i - 1} s_i' \neq 0$. The corresponding transformation matrix S consists of the eigenvector(s) $s_i = (A - \lambda_i I)^{n_i - 1} s_i'$ as well as generalized eigenvectors s_i' up to $(A - \lambda_i I)^{n_i - 2} s_i'$.

Definition 2.3. *Matrix J is said to be in the* Jordan canonical form *if it is block diagonal where the diagonal blocks, the so-called Jordan blocks J_i, contain identical eigenvalues on the diagonal, 1 on the super-diagonal, and zero everywhere else. That is,*

$$
J = \begin{bmatrix} J_1 & & \\ & \ddots & \\ & & J_l \end{bmatrix}, \quad and \quad J_i = \begin{bmatrix} \lambda_i & 1 & & \\ & \ddots & \ddots & \\ & & \ddots & 1 \\ & & & \lambda_i \end{bmatrix} \in \Re^{n_i \times n_i}, \tag{2.1}
$$

where λ_i may be identical to λ_j for some $i \neq j$ (that is, repeated eigenvalue may appear in different Jordan blocks, but eigenvalues are all the same in any given Jordan block), and $n_1 + \cdots + n_l = n$. The number of Jordan blocks with the same eigenvalue λ_i is called algebraical multiplicity *of λ_i. The order n_i of Jordan block J_i is called* geometrical multiplicity.

A generalization of distance or length in linear vector space $\mathcal{X}$, called *norm* and denoted by $\| \cdot \|$, can be defined as follows: for any $x, y \in \mathcal{X}$ and $\alpha \in \Re$, $\|x\| \geq 0$, $\|x\| = 0$ if and only if $x = 0$, $\|\alpha x\| = |\alpha| \cdot \|x\|$, and $\|x + y\| \leq \|x\| + \|y\|$. That is, a norm is a positive definite function (see Section 2.3.1) that satisfies the triangular inequality and is linear with respect to a real, positive constant multiplier. In $\Re^n$, the vector p-norm ($p \geq 1$) and ∞-norm are defined by

$$
\|x\|_p = \left[\sum_{i=1}^{n} |x_i|^p \right]^{\frac{1}{p}} \quad and \quad \|x\|_\infty = \max_{1 \leq i \leq n} |x_i|,
$$

respectively. The most commonly used are 1-norm ($p = 1$), Euclidean norm ($p = 2$) or 2-norm, and ∞-norm. All the vector norms are compatible with themselves as, for all $x, y \in \Re^n$,

$$\|x^T y\|_p \leq \|x\|_p \|y\|_p,$$

and they are also equivalent as, for any $p, q > 1$, there exist constants c_1 and c_2 such that

$$c_1 \|x\|_p \leq \|x\|_q \leq c_2 \|x\|_p.$$

The p-norm can also be defined in an infinite dimensional linear vector space. For an infinite sequence of scalars $\{x_i : i \in \aleph\} = \{x_1, x_2, \cdots\}$, p-norm $(p \geq 1)$ and ∞-norm are defined by

$$\|x\|_p = \left[\sum_{i=1}^{\infty} |x_i|^p \right]^{\frac{1}{p}} \quad \text{and} \quad \|x\|_\infty = \sup_i |x_i|,$$

respectively. A sequence $\{x_i : i \in \aleph\}$ is said to belong to l_p-space if $\|x\|_p < \infty$ and to l_∞-space if $\|x\|_\infty < \infty$.

In linear vector space $\mathcal{X} \subset \Re^n$, matrix $A \in \Re^{n \times n}$ in linear algebraic equation of $y = Ax$ represents a linear transformation, and its *induced matrix norms* are defined as

$$\|A\|_q \overset{\triangle}{=} \max_{\|x\|_q \neq 0} \frac{\|y\|_q}{\|x\|_q} = \max_{\|x\|_q = 1} \|y\|_q = \max_{\|x\|_q = 1} \|Ax\|_q,$$

where $q \geq 1$ is a positive real number including infinity. It follows that $\rho(A) \leq \|A\|$ for any induced matrix norm. According to the above definition, explicit solutions can be found for several induced matrix norms, in particular,

$$\|A\|_1 = \max_{1 \leq j \leq n} \sum_{i=1}^{n} |a_{ij}|, \quad \|A\|_2 = \sqrt{\lambda_{\max}(A^T A)}, \quad \|A\|_\infty = \|A^T\|_1,$$

where a_{ij} denotes the (i, j)th element of A, and $\lambda_{\max}(E)$ represents the operation of finding the maximum eigenvalue of E.

If a vector or a matrix is time-varying, the aforementioned vector norm or induced matrix norm can be applied pointwise at any instant of time, and hence the resulting norm value is time-varying. In this case, a functional norm defined below can be applied on top of the pointwise norm and over time: if $f : \Re^+ \to \Re$ is a Lebesgue measurable time function, then p-norm $(1 \leq p < \infty)$ and ∞-norm of $f(\cdot)$ are

$$\|f\|_p \overset{\triangle}{=} \left(\int_{t_0}^{\infty} |f(t)|^p dt \right)^{\frac{1}{p}}, \quad \text{and} \quad \|f\|_\infty \overset{\triangle}{=} \sup_{t_0 \leq t \leq \infty} |f(t)|,$$

respectively. Function $f(t)$ is said to belong to L_p-*space* if $\|f(t)\|_p < \infty$ and to L_∞-*space* if $\|f(t)\|_\infty < \infty$. Many results on stability analysis and in this chapter can be interpreted using the L_p-space or L_∞-space. For instance, any continuous and unbounded signal does not belong to either L_p-space or L_∞-space; and all uniformly bounded functions belong to L_∞-space. In fact, boundedness is always required to establish a stability result.

In a normed linear vector space, sequence $\{x_k : k \in \aleph\}$ is said to be a *Cauchy sequence* if, as $k, m \to \infty$, $\|x_k - x_m\| \to 0$. It is known and also easy to show that a convergent sequence must be a Cauchy sequence but not every Cauchy sequence is convergent. *Banach space* is a complete normed linear vector space in which every Cauchy sequence converges therein. Among the well known Banach spaces are $\Re^n$ with norm $\|\cdot\|_p$ with $1 \le p \le \infty$, and $C[a, b]$ (all continuous-time functions over interval $[a, b]$) with functional norm $\max_{t \in [a,b]} \|\cdot\|_p$.

2.2 Useful Theorems and Lemma

In this section, several mathematical theorems and lemma in systems and control theory are reviewed since they are simple yet very useful.

2.2.1 Contraction Mapping Theorem

The following theorem, the so-called *contraction mapping theorem*, is a special fixed-point theorem. It has the nice features that both existence and uniqueness of solution are ensured and that mapping T has its gain in norm less than 1 and hence is contractional. Its proof is also elementary as, by a successive application of mapping T, a convergent power series is obtained. Without $\lambda < 1$, Inequality 2.2 is referred to as the Lipschitz condition.

Theorem 2.4. *Let S be a closed sub-set of a Banach space X with norm $\|\cdot\|$ and let T be a mapping that maps S into S. If there exists constant $0 \le \lambda < 1$ such that, for all $x, y \in S$,*

$$\|T(x) - T(y)\| \le \lambda \|x - y\|, \tag{2.2}$$

then solution x^* to equation $x = T(x)$ exists and is a unique fixed-point in S.

Proof: Consider sequence $\{x_{k+1} = T(x_k), \ k \in \aleph\}$. It follows that

$$\|x_{k+l} - x_k\| \le \sum_{j=1}^{l} \|x_{k+j} - x_{k+j-1}\|$$

$$= \sum_{j=0}^{l-1} \|T(x_{k+j}) - T(x_{k+j-1})\|$$

$$\le \sum_{j=0}^{l-1} \lambda \|x_{k+j} - x_{k+j-1}\|$$

$$\le \sum_{j=0}^{l-1} \lambda^{k+j-1} \|x_2 - x_1\|$$

$$\le \frac{\lambda^{k-1}}{1 - \lambda} \|x_2 - x_1\|,$$

by which the sequence is a Cauchy sequence. By definition of $\mathcal{X}$, $x_k \to x^*$ for some x^* and hence x^* is a fixed-point since

$$\|T(x^*) - x^*\| \leq \|T(x^*) - T(x_k)\| + \|x_{k+1} - x^*\|$$
$$\leq \lambda\|x^* - x_k\| + \|x_{k+1} - x^*\| \to 0 \quad \text{as } k \to \infty.$$

Uniqueness is obvious from the fact that, if $x = T(x)$ and $y = T(y)$,

$$\|x - y\| = \|T(x) - T(y)\| < \lambda\|x - y\|,$$

which holds only if $x = y$ since $\lambda < 1$.   □

Computationally, fixed-point x^* is the limit of convergent sequence $\{x_{k+1} = T(x_k), \quad k \in \aleph\}$. Theorem 2.4 has important applications in control theory, starting with existence and uniqueness of a solution to differential equations [108, 192]. It will be shown in Section 2.3.2 that the Lyapunov direct method can be viewed as a special case of the contraction mapping theorem. Indeed, the concept of contraction mapping plays an important role in any convergence analysis.

2.2.2 Barbalat Lemma

The following lemma, known as the *Barbalat lemma* [141], is useful in convergence analysis.

Definition 2.5. *A function* $w : \Re_+ \to \Re$ *is said to be uniformly continuous if, for any* $\epsilon > 0$, *there exists* $\delta > 0$ *such that* $|t - s| < \delta$ *implies* $|w(t) - w(s)| < \epsilon$.

Lemma 2.6. *Let* $w : \Re_+ \to \Re$ *be a uniformly continuous function. Then, if* $\lim_{t \to \infty} \int_{t_0}^{t} w(\tau)d\tau$ *exists and is finite,* $\lim_{t \to \infty} w(t) = 0$.

Proof: To prove by contradiction, assume that either $w(t)$ does not have a limit or $\lim_{t \to \infty} w(t) = c \neq 0$. In the case that $\lim_{t \to \infty} w(t) = c \neq 0$, there exists some finite constant α_0 such that $\lim_{t \to \infty} \int_{t_0}^{t} w(\tau)d\tau = \alpha_0 + ct \nleq \infty$, and hence there is a contradiction. If $w(t)$ does not have a limit as $t \to \infty$, there exists an infinite time sub-sequence $\{t_i : i \in \aleph\}$ such that $\lim_{i \to \infty} t_i = +\infty$ and $|w(t_i)| \geq \epsilon_w > 0$. Since $w(t)$ is uniformly continuous, there exists interval $[t_i - \delta t_i, t_i + \delta t_i]$ within which $|w(t)| \geq 0.5|w(t_i)|$. Therefore, we have

$$\int_{t_0}^{t_i + \delta t_i} w(\tau)d\tau \geq \int_{t_0}^{t_i - \delta t_i} w(\tau)d\tau + w(t_i)\delta t_i, \quad \text{if } w(t_i) > 0,$$

and

$$\int_{t_0}^{t_i + \delta t_i} w(\tau)d\tau \leq \int_{t_0}^{t_i - \delta t_i} w(\tau)d\tau + w(t_i)\delta t_i, \quad \text{if } w(t_i) < 0.$$

Since $\{t_i\}$ is an infinite sequence and both $|w(t_i)|$ and δt_i are uniformly positive, we know by taking the limit of $t_i \to \infty$ on both sides of the above inequalities that $\int_{t_0}^t w(\tau)d\tau$ cannot have a finite limit, which contradicts the stated condition.   □

The following example illustrates the necessity of $w(t)$ being uniformly continuous for Lemma 2.6, and it also shows that imposing $w(t) \geq 0$ does not add anything. In addition, note that $\lim_{t \to \infty} w(t) = 0$ (e.g., $w(t) = 1/(1+t)$) may not imply $\lim_{t \to \infty} \int_{t_0}^t w(\tau)d\tau < \infty$ either.

Example 2.7. Consider scalar time function: for $n \in \aleph$,

$$w(t) = \begin{cases} 2^{n+1}(t - n) & \text{if } t \in [n, \ n + 2^{-n-1}] \\ 2^{n+1}(n + 2^{-n} - t) & \text{if } t \in [n + 2^{-n-1}, \ n + 2^{-n}] \ , \\ 0 & \text{everywhere else} \end{cases}$$

which is a triangle-wave sequence of constant height. It follows that $w(t)$ is continuous and that

$$\lim_{t \to \infty} \int_0^t w(\tau)d\tau = \frac{1}{2} + \frac{1}{2^2} + \frac{1}{2^3} + \cdots = 1 < \infty.$$

However, $w(t)$ does not have a limit as $t \to \infty$ as it approaches the sequence of discrete-time impulse functions in the limit.   ◇

If $w(t)$ is differentiable, $w(t)$ is uniformly continuous if $\dot{w}(t)$ is uniformly bounded. This together with Lemma 2.6 has been widely used adaptive control [232], non-linear analysis [108], and robustness analysis [192], and it will be used in Chapter 6 to analyze non-linear cooperative systems.

2.2.3 Comparison Theorem

The following theorem, known as the *comparison theorem*, can be used to find explicitly either a lower bound or an upper bound on a non-linear differential equation which itself may not have an analytical solution.

Theorem 2.8. *Consider the scalar differential equation*

$$\dot{r} = \beta(r, t), \quad r(t_0) = r_0 \tag{2.3}$$

where $\beta(r, t)$ is continuous in t and locally Lipschitz in r for all $t \geq t_0$ and $r \in \Omega \subset \Re$. Let $[t_0, T)$ be the maximal interval of existence of the solution $r(t)$ such that $r(t) \in \Omega$, where T could be infinity. Suppose that, for $t \in [t_0, T)$,

$$\dot{v} \leq \beta(v, t), \quad v(t_0) \leq r_0, \quad v(t) \in \Omega. \tag{2.4}$$

Then, $v(t) \leq r(t)$ for all $t \in [t_0, T)$.

Proof: To prove by contradiction, assume that $t_1 \geq t_0$ and $\delta t_1 > 0$ exist such that

$$v(t_1) = r(t_1), \quad \text{and} \quad r(t) < v(t) \; \forall t \in (t_1, t_1 + \delta t_1]. \tag{2.5}$$

It follows that, for $0 < h < \delta t_1$,

$$\frac{r(t+h) - r(t_1)}{h} < \frac{v(t+h) - v(t_1)}{h},$$

which in turn implies

$$\dot{r}(t_1) < \dot{v}(t_1).$$

Combined with (2.3) and (2.4), the above inequality together with the equality in (2.5) leads to the contradiction

$$\beta(t_1, r(t_1)) < \beta(t_1, v(t_1)).$$

Thus, either t_1 or $\delta t_1 > 0$ does not exist, and the proof is completed.   □

Extension of Theorem 2.8 to the multi-variable case is non-trivial and requires the so-called quasi-monotone property defined below. Should such a property hold, Theorem 2.10 can be applied. Nonetheless, the condition is usually too restrictive to be satisfied, as evident from the fact that, if $F(x,t) = Ax$ and is mixed monotone, all the entries in sub-matrices A_{11} and A_{22} must be non-negative while entries of A_{12} and A_{22} are all non-positive, where

$$A = \begin{bmatrix} A_{11} & A_{12} \\ A_{21} & A_{22} \end{bmatrix}, \quad \text{and} \quad A_{11} \in \Re^{k \times k}.$$

Definition 2.9. *Function $F(x,t) : \Re_+ \times \Re^n \to \Re^n$ is said to possess a* mixed quasi-monotone *property with respect to some fixed integer $k \in \{0, 1, \cdots, n\}$ if the following conditions hold:*

(a) For all $i \in \{1, \cdots, k\}$, function $F_i(x,t)$ is non-decreasing in x_j for $j = 1, \cdots, k$ and $j \neq i$, and it is non-increasing in x_l where $k < l \leq n$.
(b) For all $i \in \{k+1, \cdots, n\}$, function $F_i(x,t)$ is non-increasing in x_j for $j = 1, \cdots, k$, and it is non-decreasing x_l where $k < l \leq n$ and $l \neq i$.

In the special cases of $k = 0$ and $k = n$, function $F(x,t)$ is said to have quasi-monotone non-decreasing *and* quasi-monotone non-increasing *properties, respectively. Furthermore, function $F(x,t)$ is said to possess* mixed monotone *(or monotone non-decreasing or monotone non-increasing) property if $j \neq i$ and $l \neq i$ are not demanded in (a) and (b).*

Theorem 2.10. *Consider the following two sets of differential inequalities: for $v, w \in \Re^n$ and for some $k \in \{0, 1, \cdots, n\}$*

$$\begin{cases} \dot{v}_i \leq F_i(v, t), & i \in \{1, \cdots, k\} \\ \dot{v}_j > F_j(v, t), & j \in \{k+1, \cdots, n\} \end{cases} \quad \begin{cases} \dot{w}_i > F_i(w, t), & i \in \{1, \cdots, k\} \\ \dot{w}_j \leq F_j(w, t), & j \in \{k+1, \cdots, n\} \end{cases}.$$

$$(2.6)$$

or

$$\begin{cases} \dot{v}_i < F_i(v, t), & i \in \{1, \cdots, k\} \\ \dot{v}_j \geq F_j(v, t), & j \in \{k+1, \cdots, n\} \end{cases} \quad \begin{cases} \dot{w}_i \geq F_i(w, t), & i \in \{1, \cdots, k\} \\ \dot{w}_j < F_j(w, t), & j \in \{k+1, \cdots, n\} \end{cases}.$$

$$(2.7)$$

Suppose that $F(x, t)$ is continuous in t and locally Lipschitz in x for all $t \geq t_0$ and $x \in \Omega \subset \Re^n$, that initial conditions satisfy the inequalities of

$$\begin{cases} v_i(t_0) < w_i(t_0), & i \in \{1, \cdots, k\} \\ v_j(t_0) > w_j(t_0), & j \in \{k+1, \cdots, n\} \end{cases}, \quad (2.8)$$

that $F(x, t)$ has the mixed quasi-monotone property with respect to k, and that $[t_0, T)$ be the maximal interval of existence of solutions $v(t)$ and $w(t)$ such that $v(t), w(t) \in \Omega$, where T could be infinity. Then, over the interval of $t \in [t_0, T)$, inequalities $v_i(t) < w_i(t)$ and $v_j(t) > w_j(t)$ hold for all $i \in \{1, \cdots, k\}$ and for all $j \in \{k+1, \cdots, n\}$.

Proof: Define

$$\begin{cases} z_i(t) = v_i(t) - w_i(t), & i \in \{1, \cdots, k\} \\ z_j(t) = w_j(t) - v_j(t), & j \in \{k+1, \cdots, n\} \end{cases}.$$

It follows from (2.8) that $z_i(t_0) < 0$ for all i. To prove by contradiction, assume that $t_1 \geq t_0$ be the first time instant such that, for some l and for some $\delta t_1 > 0$,

$$z_l(t_1) = 0, \quad \text{and} \quad z_l(t) > 0 \ \forall t \in (t_1, t_1 + \delta t_1].$$

Hence, we know that, for all $\alpha \in \{1, \cdots, n\}$,

$$z_\alpha(t_1) \leq 0, \quad (2.9)$$

and that, for $0 < h < \delta t_1$,

$$\frac{z_l(t + h) - z_l(t_1)}{h} = \frac{z_l(t + h)}{h} > 0,$$

which in turn implies

$$\dot{z}_l(t_1) > 0.$$

Combined with (2.6) or (2.7), the above inequality becomes

$$\begin{cases} 0 < \dot{z}_l(t_1) < F_l(v(t_1), t) - F_l(w(t_1), t), & \text{if } l \leq k \\ 0 < \dot{z}_l(t_1) < F_l(w(t_1), t) - F_l(v(t_1), t), & \text{if } l > k \end{cases},$$

and, since $F(x, t)$ possesses the mixed quasi-monotone property with respect to k, the above set of inequality contradicts with (2.9) or simply

$$\begin{cases} v_i(t_1) \leq w_i(t_1), & i \in \{1, \cdots, k\} \\ v_j(t_1) \geq w_j(t_1), & j \in \{k+1, \cdots, n\} \end{cases}.$$

Hence, t_1 and $\delta t_1 > 0$ cannot co-exist, and the proof is completed. □

In Section 2.3.2, Theorem 2.8 is used to facilitate stability analysis of non-linear systems in terms of a Lyapunov function and to determine an explicit upper bound on state trajectory. In such applications as stability of large-scale interconnected systems, dynamic coupling among different sub-systems may have certain properties that render the vector inequalities of (2.6) and (2.7) in terms of a vector of Lyapunov functions v_k and, if so, Theorem 2.10 can be applied [118, 119, 156, 268]. Later in Section 6.2.1, another comparison theorem is developed for cooperative systems, and it is different from Theorem 2.10 because it does not require the mixed quasi-monotone property.

2.3 Lyapunov Stability Analysis

Dynamical systems can be described in general by the following vector differential equation:

$$\dot{x} = F'(x, u, t), \quad y = H(x, t), \tag{2.10}$$

where $x \in \Re^n$ is the state, $y \in \Re^l$ is the output, and $u \in \Re^m$ is the input. Functions $F'(\cdot)$ and $H(\cdot)$ provide detailed dynamics of the system. System 2.10 is said to be *affine* if its differential equation can be written as

$$\dot{x} = f(x) + g(x)u. \tag{2.11}$$

The control problem is to choose a feedback law $u = u(x, t)$ as the input such that the controlled system has desired the properties including stability, performance and robustness. Upon choosing the control, the closed-loop system corresponding to (2.10) becomes

$$\dot{x} = F'(x, u(x, t), t) \overset{\triangle}{=} F(x, t), \quad y = H(x, t). \tag{2.12}$$

The analysis problem is to study qualitative properties of System 2.12 for any given set of initial condition $x(t_0)$. Obviously, control design must be an integrated part of stability analysis, and quite often control $u = u(x, t)$ is chosen through stability analysis. Analysis becomes somewhat easier if functions $F(x, t)$ and $H(x, t)$ in (2.12) are independent of time. In this case, the system in the form of

$$\dot{x} = f(x), \quad y = h(x), \tag{2.13}$$

is called *autonomous*. In comparison, a system in the form of (2.12) is said to be *non-autonomous*.

As the first step of analysis, stationary points of the system, called *equilibrium points*, should be found by solving the algebraic equation: for all $t \geq t_0$,

$$F(0, t) = 0.$$

A system may have a unique equilibrium point (*e.g.*, all linear systems of $\dot{x} = Ax$ with invertible matrix A), or finite equilibrium points (*e.g.*, certain non-linear systems), or an infinite number of equilibrium points (*e.g.*, cooperative systems to be introduced later). In standard stability analysis, $x = 0$ is assumed since, if not, a constant translational transformation can be applied so that any specific equilibrium point of interest is moved to be the origin in the transformed state space. As will be shown in Chapter 5, a cooperative system typically has an infinite set of equilibrium points, its analysis needs to be done with respect to the whole set and hence should be handled differently. Should the system have periodic or chaotic solutions, stability analysis could also be done with respect to the solutions [2, 101]. In this section, the analysis is done only with respect to the equilibrium point of $x = 0$.

Qualitative properties of System 2.12 can be measured using the following *Lyapunov stability* concepts [141]. The various stability concepts provide different characteristics on how close the solution of System 2.12 will remain around or approach the origin if the solution starts in some neighborhood of the origin.

Definition 2.11. *If $x = 0$ is the equilibrium point for System 2.12, then $x = 0$ is*

(a) Lyapunov stable *(or stable) if, for every pair of $\epsilon > 0$ and $t_0 > 0$, there exists a constant $\delta = \delta(\epsilon, t_0)$ such that $\|x(t_0)\| \leq \delta$ implies $\|x(t)\| \leq \epsilon$ for $t \geq t_0$.*
(b) Unstable *if it is not Lyapunov stable.*
(c) Uniformly stable *if it is Lyapunov stable and if $\delta(\epsilon, t_0) = \delta(\epsilon)$.*
(d) Asymptotically stable *if it is Lyapunov stable and if there exists a constant $\delta'(t_0)$ such that $\|x(t_0)\| \leq \delta'$ implies $\lim_{t \to \infty} \|x(t)\| = 0$.*
(e) Uniformly asymptotically stable *if it is uniformly stable and if $\delta'(t_0) = \delta'$.*
(f) Exponentially stable *if there are constants $\delta, \alpha, \beta > 0$ such that $\|x(t_0)\| \leq \delta$ implies*

$$\|x(t)\| \leq \alpha \|x(t_0)\| e^{-\beta(t - t_0)}.$$

Lyapunov stability states that, if the initial condition of the state is arbitrarily close to the origin, the state can remain within any small hyperball centered at the origin. The relevant quantity δ represents an estimate on the radius of *stability regions*. If δ is infinite, the stability results are *global*; otherwise, the stability results are *local*. Asymptotical stability requires additionally that the solution converges to the origin, and quantity δ' is an estimate on the radius of *convergence region*. Complementary to the above Lyapunov stability concepts are the concepts of *uniform boundedness, uniform ultimate boundedness*, *asymptotic convergence*, and *input-to-state* stability (ISS) [235].

Definition 2.12. *System 2.12 is said to be*

(a) Uniformly **bounded** *if, for any $c' > 0$ and for any initial condition $\|x(t_0)\| \leq c'$, there exists a constant c such that $\|x(t)\| \leq c$ for $t \geq t_0$.*
(b) Uniformly ultimately **bounded** *with respect to $\epsilon > 0$ if it is uniformly bounded and if there exists a finite time period τ such that $\|x(t)\| \leq \epsilon$ for $t \geq t_0 + \tau$.*
(c) Asymptotically **convergent** *if it is uniformly bounded and if $x(t) \to 0$ as $t \to \infty$.*

System 2.10 is input-to-state stable if, under $u = 0$, it is uniformly asymptotically stable at $x = 0$ and if, under any bounded input u, the state is also uniformly bounded.

For a control system, uniform boundedness is the minimum requirement. It is obvious that Lyapunov stability implies uniform boundedness, that asymptotic stability implies asymptotic convergence, and that asymptotic convergence implies uniform ultimate boundedness. But, their reverse statements are not true in general. Concepts (a), (b), and (c) in Definition 2.12 are often used in robustness analysis [192] because they do not require $x = 0$ be an equilibrium point.

Several of the above concepts are illustrated by the first-order systems in the following example.

Example 2.13. (1) The first-order, time-varying, linear system of $\dot{x} = a(t)x$ has the following solution:

$$x(t) = x(t_0)e^{\int_{t_0}^{t} a(\tau)d\tau}.$$

Using this solution, one can verify the following stability results:

(1a) System $\dot{x} = -x/(1 + t)^2$ is stable but not asymptotically stable.
(1b) System $\dot{x} = (6t \sin t - 2t)x$ is stable but not uniformly stable.
(1c) System $\dot{x} = -x/(1 + t)$ is uniformly stable and asymptotically stable, but not uniformly asymptotically stable.

(2) System $\dot{x} = -x^3$ has the solution of $x(t) = 1/\sqrt{2t + 1/x^2(0)}$. Hence, the system is asymptotically stable but not exponentially stable.
(3) System $\dot{x} = -x + \text{sign}(x)e^{-t}$ has the solution $x(t) = e^{-(t-t_0)}x(t_0) + \text{sign}(x(t_0))(e^{-t_0} - e^{-t})$, where $\text{sign}(\cdot)$ is the standard sign function with $\text{sign}(0) = 0$. Thus, the system is asymptotically convergent, but it is not stable.
(4) System $\dot{x} = -x + xu$ is exponentially stable with $u = 0$, but it is not input-to-state stable. ◇

In the above example, stability is determined by analytically solving differential equations. For time-invariant linear systems, a closed-form solution is found and used in Section 2.4 to determine stability conditions. For time-varying linear systems or for non-linear systems, there is in general no closed-form solution. In these cases, stability should be studied using the Lyapunov direct method.

2.3.1 Lyapunov Direct Method

The Lyapunov direct method can be utilized to conclude various stability results without the explicit knowledge of system trajectories. It is based on the simple mathematical fact that, if a scalar function is both bounded from below and decreasing, the function has a limit as time t approaches infinity. For stability analysis of System 2.12, we introduce the following definition.

Definition 2.14. *A time function $\gamma(s)$ is said to be* strictly monotone increasing *(or* strictly monotone decreasing*) if $\gamma(s_1) < \gamma(s_2)$ (or $\gamma(s_1) > \gamma(s_2)$) for any $s_1 < s_2$.*

Definition 2.15. *A scalar function $V(x,t)$ is said to be*

(a) Positive definite *(p.d.) if $V(0,t) = 0$ and if $V(x,t) \geq \gamma_1(\|x\|)$ for some scalar, strictly monotone increasing function $\gamma_1(\cdot)$ with $\gamma_1(0) = 0$. A positive definite function $V(x,t)$ is said to be* radially unbounded *if its associated lower bounding function $\gamma_1(\cdot)$ has the property that $\gamma_1(r) \to \infty$ as $r \to +\infty$.*

(b) Positive semi-definite *(p.s.d.) if $V(x,t) \geq 0$ for all t and x.*

(c) Negative definite or negative semi-definite *(n.d. or n.s.d.) if $-V(x,t)$ is positive definite or positive semi-definite, respectively.*

(d) Decrescent *if $V(x,t) \leq \gamma_2(\|x\|)$ for some scalar, strictly monotone increasing function $\gamma_2(\cdot)$ with $\gamma_2(0) = 0$.*

Clearly, $V(x,t)$ being p.d. ensures that scalar function $V(x,t)$ is bounded from below. To study stability of System 2.12, its time derivative along any of system trajectories can be evaluated by

$$\dot{V}(x,t) = \frac{\partial V}{\partial t} + \left(\frac{\partial V}{\partial x}\right)^T \dot{x} = \frac{\partial V}{\partial t} + \left(\frac{\partial V}{\partial x}\right)^T F(x,t).$$

If $\dot{V}$ is n.d. (or n.s.d.), $V(x,t)$ keeps decreasing (or is non-increasing), asymptotic stability (or Lyapunov stability) can be concluded as stated in the following theorem, and $V(x,t)$ is referred to as a *Lyapunov function*.

Theorem 2.16. *System 2.12 is*

(a) Locally Lyapunov stable *if, in a neighborhood around the origin, $V(x,t)$ is p.d. and $\dot{V}(x,t)$ is n.s.d.*

(b) *Locally uniformly stable as* $\|x(t)\| \leq \gamma_1^{-1} \circ \gamma_2(\|x(t_0)\|)$ *if, for* $x \in \{x \in \Re^n : \|x\| < \eta\}$ *with* $\eta \geq \gamma_2(\|x(t_0)\|)$, $V(x,t)$ *is p.d. and decrescent and* $\dot{V}(x,t)$ *is negative semi-definite.*

(c) *Uniformly bounded and uniformly ultimately bounded with respect to* ϵ *if* $V(x,t)$ *is p.d. and decrescent and if, for* $x \in \{x \in \Re^n : \|x\| \geq \gamma_1^{-1} \circ \gamma_2(\epsilon)\}$, $\dot{V}(x,t)$ *is negative semi-definite.*

(d) *Locally uniformly asymptotically stable in the region of* $\{x \in \Re^n : \|x\| < \gamma_2^{-1}(\gamma_1(\eta))\}$ *if, for* $\|x\| < \eta$, $V(x,t)$ *is p.d. and decrescent and if* $\dot{V}(x,t)$ *is n.d.*

(e) *Globally uniformly asymptotically stable if* $V(x,t)$ *is p.d., radially unbounded and decrescent and if* $\dot{V}(x,t)$ *is n.d. everywhere.*

(f) *Exponentially stable if* $\gamma_1\|x\|^2 \leq V(x,t) \leq \gamma_2\|x\|^2$ *and* $\dot{V}(x,t) \leq -\gamma_3\|x\|^2$ *for positive constants* γ_i, $i = 1,2,3$.

(g) *Unstable if, in every small neighborhood around the origin,* $\dot{V}$ *is n.d. and* $V(x,t)$ *assumes a strictly negative value for some* x *therein.*

System 2.10 is ISS if $V(x,t)$ *is p.d., radially unbounded and decrescent and if* $\dot{V}(x,t) \leq -\gamma_3(\|x\|) + \gamma_3^{\beta}(\|x\|)\gamma_4(\|u\|)$, *where* $0 \leq \beta < 1$ *is a constant, and* $\gamma_i(\cdot)$ *with* $i = 3,4$ *are scalar strictly monotone increasing functions with* $\gamma_i(0) = 0$.

A useful observation is that, if $\dot{V}(x,t) \leq 0$ for $\|x\| \leq \eta$ with $\eta \geq \gamma_2(\|x(t_0)\|)$ and if $\|x(t_0)\| < \gamma_2^{-1}(\gamma_1(\eta))$, $\dot{V}(x(t_0),t_0) \leq 0$ and, for any sufficiently small $\delta t > 0$, $V(x(t),t) \leq V(x(t_0),t_0) \leq \gamma_2(\|x(t_0)\|)$ and hence $\dot{V}(x,t + \delta t) \leq 0$. By applying the observation and repeating the argument inductively, all the statements in Theorem 2.16 except for (vi) can be proven. Proof of (vi) of Theorem 2.16 will be pursued in the next subsection. Counterexamples are provided in [108] to show that the condition of radial unboundedness is needed to conclude global stability and that the decrescent condition is required for both asymptotic stability and uniform stability.

The key of applying Lyapunov direct method is to find Lyapunov function $V(x,t)$. In searching of Lyapunov function, we should use the backward process of first finding $\dot{V}$ symbolically in terms of $\partial V/\partial x$ and then selecting $\partial V/\partial x$ so that V is found and $\dot{V}$ has one of the properties listed in Theorem 2.16. The process is illustrated by the following example.

Example 2.17. Consider the second-order system:

$$\dot{x}_1 = x_2, \qquad \dot{x}_2 = -x_1^3 - x_2.$$

Let $V(x)$ be the Lyapunov function to be found. It follows that

$$\dot{V} = \frac{\partial V}{\partial x_1}x_2 + \frac{\partial V}{\partial x_2}(-x_1^3 - x_2).$$

In order to conclude Lyapunov stability, we need to show that $\dot{V}$ is at least n.s.d. To this end, we need to eliminate such sign-indefinite terms as the cross product terms of x_1 and x_2. One such choice is that

$$\frac{\partial V}{\partial x_1} x_2 = \frac{\partial V}{\partial x_2} x_1^3, \tag{2.14}$$

which holds if $V = x_1^4 + 2x_2^2$ is chosen. Consequently, we know that

$$\dot{V} = -4x_2^2$$

is n.s.d. and hence the system is globally (uniformly) Lyapunov stable.

Next, let us determine whether Lyapunov function V exists to make $\dot{V}$ n.d. It follows from the expression for $\dot{V}$ that $\dot{V}$ may become n.d. only if $\partial V/\partial x_2$ contains such a term as x_1. Accordingly, we introduce an additional term of $2ax_1$ into the previous expression for $\partial V/\partial x_2$, that is,

$$\frac{\partial V}{\partial x_2} = 2ax_1 + 4x_2$$

for some constant $a > 0$. Hence, it follows that

$$V(x) = 2ax_1x_2 + 2x_2^2 + h(x_1),$$

where function $h(x_1) > 0$ is desired. To ensure (2.14), $h(x_1)$ must contain x_1^4. On the other hand, $V(x)$ should be positive definite, and this can be made possible by including bx_1^2 in $h(x_1)$ for some constant $b > 0$. In summary, the simplest form of $V(x)$ should be

$$V(x) = 2ax_1x_2 + 2x_2^2 + x_1^4 + bx_1^2,$$

which is positive definite if $a < 2\sqrt{2b}$. Consequently, we have

$$\dot{V} = (2ax_2 + 4x_1^3 + 2bx_1)x_2 + (2ax_1 + 4x_2)(-x_1^3 - x_2)$$
$$= -2ax_1^4 - (4 - 2a)x_2^2 + (2b - 2a)x_1x_2,$$

which is n.d. under many choices of a and b (e.g., $a = b = 1$). Since V is p.d. and $\dot{V}$ is n.d., the system is shown to be asymptotically stable.

While the system is stable, many choices of $V(x)$ (such as $V(x) = x_1^2 + x_2^2$) would yield a sign-indefinite expression of $\dot{V}$.   ◇

The above example shows that the conditions in Theorem 2.16 are sufficient, Lyapunov function is not unique and, unless the backward process yields a Lyapunov function, stability analysis is inconclusive. Nonetheless, existence of a Lyapunov function is guaranteed for all stable systems. This result is stated as the following theorem, referred to *Lyapunov converse theorem*, and its proof based on qualitative properties of system trajectory can be found in [108].

Theorem 2.18. *Consider System 2.12. Then,*

(a) If the system is stable, there is a p.d. function $V(x,t)$ whose time derivative along the system trajectory is n.s.d.

(b) If the system is (globally) uniformly asymptotically stable, there is a p.d. decrescent (and radially unbounded) function $V(x,t)$ such that $\dot{V}$ is n.d.

(c) If the system is exponentially stable, there are Lyapunov function $V(x,t)$ and positive constants γ_i such that $\gamma_1\|x\|^2 \le V(x,t) \le \gamma_2\|x\|^2$, $\dot{V}(x,t) \le -\gamma_3\|x\|^2$, and $\|\partial V/\partial x\| \le \gamma_4\|x\|$.

It is the converse theorem that guarantees existence of the Lyapunov functions and makes Lyapunov direct method a universal approach for non-linear systems. Nonetheless, the theorem provides a promise rather than a recipe for finding a Lyapunov function. As a result, Lyapunov function has to be found for specific classes of non-linear systems. In the subsequent subsections, system properties are explored to search for Lyapunov functions.

2.3.2 Explanations and Enhancements

A Lyapunov function can be sought by exploiting either physical, mathematical, or structural properties of system dynamics. In what follows, physical and mathematical features of the Lyapunov direct method are explained, while structural properties of dynamics will be explored in Section 2.6.1 to construct the Lyapunov function. In Section 2.6.4, the Lyapunov function is related to a performance function in an optimal control design.

Interpretation as an Energy Function

In essence, stability describes whether and how the system trajectory moves toward its equilibrium point, and hence the motion can be analyzed using a physical measure such as an energy function or its generalization. Intuitively, if the total energy of a system keeps dissipating over time, the system will eventually lose all of its initial energy and consequently settle down to an equilibrium point. Indeed, Lyapunov function is a measure of generalized energy, and the Lyapunov direct method formalizes the dissipative argument. As an illustration, consider an one-dimensional rigid-body motion for which kinetic and potential energies are assumed to be

$$K = \frac{1}{2}m(x)\dot{x}^2, \quad P = P(x),$$

respectively, where $m(x)$ is a positive number or function. It follows from (1.16) that the dynamic equation of motion is

$$m(x)\ddot{x} + \frac{1}{2}\frac{dm(x)}{dx}\dot{x}^2 + \frac{dP(x)}{dx} = \tau. \tag{2.15}$$

Obviously, the *energy function* of the system is

$$\mathcal{E} = K + P = \frac{1}{2}m(x)\dot{x}^2 + P(x). \tag{2.16}$$

It follows that the time derivative of $\mathcal{E}$ along trajectories of System 2.15 is

$$\dot{\mathcal{E}} = \dot{x}\tau.$$

If the net external force is zero, $\dot{\mathcal{E}} = 0$, the total energy of systems is conservative, and hence System 2.15 with $\tau = 0$ is called a *conservative system*. On the other hand, if $\tau = -k\dot{x}$ for some $k > 0$, the external input is a dynamic friction force, and System 2.15 is dissipative as

$$\dot{\mathcal{E}} = -k\dot{x}^2,$$

which is n.s.d. Depending upon the property of potential energy $\mathcal{P}(x)$, equilibrium point(s) can be found, and asymptotic stability could be concluded by following Example 2.17 and finding an appropriate Lyapunov function. In other words, an energy function such as the one in (2.16) can be used as the starting point to search for the Lyapunov function.

Interpretation as a Contraction Mapping

Suppose that, for a given system $\dot{x} = F(x, t)$, Lyapunov function $V(x)$ is found such that $V(x)$ is positive definite and

$$\dot{V}(x) = \left[\frac{\partial V(x)}{\partial x}\right]^T F(x, t)$$

is negative definite. It follows that, for any infinite time sequence $\{t_k : k \in \aleph\}$, $V(x(t)) \leq V(x(t_i))$ for all $t \in [t_i, t_{i+1})$ and

$$V(x(t_{i+1})) = V(x(t_i)) + \int_{t_i}^{t_{i+1}} \dot{V}(x)d\tau \leq -\lambda V(x(t_i)),$$

for some $\lambda \in [0, 1)$. It follows from Theorem 2.4 that Lyapunov function $V(x(t_i))$ itself is a contraction mapping from which asymptotic stability can be concluded.

Enhancement by Comparison Theorem

The comparison theorem, Theorem 2.8, can be used to facilitate a generalized Lyapunov argument if the Lyapunov function and its time derivative render a solvable inequality. The following lemma is such a result in which $\dot{V}$ is not negative definite in the neighborhood around the origin.

Lemma 2.19. *Let V be a (generalized) Lyapunov function for System 2.12 such that*

$$\gamma_1(\|x(t)\|) \leq V(x, t) \leq \gamma_2(\|x\|),$$

and

$$\dot{V}(x, t) \leq -\lambda\gamma_2(\|x\|) + \epsilon\varphi(t),$$

where $\gamma_i(\cdot)$ are strictly monotone increasing functions with $\gamma_i(0) = 0$, $\lambda, \epsilon > 0$ are constants, and $0 \leq \varphi(t) \leq 1$. Then, the system is

(a) Uniformly ultimately bounded with respect to $\gamma_1^{-1}(\epsilon/\lambda)$.
(b) Asymptotically convergent if $\varphi(t)$ converges to zero.
(c) Exponentially convergent if $\varphi(t) = e^{-\beta t}$ for some $\beta > 0$.

Proof: It follows that $\dot{V}(x,t) \leq -\lambda V(x,t) + \epsilon\varphi(t)$ and hence, by Theorem 2.8, $V(x,t) \leq w(t)$, where

$$\dot{w} = -\lambda w + \epsilon\varphi(t), \quad w(t_0) = V(x(t_0), t_0).$$

Solving the scalar differential equation yields

$$V(x,t) \leq e^{-\lambda(t-t_0)}V(x(t_0), t_0) + \epsilon\int_{t_0}^{t} e^{-\lambda(t-s)}\varphi(s)ds,$$

from which the statements become obvious.   □

If $\epsilon = 0$ in the statement, Lemma 2.19 reduces to some of the stability results in Theorem 2.16.

Enhancement by Barbalat Lemma

As a useful tool in Lyapunov stability analysis, the Barbalat lemma, Lemma 2.6, can be used to conclude convergence for the cases that $\dot{V}$ is merely n.s.d. with respect to state x and may also contain an L_1-space time function. The following lemma illustrates such a result.

Lemma 2.20. *Consider System 2.12 in which function $F(x,t)$ is locally uniformly bounded with respect to x and uniformly bounded with respect to t. Let V be its Lyapunov function such that*

$$\gamma_1(\|x(t)\|) \leq V(x,t) \leq \gamma_2(\|x\|),$$

and

$$\dot{V}(x,t) \leq -\gamma_3(\|z(t)\|) + \varphi(t),$$

where $\gamma_i(\cdot)$ are strictly monotone increasing functions with $\gamma_i(0) = 0$, $\gamma_3(\cdot)$ is locally Lipschitz, $z(t)$ is a sub-vector of $x(t)$, and $\varphi(t)$ belongs to L_1-space. Then, the system is uniformly bounded and the sub-state $z(t)$ is asymptotically convergent.

Proof: It follows from the expression of $\dot{V}$ that

$$V(x(t), t) + \int_{t_0}^{t} \gamma_3(\|z(\tau)\|)d\tau \leq V(x_0, t_0) + \int_{t_0}^{t} \varphi(s)ds$$

$$\leq V(x_0, t_0) + \int_{t_0}^{\infty} \varphi(s)ds$$

$$< \infty.$$

The above inequality implies that $V(x(t), t)$ and hence $x(t)$ are uniformly bounded. Recalling properties of $F(x, t)$, we know from (2.12) that $\dot{x}$ is uniformly bounded and thus $x(t)$ as well as $z(t)$ is uniformly continuous. On the other hand, the above inequality also shows that $\gamma_3(\|z(\tau)\|)$ belongs to L_1-space. Thus, by Lemma 2.6, $z(t)$ is asymptotically convergent. $\qquad\square$

For the autonomous system in (2.13), Lemma 2.20 reduces to the famous LaSalle's *invariant set theorem*, given below. Set Ω is said to be *invariant* for a dynamic system if, by starting within Ω, its trajectory remains there for all future time.

Theorem 2.21. *Suppose that $V(x)$ is p.s.d. for $\|x\| < \eta$ and, along any trajectory of System 2.13, $\dot{V}$ is n.s.d. Then, state $x(t)$ converges either to a periodic trajectory or an equilibrium point in set $\Omega \overset{\triangle}{=} \{x \in \Re^n : \dot{V}(x) = 0, \|x\| < \eta\}$. System 2.13 is asymptotically stable if set Ω contains no periodic trajectory but only equilibrium point $x \equiv 0$. Moreover, asymptotic stability becomes global if $\eta = \infty$ and $V(x)$ is radially unbounded.*

It must be emphasized that Theorem 2.21 only holds for autonomous systems. It holds even without $V(x)$ being positive semi-definite as long as, for any constant $l > 0$, the set defined by $V(x) < l$ is closed and bounded.

Directional Derivative

Consider the autonomous system in (2.13). The corresponding Lyapunov function is time-invariant as $V = V(x)$, and its time derivative along trajectories of System 2.13 is

$$\dot{V} = \left(\frac{\partial V}{\partial x}\right)^T f(x), \quad \text{or} \quad \dot{V} = (\nabla_x V)^T f(x),$$

which is the dot product of system dynamics and gradient of Lyapunov function. Hence, as the projection of the gradient along the direction of motion, $\dot{V}$ is called directional derivative.

To simplify the notations in the subsequent discussions, the so-called Lie derivative is introduced. *Lie derivative* of scalar function $\xi(x)$ with respect to vector function $f(x)$, denoted by $L_f\xi$, is a scalar function defined by

$$L_f\xi = (\nabla_x^T \xi)f.$$

High-order Lie derivatives can be defined recursively as, for $i = 1, \cdots$,

$$L_f^0\xi = \xi, \quad L_f^i\xi = L_f(L_f^{i-1}\xi) = [\nabla_x^T(L_f^{i-1}\xi)]f, \quad \text{and} \quad L_gL_f\xi = [\nabla_x^T(L_f\xi)]g,$$

where $g(x)$ is another vector field of the same dimension. It is obvious that, if $y_j = h_j(x)$ is a scalar output of System 2.13, the ith-order time derivative of this output is simply $y_j^{(i)} = L_f^i h_j$ and that, if $V(x)$ is the Lyapunov function for System 2.13, $\dot{V} = L_f V$.

2.3.3 Control Lyapunov Function

Consider the special case that dynamics of Control System 2.10 do not explicitly depend on time, that is,

$$\dot{x} = \mathcal{F}'(x, u), \tag{2.17}$$

where $\mathcal{F}'(0,0) = 0$. For System 2.17, existence of both a stabilizing control and its corresponding Lyapunov function is captured by the following concept of control Lyapunov function.

Definition 2.22. *A smooth and positive definite function $V(x)$ is said to be a control Lyapunov function for System 2.17 if, for any $x \neq 0$,*

$$\inf_{u \in \mathbb{R}^m} L_{\mathcal{F}'(x,u)} V(x) < 0. \tag{2.18}$$

Control Lyapunov function $V(\cdot)$ is said to satisfy the small control property *if, for every $\epsilon > 0$, there exists $\delta > 0$ such that Inequality 2.18 holds for any x with $0 < \|x\| < \delta$ and for some $u(x)$ with $\|u(x)\| < \epsilon$.*

Clearly, in light of (2.18), the time derivative of the control Lyapunov function along the system trajectory can always be made negative definite by properly choosing a feedback control $u(x)$, which is sufficient for concluding at least local asymptotic stability. The converse Lyapunov theorem, Theorem 2.18, also ensures the existence of a control Lyapunov function if System 2.17 is asymptotically stabilized. The following lemma due to [8, 165] provides a sufficient condition for constructing control and concluding stability.

Lemma 2.23. *Suppose that System 2.17 has a control Lyapunov function $V(x)$. If the mapping $u \rightarrow L_{\mathcal{F}'(x,u)} V(x)$ is convex for all $x \neq 0$, then the system is globally asymptotically stable under a feedback control $u(x)$ which is continuous for all $x \neq 0$. In addition, if $V(x)$ satisfies the small control property, the control $u(x)$ is continuous everywhere.*

Given a control Lyapunov function, construction of a stabilizing control for System 2.17 is generally non-trivial. In addition, the convex condition required by the above lemma may not be valid in general. For the affine nonlinear control system in (2.11), existence of control Lyapunov function $V(x)$ is equivalent to the requirement that $L_{g(x)} V(x) = 0$ implies $L_{f(x)} V(x) < 0$. Should the control Lyapunov function be known for Affine System 2.11, a universal feedback controller is available in terms of the following Sontag formula [234]:

$$u(x) = \begin{cases} -\dfrac{L_f V + \sqrt{(L_f V)^2 + \|L_g V\|^2 \alpha(x)}}{\|L_g V\|^2} (L_g V)^T & \text{if } L_g V \neq 0 \\ 0 & \text{if } L_g V = 0 \end{cases}, \tag{2.19}$$

where $\alpha(x) \geq 0$ is a scalar function. It is straightforward to show that $u(x)$ is stabilizing as inequality $L_f V + [L_g V] u(x) < 0$ holds everywhere, that $u(x)$ is continuous except at those points satisfying $L_g V = 0$, and that $u(x)$ becomes continuous everywhere under the choice of $\alpha(x) = \xi(\|L_g V\|)$, where $\xi(\cdot)$ is a scalar function satisfying $\xi(0) = 0$ and $\xi(a) > 0$ for $a > 0$.

The problem of finding a control Lyapunov function for System 2.17 can be studied by imposing certain conditions on function $F(x, u)$, either analytical properties or special structures. In the first approach, one typically assumes that a positive definite function $V_0(x)$ (as a weak version of Lyapunov function) is already known to yield $L_{f(x)} V_0(x) \leq 0$ and that, for controllability (which will be discussed in Section 2.5), vector fields $\{f, ad_f g_k, ad_f^2 g_k, \cdots\}$ satisfy certain rank conditions, where $f(x) = \mathcal{F}'(x, 0)$ and $g_k(x) = \partial \mathcal{F}'(x, u) / \partial u_k |_{u=0}$. Based on the knowledge of $V_0(x)$ as well as f and g_k, a control Lyapunov function $V(x)$ can be constructed [152]. In the second approach, a special structural property of the system is utilized to explicitly search for control Lyapunov function, which will be carried out in Section 2.6.

2.3.4 Lyapunov Analysis of Switching Systems

Among non-autonomous systems of form (2.12), there are *switching systems* whose dynamics experience instantaneous changes at certain time instants and are described by

$$\dot{x} = F(x, s(t)), \quad s(t) \in \mathcal{I}, \tag{2.20}$$

where $s(\cdot)$ is the switching function (or selection function), and $\mathcal{I}$ is the corresponding value set (or index set). In the discussion of this subsection, it is assumed that set $\mathcal{I}$ be a finite sub-set of $\aleph$ and that, uniformly with respect to all $s(t) \in \mathcal{I}$, function $F(x, s(t))$ be bounded for each x and also be locally Lipschitz in x.

In principle, analysis of Switching System 2.20 is not much different from that of Non-autonomous System 2.12. Specifically, Lyapunov direct method can readily be applied to System 2.20, and the successful Lyapunov function is usually time dependent as $V(x, s(t))$. The dependence of V on discontinuous function $s(t)$ introduces the technical difficulty that time derivative $\dot{V}$ contains singular points. Consequently, non-smooth analysis such as semi-continuity, set valued map, generalized solution, and differential inclusions [11, 65, 110, 219, 220] need be used in stability analysis.

One way to avoid the complication of non-smooth analysis is to find a *common Lyapunov function*, that is, Lyapunov function $V(x)$ that is positive definite and whose time derivative $L_{F(x, s(t))} V(x)$ is negative definite no matter what *fixed* value $s(t)$ assumes in set $\mathcal{I}$. It is obvious that, if a (radially-unbounded) common Lyapunov function exists, System 2.20 is (globally) asymptotically stable and so are the family of autonomous systems

$$\dot{x}_s = F(x_s, s) \stackrel{\triangle}{=} \mathcal{F}_s(x_s), \quad s \in \mathcal{I}. \tag{2.21}$$

However, stability of all the autonomous systems in (2.21) is not necessary or sufficient for concluding stability of Switching System 2.20. It will be shown in Section 2.4.3 that a system switching between two stable linear autonomous systems may be unstable. Similarly, a system switching between two unstable linear autonomous systems may be stable. In other words, an asymptotically stable switching system in the form of (2.20) may not have a common Lyapunov function, nor do its induced family of autonomous systems in (2.21) have to be asymptotically stable. On the other hand, if the switching is arbitrary in the sense that switching function $s(t)$ is allowed to assume any value in index set $\mathcal{I}$ at any time, stability of Switching System 2.20 implies stability of all the autonomous systems in (2.21). Indeed, under the assumption of arbitrary switching, the corresponding Lyapunov converse theorem is available [49, 142, 155] to ensure the existence of a common Lyapunov function for an asymptotically stable switching system of (2.20) and for all the autonomous systems in (2.21). In short, stability analysis under arbitrary switching can and should be done in terms of a common Lyapunov function.

In most cases, switching occurs according to a sequence of time instants that are either fixed or unknown *a priori* and, if any, the unknown switching time instants are not arbitrary because they are determined by uncertain exogenous dynamics. In these cases, System 2.20 is piecewise-autonomous as, for a finite family of autonomous functions $\{\mathcal{F}_s(x) : s \in \mathcal{I}\}$, and for some sequence of time instances $\{t_i : i \in \aleph\}$,

$$\dot{x} = F(x, s(t_i)) \triangleq \mathcal{F}_{s(t_i)}(x), \quad \forall t \in [t_i, t_{i+1}). \tag{2.22}$$

System 2.22 may not have a common Lyapunov function, nor may its induced family of autonomous systems in (2.21). The following theorem provides the condition under which stability can be concluded using a family of Lyapunov functions defined for Autonomous Systems 2.21 and invoked over consecutive time intervals.

Theorem 2.24. *Consider the piecewise-autonomous system in (2.22). Suppose that the corresponding autonomous systems in (2.21) are all globally asymptotically stable and hence have radially unbounded positive definite Lyapunov functions $V_s(x)$ and that, along any trajectory of System 2.22, the following inequality holds: for every pair of switching times (t_i, t_j) satisfying $t_i < t_j$, $s(t_i) = s(t_j) \in \mathcal{I}$, and $s(t_k) \neq s(t_j)$ for any t_k of $t_i < t_k < t_j$,*

$$V_{s(t_j)}(x(t_j)) - V_{s(t_i)}(x(t_i)) \leq -W_{s(t_i)}(x(t_i)), \tag{2.23}$$

where $W_s(x)$ with $s \in \mathcal{I}$ are a family of positive definite continuous functions. Then, Switched System 2.22 is globally asymptotically stable.

Proof: We first show global Lyapunov stability. Suppose $\|x(t_0)\| < \delta$ for some $\delta > 0$. Since $V_s(x)$ is radially unbounded and positive definite, there exist scalar strictly monotone increasing functions $\gamma_{s1}(\cdot)$ and $\gamma_{s2}(\cdot)$ such that

$\gamma_{s1}(\|x\|) \leq V_s(x) \leq \gamma_{s2}(\|x\|)$. Therefore, it follows from (2.23) that, for any t_j,

$$V_{s(t_j)}(x(t_j)) \leq \max_{s \in \mathcal{I}} \gamma_{s2}(\delta),$$

which yields

$$\|x(t_j)\| \leq \max_{\sigma \in \mathcal{I}} \gamma_{\sigma 1}^{-1} \circ \max_{s \in \mathcal{I}} \gamma_{s2}(\delta) \triangleq \epsilon.$$

The above inequality together with asymptotic stability of autonomous systems in (2.21) imply that System 2.22 is globally Lyapunov stable.

Asymptotic stability is obvious if time sequence $\{t_i : i \in \aleph\}$ is finite. If the sequence is infinite, let $\{t_{\sigma_j} : \sigma_j \in \aleph, j \in \aleph\}$ denote its infinite sub-sequence containing all the entries of $s(t_i) = s(t_{\sigma_j}) = \sigma \in \mathcal{I}$. It follows from (2.23) that Lyapunov sequence $V_\sigma(x(t_{\sigma_j}))$ is monotone decreasing and hence has a limit c and that

$$0 = c - c = \lim_{j \to \infty} [V_\sigma(x(t_{\sigma_j})) - V_\sigma(x(t_{\sigma_{j+1}}))] \leq - \lim_{j \to \infty} W_\sigma(x(t_{\sigma_j})) \leq 0.$$

Since $W_s(\cdot)$ is positive definite and $\sigma \in \mathcal{I}$ is arbitrary, $x(t_i)$ converges to zero. The proof is completed by recalling asymptotic stability of autonomous systems in (2.21). □

In stability analysis of switching systems using multiple Lyapunov functions over time, Theorem 2.24 is representative among the available results [28, 133, 185]. Extensions such as that in [91] can be made so that some Lyapunov functions are allowed to increase during their active time intervals as long as these increases are bounded by positive definite functions properly incorporated into Inequality 2.23. Similarly, some of the systems in (2.21) may not have to be asymptotically stable provided that an inequality in the form of (2.23) holds over time. As will be illustrated in Section 2.4.3 for linear switching systems, Inequality 2.23 implies monotone decreasing over sequences of consecutive intervals, while System 2.22 can have transient increases such as overshoots during some of the intervals. To satisfy Inequality 2.23, either the solution to (2.22) or a quantitative bound on the solution should be found, which is the main difficulty in an application of Theorem 2.24. In general, Inequality 2.23 can always be ensured if the so-called *dwell time*, the length of the intervals for System 2.22 to stay as one of the systems in (2.21), is long if the corresponding system in (2.21) is asymptotically stable and sufficiently short if otherwise. Further discussions on dwell time can be found in Section 2.4.3 and in [133].

It is worth noting that, while multiple Lyapunov functions are used in Theorem 2.24 to establish stability, only one of the Lyapunov functions is actively used at every instant of time to capture the instantaneous system behavior and their cumulative effect determines the stability outcome. This is different from Comparison Theorem 2.10 by which a vector of Lyapunov functions is used simultaneously to determine qualitative behavior of a system.

2.4 Stability Analysis of Linear Systems

State space model of a linear dynamical system is

$$\dot{x} = A(t)x + B(t)u, \quad y = C(t)x + D(t)u, \tag{2.24}$$

where $x \in \Re^n$ is the state, $u \in \Re^m$ is the input, and $y \in \Re^p$ is the output. Matrices A, B, C, D are called system matrix, input matrix, output matrix, and direct coupling matrix, respectively. System 2.24 is called time-invariant if these matrices are all constant. In what follows, analysis tools of linear systems are reviewed.

2.4.1 Eigenvalue Analysis of Linear Time-invariant Systems

Eigenvalue analysis is both fundamental and the easiest to understanding stability of the linear time-invariant system:

$$\dot{x} = Ax + Bu. \tag{2.25}$$

Defining *matrix exponential function* e^{At} as

$$e^{At} = \sum_{j=0}^{\infty} \frac{1}{j!} A^j t^j, \tag{2.26}$$

we know that e^{At} satisfies the property of

$$\frac{d}{dt} e^{At} = A e^{At} = e^{At} A,$$

that the solution to $\dot{x} = Ax$ is $x(t) = e^{A(t-t_0)}x(t_0)$ and hence $e^{A(t-t_0)}$ is called *state transition matrix*, and that the solution to (2.25) is

$$x(t) = e^{At}x(0) + \int_0^t e^{A(t-\tau)} Bu(\tau)d\tau. \tag{2.27}$$

The application of a similarity transformation, reviewed in Section 2.1, is one of the time-domain approaches for solving state space equations and revealing stability properties. Consider System 2.25 and assume that, under similarity transformation $z = S^{-1}x$, $\dot{z} = S^{-1}ASz = Jz$ where J is the Jordan canonical form in (2.1). It follows from the structure of J and the Taylor series expansion in (2.26) that

$$e^{At} = S \begin{bmatrix} e^{J_1 t} & & \\ & \ddots & \\ & & e^{J_l t} \end{bmatrix} S^{-1},$$

in which

$$e^{J_i t} = \begin{bmatrix} e^{\lambda_i t} & te^{\lambda_i t} & \cdots & \frac{t^{n_i-1}}{(n_i-1)!}e^{\lambda_i t} \\ 0 & \ddots & \ddots & \vdots \\ \vdots & \vdots & \ddots & te^{\lambda_i t} \\ 0 & 0 & \cdots & e^{\lambda_i t} \end{bmatrix}. \tag{2.28}$$

Therefore, the following necessary and sufficient conditions [102, 176] can be concluded from the solutions of (2.27) and (2.28):

(a) System 2.25 with $u = 0$ is Lyapunov stable if and only if its eigenvalues are not in the right open half plane and those on the imaginary axis are of geometrical multiplicity one.

(b) System 2.25 with $u = 0$ is asymptotically stable if and only if matrix A is Hurwitz (i.e., its eigenvalues are all in the left open half plane).

(c) System 2.25 is input-to-state stable if and only if it is asymptotically stable.

(d) System 2.25 with $u = 0$ is exponentially stable if and only if it is asymptotically stable.

2.4.2 Stability of Linear Time-varying Systems

Consider first the continuous-time linear time-varying system

$$\dot{x} = A(t)x, \quad x \in \Re^n. \tag{2.29}$$

Its solution can be expressed as

$$x(t) = \Phi(t, t_0)x(t_0), \tag{2.30}$$

where state transition matrix $\Phi(\cdot)$ has the properties that $\Phi(t_0, t_0) = I$, $\Phi(t, s) = \Phi^{-1}(s, t)$, and $\partial\Phi(t, t_0)/\partial t = A(t)\Phi(t, t_0)$. It follows that the above system is asymptotically stable if and only if

$$\lim_{t \to \infty} \Phi(t, t_0) = 0.$$

However, an analytical solution of $\Phi(t, t_0)$ is generally difficult to find except for simple systems. The following simple example shows that eigenvalue analysis does not generally reveal stability of linear time-varying systems.

Example 2.25. Consider System 2.29 with

$$A(t) = \begin{bmatrix} -1 + \sqrt{2}\cos^2 t & 1 - \sqrt{2}\sin t \cos t \\ -1 - \sqrt{2}\sin t \cos t & -1 + \sqrt{2}\sin^2 t \end{bmatrix}.$$

Matrix $A(t)$ is continuous and uniformly bounded. By direct computation, one can verify that the corresponding state transition matrix is

$$\Phi(t,0) = \begin{bmatrix} e^{(\sqrt{2}-1)t} & e^{-t}\sin t \\ -e^{(\sqrt{2}-1)t}\sin t & e^{-t}\cos t \end{bmatrix}.$$

Hence, the system is unstable. Nonetheless, eigenvalues of matrix $A(t)$ are at $[-(2-\sqrt{2})\pm\sqrt{2}j]/2$, both of which are time-invariant and in the left open half plan.   ◇

Piecewise-constant systems are a special class of time-varying systems. Consider a continuous-time switching system which is in the form of (2.29) and whose system matrix $A(t)$ switches between two constant matrices as, for some $\tau > 0$ and for $k \in \aleph$,

$$A(t) = \begin{cases} A_1 \text{ if } t \in [2k\tau, 2k\tau + \tau) \\ A_2 \text{ if } t \in [2k\tau + \tau, 2k\tau + 2\tau) \end{cases}. \tag{2.31}$$

Since the system is piecewise-constant, the state transition matrix is known to be

$$\Phi(t,0) = \begin{cases} e^{A_1(t-2k\tau)}H^k & \text{for } [2k\tau, 2k\tau + \tau) \\ e^{A_2(t-2k\tau-\tau)}e^{A_1\tau}H^k & \text{for } [2k\tau + \tau, 2k\tau + 2\tau) \end{cases}, \quad H = e^{A_2\tau}e^{A_1\tau},$$

whose convergence depends upon the property of matrix exponential H^k. Indeed, for System 2.29 satisfying (2.31), we can define the so-called average system as

$$\dot{x}_a = A_a x_a, \quad x \in \Re^n. \tag{2.32}$$

where $x_a(2k\tau) = x(2k\tau)$ for all $k \in \aleph$, $e^{2A_a\tau} = H$, and matrix A_a can be expressed in terms of Baker-Campbell-Hausdorff formula [213] as

$$A_a = A_1 + A_2 + \frac{1}{2}ad_{A_1}(A_2) + \frac{1}{12}\{ad_{A_1}(ad_{A_1}A_2) + ad_{A_2}(ad_{A_1}A_2)\}$$
$$+ \cdots, \tag{2.33}$$

and $ad_A B = AB - BA$ is the so-called Lie bracket (see Section 2.5 for more details). Note that A_a in (2.33) is an infinite convergent sequence in which the generic expression of ith term is not available. If $ad_A B = 0$, the system given by (2.31) is said to have commuting matrices A_i and, by invoking (2.32) and (2.33), it has the following solution:

$$x(t) = e^{A_1(\tau+\tau+\cdots)}e^{A_2(\tau+\tau+\cdots)}x(0),$$

which is exponentially stable if both A_1 and A_2 are Hurwitz. By induction, a linear system whose system matrix switches among a finite set of commuting Hurwitz matrices is exponentially stable. As a relaxation, the set of matrices are said to have *nilpotent Lie algebra* if all Lie brackets of sufficiently high-order are zero, and it is shown in [84] that a linear system whose system matrix switches among a finite set of Hurwitz matrices of nilpotent Lie algebra is also exponentially stable.

Similarly, consider a discrete-time linear time-varying system

$$x_{k+1} = A_k x_k, \quad x_k \in \Re^n. \tag{2.34}$$

It is obvious that the above system is asymptotically stable if and only if

$$\lim_{k \to \infty} A_k A_{k-1} \cdots A_2 A_1 \triangleq \lim_{k \to \infty} \prod_{j=1}^{k} A_j = 0. \tag{2.35}$$

In the case that A_k arbitrarily switches among a finite number matrices $\{D_1, \cdots, D_l\}$, The sequence convergence of (2.35) can be embedded into stability analysis of the switching system:

$$z_{k+1} = \overline{A}_k z_k, \tag{2.36}$$

where $\overline{A}_k$ switches between two constant matrices $\overline{D}_1$ and $\overline{D}_2$,

$$\overline{D}_1 = \mathrm{diag}\{D_1, \cdots, D_l\}, \quad \overline{D}_2 = T \otimes I,$$

and $T \in \Re^{l \times l}$ is any of the cyclic permutation matrices (see Sections 4.1.1 and 4.1.3 for more details). Only in the trivial case of $n = 1$ do we know that the standard time-invariant stability condition, $|D_i| < 1$, is both necessary and sufficient for stability of System 2.36. The following example shows that, even in the simple case of $n = 2$, the resulting stability condition on System 2.36 is not general or consistent enough to become satisfactory. That is, there is no simple stability test for switching systems in general.

Example 2.26. Consider the pair of matrices: for constants $\alpha, \beta > 0$,

$$D_1 = \sqrt{\beta} \begin{bmatrix} 1 & \alpha \\ 0 & 1 \end{bmatrix}, \quad D_2 = D_1^T.$$

It follows that, if the system matrix switches among D_1 and D_2, product $\overline{A}_k \overline{A}_{k-1}$ has four distinct choices: D_1^2, D_2^2, $D_1 D_2$, and $D_2 D_1$. Obviously, both D_1^2 and D_2^2 are Schur (*i.e.*, asymptotically stable) if and only if $\beta < 1$. On the other hand, matrix

$$D_1 D_2 = \beta \begin{bmatrix} 1 + \alpha^2 & \alpha \\ \alpha & 1 \end{bmatrix}$$

is Schur if and only if $\beta < 2/[2 + \alpha^2 + \sqrt{(2 + \alpha^2)^2 - 4}]$, in which β has to approach zero as α becomes sufficiently large.

Such a matrix pair of $\{D_1, D_2\}$ may arise from System 2.29. For instance, consider the case of (2.31) with $c > 0$:

$$A_1 = \begin{bmatrix} -1 & c \\ 0 & -1 \end{bmatrix}, \quad A_2 = A_1^T, \quad \text{and} \quad e^{A_1 \tau} = e^{-\tau} \begin{bmatrix} 1 & c\tau \\ 0 & 1 \end{bmatrix}.$$

Naturally, System 2.34 can exhibit similar properties.   ◇

The above discussions demonstrate that eigenvalue analysis does not generally apply to time-varying systems and that stability depends explicitly upon changes of the dynamics. In the case that all the changes are known *a priori*, eigenvalue analysis could be applied using the equivalent time-invariant system of (2.32) (if found). In general, stability of time-varying systems should be analyzed using such tools as the Lyapunov direct method.

2.4.3 Lyapunov Analysis of Linear Systems

The Lyapunov direct method can be used to handle linear and non-linear systems in a unified and systematic manner. Even for linear time-invariant systems, it provides different perspectives and insights than linear tools such as impulse response and eigenvalues. For Linear System 2.29, the Lyapunov function can always be chosen to be a *quadratic function* of form

$$V(x,t) = x^T P(t)x,$$

where $P(t)$ is a symmetric matrix and is called *Lyapunov function matrix*. Its time derivative along trajectories of System 2.29 is also quadratic as

$$\dot{V}(x,t) = -x^T Q(t)x,$$

where matrices $P(t)$ and $Q(t)$ are related by the so-called differential Lyapunov equation

$$\dot{P}(t) = -A^T(t)P(t) - P(t)A(t) - Q(t). \tag{2.37}$$

In the case that the system is time-invariant, $Q(t)$ and hence $P(t)$ can be selected to be constant, and (2.37) becomes algebraic.

To conclude asymptotic stability, we need to determine whether the two quadratic functions $x^T P(t)x$ and $x^T Q(t)x$ are positive definite. To this end, the following Rayleigh-Ritz inequality should be used. For any symmetric matrix H:

$$\lambda_{min}(H)\|x\|^2 \le x^T H x \le \lambda_{max}(H)\|x\|^2,$$

where $\lambda_{min}(H)$ and $\lambda_{max}(H)$ are the minimum and maximum eigenvalues of H, respectively. Thus, stability of linear systems can be determined by checking whether matrices P and Q are positive definite (p.d.). Positive definiteness of a symmetric matrix can be checked using one of the following tests:

(a) Eigenvalue test: a symmetric matrix P is p.d. (or p.s.d.) if and only if all its eigenvalues are positive (or non-negative).
(b) Principal minor test: a symmetric matrix P is p.d. (or p.s.d.) if and only if all its leading principal minors are positive (or non-negative).
(c) Factorization test: a symmetric matrix P is p.s.d. (or p.d.) if and only if $P = WW^T$ for some (invertible) matrix W.

(d) Gershgorin test: a symmetric matrix P is p.d. or p.s.d. if, for all $i \in \{1, \cdots, n\}$,

$$p_{ii} > \sum_{j=1, j \neq i}^{n} |p_{ij}| \quad \text{or} \quad p_{ii} \geq \sum_{j=1, j \neq i}^{n} |p_{ij}|,$$

respectively.

For time-varying matrices $P(t)$ and $Q(t)$, the above tests can be applied but should be strengthened to be uniform with respect to t.

As discussed in Section 2.3.1, Lyapunov function $V(x, t)$ and its matrix $P(t)$ should be solved using the backward process. The following theorem provides such a result.

Lemma 2.27. *Consider System 2.29 with uniformly bounded matrix $A(t)$. Then, it is uniformly asymptotically stable and exponentially stable if and only if, for every uniformly bounded and p.d. matrix $Q(t)$, solution $P(t)$ to (2.37) is positive definite and uniformly bounded.*

Proof: Sufficiency follows directly from a Lyapunov argument with Lyapunov function $V(x, t) = x^T P(t)x$. To show necessity, choose

$$P(t) = \int_t^\infty \Phi^T(\tau, t) Q(\tau) \Phi(\tau, t) d\tau, \tag{2.38}$$

where Q be p.d. and uniformly bounded, and $\Phi(\cdot, \cdot)$ is the state transition matrix in (2.30). It follows from the factorization test that $P(t)$ is positive definite. Since matrix $A(t)$ is uniformly bounded, $\Phi^T(t, \tau)$ is exponentially convergent and uniformly bounded if and only if the system is exponentially stable. Thus, $P(t)$ defined above exists and is also uniformly bounded if and only if the system is exponentially stable. In addition, it follows that

$$\dot{P}(t) = \int_t^\infty \frac{\partial \Phi^T(\tau, t)}{\partial t} Q(\tau) \Phi(\tau, t) d\tau + \int_t^\infty \Phi^T(\tau, t) Q(\tau) \frac{\partial \Phi(\tau, t)}{\partial t} d\tau - Q(t)$$
$$= -A^T(t)P(t) - P(t)A(t) - Q(t),$$

which is (2.37). This completes the proof.   □

To solve for Lyapunov function $P(t)$ from Lyapunov Equation 2.37, $A(t)$ needs to be known. Since finding $P(t)$ is computationally similar to solving for state transition matrix $\Phi(\cdot, \cdot)$, finding a Lyapunov function may be quite difficult for some linear time-varying systems. For linear time-invariant systems, Lyapunov function matrix P is constant as the solution to either the algebraic Lyapunov equation or the integral given below:

$$A^T P + PA = -Q, \quad P = \int_0^\infty e^{A^T \tau} Q e^{A \tau} d\tau, \tag{2.39}$$

where Q is any positive definite matrix.

The Lyapunov direct method can be applied to analyze stability of linear piecewise-constant systems. If System 2.20 is linear and has arbitrary switching, a common Lyapunov function exists, but may not be quadratic as indicated by the counterexample in [49]. Nonetheless, it is shown in [158] that the common Lyapunov function can be chosen to be of the following homogeneous form of degree 2: for some vectors $c_i \in \Re^n$,

$$V(x) = \max_{1 < i < k} V_i(x), \quad V_i(x) = x^T c_i c_i^T x.$$

Using the notations in (2.22), a linear piecewise-constant system can be expressed as

$$\dot{x} = A_{s(t_i)} x, \quad s(t_i) \in \mathcal{I}. \tag{2.40}$$

Assuming that A_s be Hurwitz for all $s \in \mathcal{I}$, we have $V_s(x) = x^T P_s x$ where P_s is the solution to algebraic Lyapunov equation

$$A_s^T P_s + P_s A_s = -I.$$

Suppose that $t_i < t_{i+1} < t_j$ and that $s(t_i) = s(t_j) = p$ and $s(t_{i+1}) = q \neq p$. It follows that, for $t \in [t_i, t_{i+1})$,

$$\dot{V}_p = -x^T x \leq -\sigma_p V_p,$$

and hence

$$V_p(x(t_{i+1})) \leq e^{-\sigma_p \tau_{i+1}} V_p(x(t_i)),$$

where $\tau_{i+1} = t_{i+1} - t_i$ is the dwell time, and $\sigma_p = 1/\lambda_{\max}(P_p)$ is the time constant. Similarly, it follows that, over the interval $[t_{i+1}, t_{i+2})$

$$V_q(x(t_{i+2})) \leq e^{-\sigma_q \tau_{i+2}} V_q(x(t_{i+1})).$$

Therefore, we have

$$
\begin{aligned}
&V_p(x(t_{i+2})) - V_p(x(t_i)) \\
&\leq \lambda_{max}(P_p) \| x(t_{i+2}) \|^2 - V_p(x(t_i)) \\
&\leq \frac{\lambda_{max}(P_p)}{\lambda_{min}(P_q)} V_q(x(t_{i+2})) - V_p(x(t_i)) \\
&\leq \frac{\lambda_{max}(P_p)}{\lambda_{min}(P_q)} e^{-\sigma_q \tau_{i+2}} V_q(x(t_{i+1})) - V_p(x(t_i)) \\
&\leq \frac{\lambda_{max}(P_p)}{\lambda_{min}(P_p)} \frac{\lambda_{max}(P_q)}{\lambda_{min}(P_q)} e^{-\sigma_q \tau_{i+2}} V_p(x(t_{i+1})) - V_p(x(t_i)) \\
&\leq -\left[1 - \frac{\lambda_{max}(P_p)}{\lambda_{min}(P_p)} \frac{\lambda_{max}(P_q)}{\lambda_{min}(P_q)} e^{-\sigma_q \tau_{i+2}} e^{-\sigma_p \tau_{i+1}} \right] V_p(x(t_i)),
\end{aligned}
$$

which is always negative definite if either time constants σ_s are sufficiently small or dwell times τ_k are sufficiently long or both, while the ratios of

$\lambda_{max}(P_s)/\lambda_{min}(P_s)$ remain bounded. By induction and by applying Theorem 2.24, we know that asymptotic stability is maintained for a linear system whose dynamics switch relatively slowly among a finite number of relatively fast dynamics of exponentially stable autonomous systems. This result illustrates robustness of asymptotic stability with respect to switching, but it is quite conservative (since the above inequality provides the worst estimate on transient overshoots of the switching system). Similar analysis can also be done for Discrete-time System 2.36.

In the special case that matrices A_s are all commuting Hurwitz matrices, it has been shown in Section 2.4.2 that System 2.40 is always exponentially stable for all values of time constants and dwell times. Furthermore, System 2.40 has a quadratic common Lyapunov function in this case. For instance, consider the case that $\mathcal{I} = \{1,2\}$ and let P_1 and P_2 be the solutions to Lyapunov equations

$$A_1^T P_1 + P_1 A_1 = -I, \quad A_2^T P_2 + P_2 A_2 = -P_1.$$

Then, P_2 is the common Lyapunov function matrix since $e^{A_1 \tau_1}$ and $e^{A_2 \tau_2}$ commute and, by (2.39),

$$P_2 = \int_0^\infty e^{A_2^T \tau_2} P_1 e^{A_2 \tau_2} d\tau_2 = \int_0^\infty e^{A_1^T \tau_1} W_2 e^{A_1 \tau_1} d\tau_1,$$

where

$$W_2 = \int_0^\infty e^{A_2^T \tau_2} e^{A_2 \tau_2} d\tau_2$$

is positive definite. Given any finite index set $\mathcal{I}$, the above observation leads to a recursive process of finding the quadratic common Lyapunov function [173].

2.5 Controllability

Roughly speaking, a control system is locally controllable if a steering law for its control can be found to move the state to any specified point in a neighborhood of its initial condition. On the other hand, Affine System 2.11 can be rewritten as

$$\dot{x} = f(x) + \sum_{i=1}^m g_i(x)u_i = \sum_{i=0}^m g_i(x)u_i, \tag{2.41}$$

where $x \in \Re^n$, $u_0 = 1$, $g_0(x) \triangleq f(x)$, and vector fields $g_i(x)$ are assumed to be analytic and linearly independent. Thus, controllability should be determined by investigating time evolution of the state trajectory of (2.41) under all possible control actions.

To simplify the notation, we introduce the so-called *Lie bracket*: for any pair of $f(x), g(x) \in \Re^n$, $[f, g]$ or $ad_f g$ (where ad stands for "adjoint") is a vector function of the two vector fields and is defined by

$$[f, g] = (\nabla_x^T g)f - (\nabla_x^T f)g.$$

High-order Lie brackets can be defined recursively as, for $i = 1, \cdots$,

$$ad_f^0 g = g, \quad \text{and} \quad ad_f^i g = [f, ad_f^{i-1} g].$$

An important property associated with the Lie derivative and Lie bracket is the so-called *Jacobi identity*: for any $f(x), g(x) \in \Re^n$ and $\xi(x) \in \Re$,

$$L_{ad_f g} \xi = L_f L_g \xi - L_g L_f \xi, \tag{2.42}$$

and it can easily be verified by definition. The following are the concepts associated with Lie bracket.

Definition 2.28. *A set of linearly independent vector fields $\{\xi_l(x) : l = 1, \cdots, k\}$ is said to be* involutive *or* closed under Lie bracket *if, for all $i, j \in \{1, \cdots, k\}$, the Lie bracket $[\xi_i, \xi_j]$ can be expressed as a linear combination of ξ_1 up to ξ_k.*

Definition 2.29. *Given smooth linearly independent vector fields $\{\xi_l(x) : l = 1, \cdots, k\}$, the tangent-space distribution is defined as*

$$\Delta(x) = span\{\xi_1(x), \xi_2(x), \cdots, \xi_k(x)\}.$$

$\Delta(x)$ *is* regular *if the dimension of $\Delta(x)$ does not vary with x. $\Delta(x)$ is* involutive *if $[f, g] \in \Delta$ for all $f, g \in \Delta$. Involutive closure $\overline{\Delta}$ of $\Delta(x)$, also called* Lie algebra, *is the smallest distribution that contains Δ and is closed under Lie bracket.*

While Lie brackets of $g_i(x)$ do not have an explicit physical meaning, the span of their values consists of all the incremental movements achievable for System 2.41 under modulated inputs. For instance, consider the following two-input system:

$$\dot{x} = g_1(x)u_1 + g_2(x)u_2, \quad x(0) = x_0.$$

Should piecewise-constant inputs be used, there are only two linearly independent values for the input vector at any fixed instant of time, and they can be used to construct any steering control input over time. For instance, the following is a piecewise-constant steering control input:

$$\left[\, u_1(t)\ u_2(t)\,\right]^T = \begin{cases} \left[1\ 0\right]^T & t \in (0, \varepsilon] \\ \left[0\ 1\right]^T & t \in (\varepsilon, 2\varepsilon] \\ \left[-1\ 0\right]^T & t \in (2\varepsilon, 3\varepsilon] \\ \left[0\ -1\right]^T & t \in (3\varepsilon, 4\varepsilon] \end{cases},$$

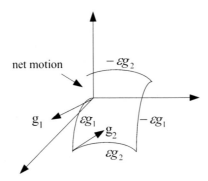

Fig. 2.1. Incremental motion in terms of Lie bracket

where $\varepsilon > 0$ is some sufficiently small constant. One can calculate the second-order Taylor series expansions of $x(i\varepsilon)$ recursively for $i = 1, 2, 3, 4$ and show that

$$[g_1, g_2](x_0) = \lim_{h \to 0} \frac{x(4\varepsilon) - x_0}{\varepsilon^2}.$$

As shown by Fig. 2.1, Lie bracket $[g_1, g_2](x_0)$ is the net motion generated under the steering control. In general, controllability of Affine System 2.11 can be determined in general by a rank condition on its Lie algebra, as stated by the following theorem [40, 114].

Definition 2.30. *A system is* controllable *if, for any two points* $x_0, x_f \in \Re^n$, *there exist a finite time* T *and a control* $u(x, t)$ *such that the system solution satisfies* $x(t_0) = x_0$ *and* $x(t_0 + T) = x_f$. *The system is* small-time locally controllable *at* x_1 *if* $u(x, t)$ *can be found such that* $x(t_0 + \delta t) = x_1$ *for sufficiently small* $\delta t > 0$ *and that* $x(t)$ *with* $x(t_0) = x_0$ *stays near* x_1 *at all times.*

Theorem 2.31. *Affine System 2.41 is small-time locally controllable at* x *if the involutive closure of*

$$\Delta(x) = \{f(x),\ g_1(x),\ \cdots,\ g_m(x)\}$$

is of dimension n, *that is, the rank of the controllability Lie algebra is* n.

If an affine system is not controllable, controllability decomposition can be applied to separate controllable dynamics from uncontrollable dynamics. For non-affine systems, conditions could be obtained using local linearization with respect to control u. More details on non-linear controllability can be found in texts [12, 96, 175, 223]. For Linear System 2.25, the controllability Lie algebra reduces to

$$\{g, ad_f g, \cdots, ad_f^{n-1} g\} = \{B, AB, \cdots, A^{n-1}B\},$$

which is the linear controllability matrix $\mathcal{C} = \begin{bmatrix} B & AB & A^2 B & \cdots & A^{n-1}B \end{bmatrix}$. And, System 2.25 is said to be in the controllable canonical form if the matrix pair (A, B) is given by

$$A_c \triangleq \begin{bmatrix} 0 & 1 & 0 & \cdots & 0 \\ 0 & 0 & 1 & \cdots & 0 \\ \vdots & \vdots & \vdots & \ddots & \vdots \\ 0 & 0 & 0 & \cdots & 1 \\ 0 & 0 & 0 & \cdots & 0 \end{bmatrix}, \quad B_c \triangleq \begin{bmatrix} 0 \\ 0 \\ \vdots \\ 0 \\ 1 \end{bmatrix}. \tag{2.43}$$

Consider the linear time-varying system:

$$\dot{x} = A(t)x + B(t)u, , \tag{2.44}$$

where matrices $A(t)$ and $B(t)$ are uniformly bounded and of proper dimension. Their controllability can directly be studied using its solution

$$x(t) = \Phi(t, t_0) x(t_0) + \int_{t_0}^t \Phi(t, \tau) B(\tau) u(\tau) d\tau$$

$$= \Phi(t, t_0) \left[x(t_0) + \int_{t_0}^t \Phi(t_0, \tau) B(\tau) u(\tau) d\tau \right].$$

It is straightforward to verify that, if the matrix

$$W_c(t_0, t_f) = \int_{t_0}^{t_f} \Phi(t_0, \tau) B(\tau) B^T(\tau) \Phi^T(t_0, \tau) d\tau \tag{2.45}$$

is invertible, the state can be moved from any x_0 to any x_f under control

$$u(t) = -B^T(t) \Phi^T(t_0, t) W_c^{-1}(t_0, t_f)[x_0 - \Phi(t_0, t_f) x_f].$$

Matrix $W_c(t_0, t_f)$ is the so-called *controllability Gramian*, and it is easy to show [39] that $W_c(t_0, t_f)$ is invertible and hence System 2.44 is controllable if and only if the n rows of n-by-m matrix function $\Phi(t_0, t)B(t)$ are linear independent over time interval $[t_0, t_f]$ for some $t_f > t_0$. To develop a controllability condition without requiring the solution of state transition matrix $\Phi(t_0, t)$, it follows that

$$\frac{\partial^k}{\partial t^k} \Phi(t_0, t) B(t) = \Phi(t_0, t) E_k(t), \quad k = 0, 1, \cdots, (n-1),$$

where $E_0(t) = B(t)$ and $E_{k+1}(t) = -A(t)E_k(t) + \dot{E}_k(t)$ for $k = 0, 1, \cdots, (n-1)$. Hence, controllability of System 2.44 is equivalent to the Kalman rank condition of

$$\text{rank} \begin{bmatrix} E_0(t) & E_1(t) & \cdots & E_{n-1}(t) \end{bmatrix} = n,$$

Comparing (2.45) and (2.38), we know that, under the choice of positive semi-definite matrix $Q(t) = B(t)B^T(t)$, Lyapunov Equation 2.37 still yields positive definite Lyapunov function $P(t)$ provided that System 2.44 is controllable and that its uncontrolled dynamics in (2.29) are asymptotically stable. To design a Lyapunov-based control and to ensure global and uniformly asymptotic stability, it is necessary that the system solution is uniformly bounded and that Lyapunov function matrix $P(t)$ exists and is also uniformly bounded, which leads to the following definition and theorem [103]. Later in Section 2.6.3, Control 2.46 below is shown to be optimal as well.

Definition 2.32. *System 2.44 is said to be* uniformly completely controllable *if the following two inequalities hold for all t:*

$$0 < \alpha_{c1}(\delta)I \le W_c(t, t+\delta) \le \alpha_{c2}(\delta)I, \quad \|\Phi(t, t+\delta)\| \le \alpha_{c3}(\delta),$$

where $W_c(t_0, t_f)$ is defined by (2.45), $\delta > 0$ is a fixed constant, and $\alpha_{ci}(\cdot)$ are fixed positively-valued functions.

Theorem 2.33. *Consider System 2.44 under control*

$$u = -R^{-1}(t)B^T(t)P(t)x, \tag{2.46}$$

where matrices $Q(t)$ and $R(t)$ are chosen to be positive definite and uniformly bounded, and matrix $P(t)$ is the solution to the differential Riccati *equation*

$$\dot{P} + [PA + A^T P - PBR^{-1}B^T P + Q] = 0 \tag{2.47}$$

with terminal condition of $P(\infty)$ being positive definite. Then, if System 2.44 is uniformly completely controllable, Lyapunov function $V(x,t) = x^T P(t)x$ is positive definite and decrescent, and Control 2.46 is asymptotically stabilizing.

2.6 Non-linear Design Approaches

In this section, three popular design methods are outlined for non-linear systems; they are backstepping design, feedback linearization, and optimal control.

2.6.1 Recursive Design

In this section, we focus upon the following class of *feedback systems*:

$$\begin{cases} \dot{x}_1 = f_1(x_1, t) + x_2, \\ \dot{x}_2 = f_2(x_1, x_2, t) + x_3, \\ \vdots \\ \dot{x}_n = f_n(x_1, \cdots, x_n, t) + u, \end{cases} \tag{2.48}$$

where $x_i \in \Re^l$, x_1 is the output, and u is the input. Letting x_n up to x_1 be the outputs of a chain of pure integrators, we can graphically connect the inputs to the integrators by using dynamic equations in (2.48). In the resulting block diagram, there is a feedforward chain of integrators, and all other connections are feedback. In particular, the ith integrator has x_{i+1} and x_i as its input and output, respectively. Should $f_i(x_1, \cdots, x_i, t) = f_i(x_i, t)$, System 2.48 would become a cascaded chain of first-order non-linear sub-systems. As such, System 2.48 is also referred to as a *cascaded system*. Several classes of physical systems, especially electromechanical systems [130, 195, 242], have by nature this structural property on their dynamics, and their controllability is guaranteed.

The aforementioned structural property of cascaded dynamics provides a natural and intuitive way for finding both Lyapunov function and a stabilizing control. The systematic procedure, called *backstepping* or *backward recursive design* [72, 116, 192], is a step-by-step design process in which the first-order sub-systems are handled one-by-one and backwards from the output x_1 back to the input u. Specifically, let us consider the case of $n = 2$ and begin with the first sub-system in (2.48), that is,

$$\dot{x}_1 = f_1(x_1, t) + x_2.$$

If x_2 were a control variable free to be selected, a choice of x_2 would easily be found to stabilize the first sub-system. Because x_2 is a state variable rather than a control, we rewrite the first sub-system as

$$\dot{x}_1 = f_1(x_1, t) + x_2^d(x_1, t) + [x_2 - x_2^d(x_1, t)],$$

and choose the fictitious control $x_2^d(x_1, t)$ (such as the most obvious choice of $x_2^d(x_1, t) = -f_1(x_1, t) - x_1$) to stabilize asymptotically fictitious system $\dot{x}_1 = f_1(x_1, t) + x_2^d(x_1, t)$ by ensuring the inequality

$$2x_1^T[f_1(x_1, t) + x_2^d(x_1, t)] \leq -\|x_1\|^2.$$

Thus, choosing Lyapunov sub-function $V_1(x_1) = \|x_1\|^2$, we have that, along the trajectory of (2.48),

$$\dot{V}_1 \leq -2\|x_1\|^2 + 2x_1^T z_2,$$

where $z_2 = x_2 - x_2^d(x_1, t)$ is a transformed state variable of x_2. As shown in Section 2.3.2, the above inequality of $\dot{V}_1$ implies that, if z_2 is asymptotically convergent, so is x_1. To ensure asymptotic convergence of z_2, we know from its definition that

$$\dot{z}_2 = u - \frac{\partial x_2^d}{\partial t} - \left(\frac{\partial x_2^d}{\partial x_1}\right)^T [f_1(x_1, t) + x_2].$$

As before, the above dynamic equation of z_2 is of first-order and hence control u can be easily found (for instance, the choice rendering $\dot{z}_2 = -z_2$) such that, with $V_2(x_1, x_2) = \|z_2\|^2$,

$$\dot{V}_2 \le -2\|z_2\|^2.$$

Therefore, we now have found Lyapunov function $V = V_1 + \alpha_2 V_2$ whose time derivative is

$$\dot{V} = -2\|x_1\|^2 + 2x_1^T z_2 - 2\alpha_2 \|z_2\|^2,$$

which is negative definite for any choice of $\alpha_2 > 1/4$. Hence, under the control u selected, both x_1 and z_2 and consequently both x_1 and x_2 are globally asymptotically stable. It is straightforward to see that, by induction, a stabilizing control and the corresponding Lyapunov function can be found recursively for System 2.48 of any finite-order n.

2.6.2 Feedback Linearization

The feedback linearization approach provides the conditions under which a pair of state and control transformations exist such that a non-linear system is mapped (either locally or globally) into the linear controllable canonical form. Specifically, the objective of feedback linearization is to map System 2.11 into the form

$$\dot{z} = A_c z + B_c v,$$
$$\dot{w} = \phi(z, w), \tag{2.49}$$

where the pair $\{A_c, B_c\}$ is that in (2.43), $z(x) \in \Re^r$ is the state of the feedback linearized sub-system, r is the so-called relative degree of the system, w is the state of so-called internal dynamics, $[z^T \ w^T] = [z^T(x) \ w^T(x)]$ is the state transformation, and $v = v(u, x)$ is the control transformation. Then, standard linear control results can be applied through the transformations to the original non-linear system provided that the internal dynamics are minimum phase (*i.e.*, the zero dynamics of $\dot{w} = \phi(0, w)$ are asymptotically stable).

By (2.49), System 2.11 with $m = 1$ is feedback linearizable if function $h(x) \in \Re$ exists such that

$$z_1 = h(x); \quad \dot{z}_i = z_{i+1}, \ i = 1, \cdots, r-1; \quad \dot{z}_r = v; \quad \dot{w} = \phi(z, w) \tag{2.50}$$

for some vector functions $w(x)$ and $\phi(\cdot)$ and for transformed control $v(u, x)$. It follows from dynamics of System 2.11 that $\dot{z}_i = L_f z_i + (L_g z_i)u$. Therefore, equations of (2.50) are equivalent to

$$\begin{cases} L_g w_j = 0, \ j = 1, \cdots, n-r; \\ L_g L_f^{i-1} h = 0, \ i = 1, \cdots, r-1; \\ L_g L_f^r h \ne 0, \end{cases} \tag{2.51}$$

while the control transformation and linearized state variables are defined by

$$v = L_f z_r + (L_g z_r)u; \quad z_1 = h(x), \quad z_{i+1} = L_f z_i \ i = 1, \cdots, r-1. \tag{2.52}$$

By Jacobi Identity 2.42, $L_g h = L_g L_f h = 0$ if and only if $L_g h = L_{ad_f g} h = 0$. By induction, we can rewrite the partial differential equations in (2.51) as

$$\begin{cases} L_g w_j = 0, \ j = 1, \cdots, n - r; \\ L_{ad_f^i g} h = 0, \ i = 1, \cdots, r - 1; \\ L_g L_f^r h \neq 0. \end{cases} \tag{2.53}$$

In (2.53), there are $(n - 1)$ partial differential equations that are all homogeneous and of first-order. Solutions to these equations can be found under rank-based conditions, and they are provided by the following theorem often referred to as the Frobenius theorem [96]. For control design and stability analysis, the transformation from x to $[z^T, w^T]^T$ needs to be diffeomorphic (*i.e.*, have a unique inverse), which is also ensured by Theorem 2.34 since, by implicit function theorem [87], the transformation is diffeomorphic if its Jacobian matrix is invertible and since the Jacobian matrix consists of $\triangledown q_j(x)$, the gradients of the solutions.

Theorem 2.34. *Consider $k(n-k)$ first-order homogeneous partial differential equations:*

$$L_{\xi_i} q_j = 0, \quad i = 1, \cdots, k; \quad j = 1, \cdots, n - k, \tag{2.54}$$

where $\{\xi_1(x), \xi_2(x), \cdots, \xi_k(x)\}$ is a set of linearly independent vectors in $\Re^n$, and $q_j(x)$ are the functions to be determined. Then, the solutions $q_j(x)$ to (2.54) exist if and only if distribution $\{\xi_1(x), \xi_2(x), \cdots, \xi_k(x)\}$ is involutive. Moreover, under the involutivity condition, the gradients of the solutions, $\triangledown q_j(x)$, are linearly independent.

Applying Theorem 2.34 to (2.53), we know that System 2.11 is feedback linearizable with relative degree $r = n$ (*i.e.*, full state feedback linearizable) if the set of vector fields $\{g, ad_f g, \cdots, ad_f^{n-2} g\}$ is involutive and if matrix $[g \ ad_f g \ \cdots \ ad_f^{n-1} g]$ is of rank n. Upon verifying the conditions, the first state variable $z_1 = h(x)$ can be found such that $L_{ad_f^i g} h = 0$ for $i = 1, \cdots, n - 1$ but $L_g L_f^n h \neq 0$, the rest of state variables and the control mapping are given by (2.52). The following theorem deals with the case of $m \geq 1$ and $r = n$, and a similar result can be applied for the general case of $r < n$ [96].

Theorem 2.35. *System 2.11 is feedback linearizable (i.e., can be mapped into the linear controllable canonical form) under a diffeomorphic state transformation $z = T(x)$ and a control mapping $u = \alpha(x) + \beta(x)v$ if and only if the nested distributions defined by $\mathcal{D}_0 = \text{span} \{ g_1 \cdots g_m \}$ and $\mathcal{D}_i = \mathcal{D}_{i-1} + \text{span} \{ ad_f^i g_1 \cdots ad_f^i g_m \}$ with $i = 1, \cdots, n - 1$ have the properties that $\mathcal{D}_l$ are all involutive and of constant rank for $0 \leq l \leq n - 2$ and that rank $\mathcal{D}_{n-1} = n$.*

As shown in Section 2.5, matrix $[g \ ad_f g \ \cdots \ ad_f^{n-1} g]$ having rank n is ensured by non-linear controllability. The involutivity condition on set

$\{g\ ad_f g\ \cdots\ ad_f^{n-2} g\}$ ensures the existence and diffeomorphism of state and control transformations but does not have a clear intuitive explanation. Note that the involutivity condition is always met for linear systems (since the set consists of constant vectors only).

2.6.3 Optimal Control

For simplicity, consider Affine System 2.11 and its optimal control problem over the infinite horizon. That is, our goal is to find control u such that the following performance index

$$J(x(t_0)) = \int_{t_0}^{\infty} L(x(t), u(t))dt \tag{2.55}$$

is minimized, subject to Dynamic Equation 2.11 and terminal condition of $x(\infty) = 0$. There are two approaches to solve the optimal control problem: the Euler-Lagrange method based on Pontryagin minimum principle, and the principle of optimality in dynamic programming.

By using a Lagrange multiplier, the Euler-Lagrange method converts an optimization problem with equality constraints into one without any constraint. That is, System Equation 2.11 is adjoined into performance index J as

$$J(t_0) = \int_{t_0}^{\infty} \{L(x, u) + \lambda^T [f(x) + g(x)u - \dot{x}]\}dt = \int_{t_0}^{\infty} [H(x, u, \lambda, t) - \lambda^T \dot{x}]dt,$$

where $\lambda \in \Re^n$ is the Lagrange multiplier, and

$$H(x, u, \lambda, t) = L(x, u) + \lambda^T [f(x) + g(x)u] \tag{2.56}$$

is the Hamiltonian. Based on calculus of variations [187], J is optimized locally by u if $\delta^1 J = 0$, where $\delta^1 J$ is the first-order variation of J due to variation δu in u and its resulting state variation δx in x. Integrating in part the term containing $\dot{x}$ in J and then finding the expression of its variation, one can show that $\delta^1 J = 0$ holds under the following equations:

$$\dot{\lambda} = -\frac{\partial H}{\partial x} = -\frac{\partial L}{\partial x} - \left(\frac{\partial f}{\partial x}\right)^T \lambda - \sum_{i=1}^{n} \sum_{j=1}^{n_u} \lambda_i \frac{\partial g_{ij}}{\partial x} u_j, \quad \lambda(\infty) = 0, \tag{2.57}$$

$$0 = \frac{\partial H}{\partial u} = \frac{\partial L}{\partial u} + g^T \lambda, \quad \text{or} \quad H(x, \lambda, u) = \min_u H(x, \lambda, u). \tag{2.58}$$

If instantaneous cost $L(\cdot)$ is a p.d. function, optimal control can be solved from (2.58). Equation 2.58 is referred to as the *Pontryagin minimum principle*. Equation 2.57 is the so-called costate equation; it should be solved simultaneously with State Equation 2.11. Equations 2.11 and 2.57 represent a two-point boundary-value problem, and they are necessary for optimality but generally not sufficient. To ensure that the value of J is a local minimum, its second-order variation $\delta^2 \mathcal{J}$ must be positive, that is, the following Hessian matrix of Hamiltonian H should be positive definite:

$$\mathcal{H}_H \triangleq \begin{bmatrix} \dfrac{\partial^2 H}{\partial x^2} & \dfrac{\partial^2 H}{\partial x \partial u} \\[2mm] \dfrac{\partial^2 H}{\partial u \partial x} & \dfrac{\partial^2 H}{\partial u^2} \end{bmatrix}.$$

For the linear time-varying system in (2.44), it is straightforward to verify that $\lambda = P(t)x$ and $\mathcal{J}(t_0) = V(x(t_0), t_0)$ and that Control 2.46 is optimal.

Alternatively, an optimal control can be derived by generalizing Performance Index 2.55 into

$$J^*(x(t)) = \inf_u \int_t^\infty L(x(\tau), u(\tau)) d\tau,$$

which can be solved using dynamic programming. The *principle of optimality* states that, if u^* is the optimal control under Performance Index 2.55, control u^* is also optimal with respect to the above measure for the same system but with initial condition $x(t) = x^*(t)$. Applying the principle of optimality to $J^*(x(t + \delta t))$ and invoking Taylor series expansion yield the following partial differential equation:

$$\min_u H(x, u, \lambda)\Big|_{\lambda = \nabla_x J^*(x)} = -\nabla_t J^*(x) = 0. \tag{2.59}$$

which is so-called Hamilton-Jacobi-Bellman (HJB) equation [10, 16, 33, 97]. For Affine System 2.11 and under the choice of $L(x, u) = \alpha(x) + \|u\|^2$, the HJB equation reduces to

$$L_f J^* - \frac{1}{4} \|L_g J^*\|^2 + \alpha(x) = 0, \tag{2.60}$$

where $\alpha(x) \geq 0$ is any scalar function. If the above partial differential equation has a continuously differentiable and positive definite solution $J^*(x)$, the optimal control is

$$u^*(x) = -\frac{1}{2}(L_g J^*)^T,$$

under which the closed-loop system is asymptotically stable. However, solving HJB Equation 2.60 in general is quite difficult even for Affine System 2.11.

2.6.4 Inverse Optimality and Lyapunov Function

Solving HJB Equation 2.60 is equivalent to finding control Lyapunov function $V(x)$. If J^* is known, the Lyapunov function can be set as $V(x) = J^*(x)$, and the optimal control is stabilizing since

$$L_f J^* + (L_g J^*) u^* = -\frac{1}{4} \| L_g J^* \|^2 - \alpha(x) \le 0.$$

On the other hand, given a control Lyapunov function $V(x)$, a class of stabilizing controls for System 2.11 can be found, and performance under any of these controls can be evaluated. For instance, consider Control 2.19 in Section 2.3.3. Although optimal value function $J^*(x)$ is generally different from $V(x)$, we can assume that, for some scalar function $k(x)$, $\nabla_x J^* = 2k(x) \nabla_x V$. That is, in light of stability outcome, J^* has the same level curves as those of V. Substituting the relationship into (2.60) yields

$$2k(x) L_f V - k^2(x) \| L_g V \|^2 + \alpha(x) = 0,$$

which is a quadratic equation in $k(x)$. Choosing the solution with positive square root, we know that

$$k(x) = \frac{L_f V + \sqrt{(L_f V)^2 + \| L_g V \|^2 \alpha(x)}}{\| L_g V \|^2}$$

and that optimal control $u^*(x)$ reduces to Control 2.19. That is, for any choice of $\alpha(x)$, Control 2.19 is optimal with respect to J^*, which is called inverse optimality [72, 234].

In the special case of linear time-invariant systems, the relationship between the Lyapunov-based control design and inverse optimality is more straightforward. Should System 2.25 be controllable, a stabilizing control of general form $u = -Kx$ exists such that $A - BK$ is Hurwitz. For any pair of p.d. matrices $Q \in \Re^{n \times n}$ and $R \in \Re^{m \times m}$, matrix sum $(Q + K^T RK)$ is p.d. and hence, by Lemma 2.27, solution P to Lyapunov equation

$$P(A - BK) + (A - BK)^T P = -(Q + K^T RK)$$

is p.d. Thus, it follows from System 2.25 under control $u = -Kx$ that

$$x^T(t) P x(t) = -\int_t^\infty \frac{dx^T P x}{d\tau} d\tau = \int_t^\infty (x^T Q x + u^T Ru) d\tau.$$

Thus, every stabilizing control is inversely optimal with respect to a quadratic performance index, and the optimal performance value over $[t, \infty)$ is the corresponding quadratic Lyapunov function.

2.7 Notes and Summary

The Lyapunov direct method [141] is a universal approach for both stability analysis and control synthesis of general systems, and finding a Lyapunov function is the key [108, 264]. As shown in Sections 2.3.2 and 2.6, the Lyapunov function [141] or the pair of control Lyapunov function [235, 236] and the corresponding controller can be searched for by analytically exploiting system properties, that is, a control Lyapunov function [8] can be constructed for systems of special forms [261], for systems satisfying the Jurdejevic-Quinn conditions [61, 152], for feedback linearizable systems [94, 96, 99, 175], and for all the systems to which recursive approaches are applicable [116, 192]. Numerically, a Lyapunov function can also be found by searching for its dual of density function [210] or the corresponding sum-of-square numerical representation in terms of a polynomial basis [190]. Alternative approaches are that a numerical solution to the HJB equation yields the Lyapunov function as an optimal value function [113] and that set-oriented partition of the state space renders discretization and approximation to which graph theoretical algorithms are applicable [81]. For linear control systems, the problem reduces to a set of linear matrix inequalities (LMI) which are convex and can be solved using semi-definite programming tools [27], while linear stochastic systems can be handled using Perron-Frobenius operator [121].

Under controllability and through the search of control Lyapunov function, a stabilizing control can be designed for dynamic systems. In Chapter 3, several vehicle-level controls are designed for the class of non-holonomic systems that include various vehicles as special cases. To achieve cooperative behaviors, a team of vehicles needs to be controlled through a shared sensing/communication network. In Chapter 4, a mathematical representation of the network is introduced, and a matrix-theoretical approach is developed by extending the results on piecewise-constant linear systems in Section 2.4.2. In Chapter 5, cooperative stability is defined, and the matrix-theoretical approach is used to study cooperative controllability over a network, to design a linear cooperative control, and to search for the corresponding control Lyapunov function. Due to the changes in the network, a family of control Lyapunov functions would exist over consecutive time intervals in a way similar to those in Theorem 2.24. Since the networked changes are uncertain, the control Lyapunov functions cannot be determined and their changes over time cannot be assessed. Nonetheless, all the control Lyapunov functions are always quadratic and have the same square components, and these components can be used individually and together as a vector of Lyapunov function components to study stability. This observation leads to the Lyapunov function component-based methodology which is developed in Chapter 6 to extend both Theorems 2.10 and 2.24 and to analyze and synthesize non-linear cooperative systems.

3

Control of Non-holonomic Systems

Analysis and control of non-holonomic systems are addressed in this chapter. In order to proceed with the analysis and control design in a more systematic way, the so-called chained form is introduced as the canonical form for non-holonomic systems. As examples, kinematic equations of several vehicles introduced in Chapter 1 are transformed into the chained form. Then, a chained system is shown to be non-linearly controllable but not uniformly completely controllable in general, and it is partially and dynamically feedback linearizable, but can only be stabilized under a discontinuous and/or time-varying control. Based on these properties, open-loop steering controls are synthesized to yield a trajectory for the constrained system to move either continually or from one configuration to another. To ensure that a continual and constrained trajectory is followed, feedback tracking controls can be designed and implemented with the steering control. To move the system to a specific configuration, stabilizing controls can be used. For a vehicle system, the formation control problem is investigated. To make the vehicle system comply with environmental changes, either a real-time optimized path planning algorithm or a multi-objective reactive control can be deployed.

3.1 Canonical Form and Its Properties

In this section, the chained form is introduced as the canonical form for the kinematic constraints of non-holonomic systems. It is shown that, through state and control transformations, the kinematic sub-system of a non-holonomic system can be mapped into the chained form or one of its extensions. Vehicle models introduced in Chapter 1 are used as examples to determine the state and control transformations. Properties of the chained form are studied for the purpose of systematic control designs in the subsequent sections. In particular, controllability, feedback linearizability, existence of smooth control, and uniform complete controllability are detailed for chained form systems.

3.1.1 Chained Form

The n-variable single-generator m-input chained form is defined by [169, 266]:

$$
\begin{cases}
\dot{x}_1 = u_1, \\
\dot{x}_{21} = u_1 x_{22}, \quad \cdots, \quad \dot{x}_{2(n_2-1)} = u_1 x_{2n_2}, \quad \dot{x}_{2n_2} = u_2, \\
\vdots \\
\dot{x}_{m1} = u_1 x_{m2}, \quad \cdots, \quad \dot{x}_{m(n_m-1)} = u_1 x_{mn_m}, \quad \dot{x}_{mn_m} = u_m,
\end{cases}
\tag{3.1}
$$

where $x = [x_1, x_{21}, \cdots, x_{2n_2}, \cdots, x_{m1}, \cdots, x_{mn_m}]^T \in \Re^n$ is the state, $u = [u_1, \cdots, u_m]^T \in \Re^m$ is the control, and $y = [x_1, x_{21}, \cdots, x_{m1}]^T \in \Re^m$ is the output. If $m = 2$, Chained Form 3.1 reduces to

$$
\dot{x}_1 = u_1, \quad \dot{x}_2 = u_1 x_3, \quad \cdots, \quad \dot{x}_{n-1} = u_1 x_n, \quad \dot{x}_n = u_2.
\tag{3.2}
$$

Since analysis of and control design for (3.1) and (3.2) are essentially identical, we will focus mostly upon the two-input chained form. Through reordering the state variables, (3.2) renders its alternative expressions:

$$
\dot{z}_1 = u_1, \quad \dot{z}_2 = u_2, \quad \dot{z}_3 = z_2 u_1, \quad \cdots, \quad \dot{z}_n = z_{n-1} u_1,
\tag{3.3}
$$

where $z_1 = x_1$, and $z_j = x_{n-j+2}$ for $j = 2, \cdots, n$.

Given a mechanical system subject to non-holonomic constraints, it is often possible to convert its constraints into the chained form either locally or globally by using a coordinate transformation and a control mapping. The transformation process into the chained form and the conditions are parallel to those of feedback linearization studied in Section 2.6.2. For example, consider a two-input kinematic system in the form of (1.12), that is,

$$
\dot{q} = g_1(q)v_1 + g_2(q)v_2,
\tag{3.4}
$$

where $q \in \Re^n$ with $n > 2$, vector fields g_1 and g_2 are linearly independent, and $v = [v_1, v_2]^T \in \Re^2$ is the vector of original control inputs. Should System 3.4 be mapped into Chained Form 3.2, we know from direct differentiation and Jacobi Identity 2.42 that $x_1 = h_1(q)$ and $x_2 = h_2(q)$ exist and satisfy the following conditions:

$$
\begin{cases}
\left(\dfrac{\partial h_1}{\partial q}\right)^T \Delta_1 = 0, \quad \left(\dfrac{\partial h_1}{\partial q}\right)^T g_1 = 1, \\
\left(\dfrac{\partial h_2}{\partial q}\right)^T \Delta_2 = 0, \quad \left(\dfrac{\partial h_2}{\partial q}\right)^T ad_{g_1}^{n-2} g_2 \neq 0,
\end{cases}
\tag{3.5}
$$

where $\Delta_j(q)$ with $j = 0, 1, 2$ are distributions defined by

$$
\Delta_0(q) \triangleq \operatorname{span}\{g_1, g_2, ad_{g_1} g_2, \cdots, ad_{g_1}^{n-2} g_2\},
$$

$$
\Delta_1(q) \triangleq \operatorname{span}\{g_2, ad_{g_1} g_2, \cdots, ad_{g_1}^{n-2} g_2\},
$$

$$
\Delta_2(q) \triangleq \operatorname{span}\{g_2, ad_{g_1} g_2, \cdots, ad_{g_1}^{n-3} g_2\}.
$$

Under Condition 3.5, System 3.4 can be mapped into Chained Form 3.2 under state transformation

$$x = \begin{bmatrix} h_1 & h_2 & L_{g_1} h_2 & \cdots & L_{g_1}^{n-2} h_2 \end{bmatrix}^T \triangleq T_x(q)$$

and control transformation

$$u = \begin{bmatrix} 1 & 0 \\ L_{g_1}^{n-1} h_2 & L_{g_2} L_{g_1}^{n-2} h_2 \end{bmatrix} v \triangleq T_u(q)v.$$

The state transformation corresponding to Chained Form 3.3 is

$$z = \begin{bmatrix} h_1 & L_{g_1}^{n-2} h_2 & \cdots & L_{g_1} h_2 & h_2 \end{bmatrix}^T \triangleq T_z(q).$$

Applying Theorem 2.34 to the first-order homogeneous partial differential equations in (3.5), we know that transformations $T_x(q)$ and $T_u(q)$ exist and are diffeomorphic for Systems 3.4 and 3.2 if and only if both Δ_1 and $\Delta_2 \subset \Delta_1$ are involutive, $\Delta_0(q)$ is of dimension of n and hence also involutive, and $h_1(q)$ is found to yield $(\partial h_1/\partial q)^T g_1 = 1$. Similarly, the following theorem can be concluded [266] as the sufficient conditions under which a driftless system can be transformed into Chained Form 3.1.

Theorem 3.1. *Consider the driftless non-holonomic system*

$$\dot{q} = \sum_{i=1}^{m} g_i(q)v_i, \tag{3.6}$$

where $q \in \Re^n$, $v_i \in \Re$, and vector fields of g_i are smooth and linearly independent. Then, there exist state transformation $x = T_x(q)$ and control mapping $u = T_u(q)v$ to transform System 3.6 into Chained Form 3.1 if there exist functions $\{h_1, \cdots, h_m\}$ of form

$$h_i = \begin{bmatrix} h_{i1}(q) & h_{i2}(q) & \cdots & h_{in}(q) \end{bmatrix}^T, \quad \begin{cases} h_{11} = 1 \\ h_{j1} = 0 \end{cases}, \quad \begin{cases} i = 1, \cdots, m \\ j = 2, \cdots, m \end{cases}$$

such that the distributions

$$\Delta_k \triangleq span\{ad_{h_1}^i h_2, \cdots, ad_{h_1}^i h_m : \ 0 \le i \le k\}, \ 0 \le k \le n-1 \tag{3.7}$$

have constant dimensions, are all involutive, and Δ_{n-1} is of dimension $(n-1)$.

In what follows, several non-holonomic vehicle models derived in Chapter 1 are converted into Chained Form 3.1 .

Differential-drive Vehicle

Consider Kinematic Model 1.31, and choose $h_1(q) = \theta$ and $h_2(q) = x \sin \theta - y \cos \theta$. It is straightforward to verify that Condition 3.5 holds. Therefore, the coordinate transformation is

$$z_1 = \theta, \quad z_2 = x\cos\theta + y\sin\theta, \quad z_3 = x\sin\theta - y\cos\theta \qquad (3.8)$$

and the control mapping is

$$u_1 = v_2 + z_3 v_1, \quad u_2 = v_1. \qquad (3.9)$$

Under the pair of diffeomorphic transformations, Kinematic Model 1.31 is converted into the following chained form:

$$\dot{z}_1 = v_1, \quad \dot{z}_2 = v_2, \quad \dot{z}_3 = z_2 v_1.$$

Car-like Vehicle

Consider Kinematic Model 1.35. Condition 3.5 holds under the choices of $h_1(q) = x$ and $h_2(q) = y$. It is straightforward to verify that Kinematic Equation 1.35 is converted into the following chained form

$$\dot{z}_1 = v_1, \quad \dot{z}_2 = v_2, \quad \dot{z}_3 = z_2 v_1, \quad \dot{z}_4 = z_3 v_1. \qquad (3.10)$$

under the state transformation

$$z_1 = x, \quad z_2 = \frac{\tan(\phi)}{l\cos^3(\theta)}, \quad z_3 = \tan(\theta), \quad z_4 = y, \qquad (3.11)$$

and control mapping

$$u_1 = \frac{v_1}{\rho\cos(\theta)}, \quad u_2 = -\frac{3\sin(\theta)}{l\cos^2(\theta)}\sin^2(\phi)v_1 + l\cos^3(\theta)\cos^2(\phi)v_2. \qquad (3.12)$$

Mappings 3.11 and 3.12 are diffeomorphic in the region where $\theta \in (-\pi/2,\ \pi/2)$.

Fire Truck

Consider Kinematic Model 1.40, and select the following vector fields:

$$h_1 = \begin{bmatrix} 1 \\ \tan\theta_1 \\ 0 \\ \frac{1}{l_f}\sec\theta_1\tan\phi_1 \\ 0 \\ -\frac{1}{l_b}\sec\theta_1\sec\phi_2\sin(\phi_2 - \theta_1 + \theta_2) \end{bmatrix}, \quad h_2 = \begin{bmatrix} 0 \\ 0 \\ 1 \\ 0 \\ 0 \\ 0 \end{bmatrix}, \quad h_3 = \begin{bmatrix} 0 \\ 0 \\ 0 \\ 0 \\ 1 \\ 0 \end{bmatrix}.$$

It is routine to check that, excluding all the singular hyperplanes of $\theta_1 - \theta_2 = \phi_1 = \phi_2 = \theta_1 = \pi/2$, the distributions of Δ_k defined in Theorem 3.1 all have constant dimensions and are involutive for $k = 0, 1, \cdots, 5$ and that Δ_5 is of dimension 5. Thus, Kinematic Equation 1.40 is transformed into the chained form of

$$\begin{cases} \dot{z}_1 = v_1 \\ \dot{z}_{21} = v_2, \quad \dot{z}_{22} = z_{21}v_1, \quad \dot{z}_{23} = z_{22}v_1, \\ \dot{z}_{31} = v_3, \quad \dot{z}_{32} = z_{31}v_1. \end{cases}$$

Indeed, the corresponding state and control transformations are [34]

$$\begin{cases} z_1 = x_1, \\ z_{21} = \dfrac{1}{l_f} \tan\phi_1 \sec^3\theta_1, \quad z_{22} = \tan\theta_1, \quad z_{23} = y_1, \\ z_{31} = -\dfrac{1}{l_b}\sin(\phi_2 - \theta_1 + \theta_2)\sec\phi_2\sec\theta_1, \quad z_{32} = \theta_2, \end{cases}$$

and

$$\begin{cases} v_1 = u_1\cos\theta_1, \\ v_2 = \dfrac{3}{l_f^2}\tan^2\phi_1\tan\theta_1\sec^4\theta_1 v_1 + \dfrac{1}{l_f^2}\sec^2\phi_1\sec^3\theta_1 u_2, \\ v_3 = \dfrac{1}{l_f l_b}\cos(\phi_2 + \theta_2)\tan\phi_1\sec\phi_2\sec^3\theta_1 v_1 \\ \qquad + \dfrac{1}{l_b^2}\cos(\phi_2 - \theta_1 + \theta_2)\sin(\phi_2 - \theta_1 + \theta_2)\sec^2\phi_2\sec^2\theta_1 v_1 \\ \qquad - \dfrac{1}{l_f}\cos(\theta_2 - \theta_1)\sec^2\phi_2\sec\theta_1 u_3. \end{cases}$$

3.1.2 Controllability

Controllability of a dynamic system provides a definitive answer to the question whether the state can be driven to a specific point from any (nearby) initial condition and under an appropriate choice of control. As shown in Section 2.5, there are two basic approaches to check controllability. Given a non-linear system, we can determine linear controllability by first linearizing the non-linear system at an equilibrium point (typically assumed to be the origin) or along a given trajectory and then invoking the Kalman rank condition or calculating the controllability Gramian for the resulting linear system (which is either time-invariant or time-varying). Alternatively, we can determine non-linear controllability by simply applying Theorem 2.31 which, also known as Chow's theorem, is essentially a rank condition on the Lie brackets of vector fields of the system. Note that a controllable non-linear system may not have a controllable linearization, which is illustrated by the following simple example.

Example 3.2. Consider a differential-drive vehicle whose model is

$$\dot{x}_1 = u_1\cos x_3, \quad \dot{x}_2 = u_1\sin x_3, \quad \dot{x}_3 = u_2, \tag{3.13}$$

which yields

$$g_1(x) = \begin{bmatrix} \cos x_3 & \sin x_3 & 0 \end{bmatrix}^T, \quad g_2(x) = \begin{bmatrix} 0 & 0 & 1 \end{bmatrix}^T. \tag{3.14}$$

Its linearized system at the origin is

$$\dot{x} = Ax + Bu, \quad A = 0, \quad B = \begin{bmatrix} 1 & 0 \\ 0 & 0 \\ 0 & 1 \end{bmatrix},$$

which is not controllable. On the other hand, it follows from

$$ad_{g_1} g_2 = \begin{bmatrix} \sin x_3 & -\cos x_3 & 0 \end{bmatrix}$$

that $\text{rank}\{g_1, g_2, ad_{g_1} g_2\} = 3$. That is, the system is (globally non-linearly) controllable. ◇

Physical explanation of Example 3.2 is that, due to non-holonomic constraints, the wheeled vehicle is not allowed to move sideways and hence is not linearly controllable but it can accomplish parallel parking through a series of maneuvers and thus is non-linearly controllable. As pointed out in Section 2.5, motion can always be accomplished in the sub-space spanned by Lie brackets of vector fields. As such, the basic conclusions in Example 3.2 hold in general for non-holonomic systems. To see this, consider Chained System 3.2 whose vector fields are

$$g_1(x) = \begin{bmatrix} 1 & x_3 & \cdots & x_n & 0 \end{bmatrix}^T, \quad g_2(x) = \begin{bmatrix} 0 & 0 & \cdots & 0 & 1 \end{bmatrix}. \tag{3.15}$$

Every point in the state space is an equilibrium point of System 3.2, and its linearized system (at any point) is

$$\dot{x} = Ax + Bu, \quad A = 0, \quad B = \begin{bmatrix} 1 & 0 & \cdots & 0 & 0 \\ 0 & 0 & \cdots & 0 & 1 \end{bmatrix}^T,$$

which is not controllable. Direct computation yields

$$ad_{g_1} g_2 = \begin{bmatrix} 0 \\ 0 \\ \vdots \\ 0 \\ -1 \\ 0 \end{bmatrix}, \quad ad_{g_1}^2 g_2 = \begin{bmatrix} 0 \\ 0 \\ \vdots \\ 0 \\ 1 \\ 0 \\ 0 \end{bmatrix}, \quad \cdots, \quad ad_{g_1}^{n-2} g_2 = \begin{bmatrix} 0 \\ (-1)^{n-2} \\ 0 \\ \vdots \\ 0 \\ 0 \end{bmatrix}, \tag{3.16}$$

which implies that $\text{rank}\{g_1, g_2, ad_{g_1} g_2, \cdots, ad_{g_1}^{n-2} g_2\} = n$ globally. Hence, Chained System 3.2 is (globally non-linearly) controllable. The following facts can be further argued [169]:

(a) The presence of holonomic constraints makes a constrained system not controllable due to the fact that the system cannot move to any point violating the holonomic constraints.
(b) Once the holonomic constraints are met through order reduction, the reduced-order system becomes controllable.
(c) A constrained system only with non-holonomic constraints is always controllable.

3.1.3 Feedback Linearization

Should a non-linear system be feedback linearizable, its control design becomes simple. In what follows, non-holonomic systems are shown not to be feedback linearizable, but the feedback linearization technique is still applicable under certain circumstances.

Feedback Linearization

As discussed in Section 2.6.2, the non-linear system

$$\dot{x} = f(x) + g(x)u \tag{3.17}$$

is feedback linearizable if it can be transformed into the linear controllable canonical form under a state transformation $z = T(x)$ and a static control mapping $u = \alpha(x) + \beta(x)v$. If the relative degree is less than the system dimension, the transformed system also contains non-linear internal dynamics as

$$\dot{z} = A_c z + B_c v, \quad \dot{w} = \phi(z, w), \tag{3.18}$$

where $\begin{bmatrix} z^T & w^T \end{bmatrix}^T = T(x)$ is the state transformation.

According to Theorem 2.35, Chained System 3.2 (which has no drift as $f = 0$) is feedback linearizable if and only if $\{g_1(x), g_2(x)\}$ is involutive. It follows from (3.15) and (3.16) that $ad_{g_1}g_2 \notin \mathrm{span}\{g_1, g_2\}$ and hence $\{g_1, g_2\}$ is not involutive. Thus, Chained System 3.2 is not feedback linearizable. In fact, for any driftless non-holonomic system in the form of (3.6) and with $n > m$, vector fields are never involutive because $\mathrm{rank}\{g_1, \cdots, g_m\} = m < n$ while controllability implies $\mathrm{rank}\{g_1, g_2, ad_{g_1}g_2, \cdots, ad_{g_1}^{n-2}g_2\} = n$. Therefore, it is concluded that any driftless non-holonomic system must not be feedback linearizable.

Partial Feedback Linearization over a Region

System 3.17 is said to be *partially feedback linearizable* if, under a partial state transformation $z = T'(x)$ and a static control mapping $u = \alpha(x) + \beta(x)v$, it can be mapped into the following pseudo-canonical form [37]:

$$\dot{z} = A_c z + B_c v_1, \quad \dot{w} = \phi_1(z, w) + \phi_2(z, w)v, \tag{3.19}$$

where $\begin{bmatrix} z^T & w^T \end{bmatrix}^T = T(x)$ is the state transformation, and $v = \begin{bmatrix} v_1^T & v_2^T \end{bmatrix}$ with $v_1 \in \Re^p$ and $v_2 \in \Re^{m-p}$. The key differences between feedback linearization and partial feedback linearization can be seen by comparing (3.18) and (3.19); that is, in partial feedback linearization, the linearized sub-system does not contain all the control inputs, and the non-linear internal dynamics are allowed to be driven by the transformed control v. Conditions on partial feedback linearization and the largest feedback linearizable sub-system can be found in [144].

Non-linear dynamics of Chained System 3.2 are partially feedback lineariz-able not in any neighborhood around the origin, but in a region where certain singularities can be avoided. Specifically, consider the region $\Omega = \{x \in \Re^n : x_1 \neq 0\}$ and define the following coordinate transformation: for any $x \in \Omega$,

$$\xi_1 = x_1, \ \xi_2 = \frac{x_2}{x_1^{n-2}}, \ \cdots, \ \xi_i = \frac{x_i}{x_1^{n-i}}, \ \cdots, \ \xi_n = x_n. \tag{3.20}$$

Applying Transformation 3.20 to (3.2) yields

$$\dot{\xi}_1 = u_1, \ \dot{\xi}_2 = \frac{\xi_3 - (n-2)\xi_2}{\xi_1} u_1, \ \cdots, \ \dot{\xi}_i = \frac{\xi_{i+1} - (n-i)\xi_i}{\xi_1} u_1, \ \cdots, \ \dot{\xi}_n = u_2. \tag{3.21}$$

Should $x_1(t_0) \neq 0$, $x(t) \in \Omega$ is ensured under control

$$u_1 = -kx_1, \tag{3.22}$$

where constant $k > 0$ can be arbitrarily chosen. Substituting (3.22) into (3.21), we obtain the linearized sub-system

$$\dot{\xi}_2 = k(n-2)\xi_2 - k\xi_3, \ \cdots, \ \dot{\xi}_i = k(n-i)\xi_i - k\xi_{i+1}, \ \cdots, \ \dot{\xi}_n = u_2, \tag{3.23}$$

which is time-invariant and controllable. It is straightforward to map Linear System 3.23 into the linear controllable canonical form. This partial feedback linearization over region Ω is referred to as the σ-process [9], and it will be used in Section 3.3.3 for a control design.

Dynamic Feedback Linearization

As an extension, System 3.17 is said to be *dynamically feedback linearizable* if it can be transformed into the linear controllable canonical form under a dynamic controller of form

$$u = \alpha_1(x, \xi) + \beta_1(x, \xi)v, \quad \dot{\xi} = \alpha_2(x, \xi) + \beta_2(x, \xi)v,$$

and an augmented state transformation $z = T(x, \xi)$, where $\xi \in \Re^l$ for some integer $l > 0$. Typically, an m-input affine system in the form of (3.17) is feed-back linearizable if output $z_{1j}(x)$ with $j = 1, \cdots, m$ is given (or can be found) such that u appears in $z_{ij}^{(r_j)}$ but not in $z_{ij}^{(k)}$ for $k < r_j$ and if $r_1 + \cdots + r_m = n$. In other words, a well-defined vector relative degree $[r_1 \ \cdots \ r_m]^T$ is required for feedback linearization. Typically, a system is dynamically feedback lin-earizable but not (statically) feedback linearizable because the vector relative degree of the system does not exist but can be found by considering u_j as a part of the state variables and by introducing its time derivative $u_j^{(l_j)}$ as the control variables. In other words, the partially feedback linearized system of (3.19) could be used to derive conditions on dynamic feedback linearization. Both sufficient conditions [38] and necessary condition [233] have been reported.

For many physical systems, dynamic feedback linearization can be done by appropriately choosing the system output, as illustrated by the following examples.

Example 3.3. Consider Model 3.13 of a differential-drive vehicle. It follows from (3.14) that

$$ad_{g_1}g_2 = \begin{bmatrix} \sin x_3 & -\cos x_3 & 0 \end{bmatrix}^T \quad \text{and} \quad \text{rank} \begin{bmatrix} g_1 & g_2 & ad_{g_1}g_2 \end{bmatrix} = 3,$$

which implies that $\mathcal{D}_0 = \text{span}\{g_1, g_2\}$ is not involutive and, by Theorem 2.35, Kinematic Equation 3.13 is not static feedback linearizable. On the other hand, choosing x_1, x_2 as the output variables, differentiating them twice and substituting the equations of (3.13) into the result yield

$$\begin{bmatrix} \ddot{x}_1 \\ \ddot{x}_2 \end{bmatrix} = \begin{bmatrix} \cos x_3 & -u_1 \sin x_3 \\ \sin x_3 & u_1 \cos x_3 \end{bmatrix} \begin{bmatrix} \dot{u}_1 \\ u_2 \end{bmatrix} \triangleq \begin{bmatrix} v_1 \\ v_2 \end{bmatrix},$$

which is in the linear canonical form with respect to the transformed control v. Solving $\dot{u}_1$ and u_2 in terms of v yields the dynamic control mapping:

$$u_1 = \xi, \quad \dot{\xi} = v_1 \cos x_3 + v_2 \sin x_3, \quad u_2 = \frac{1}{u_1}(v_2 \cos\theta - v_1 \sin\theta).$$

Note that the above transformation becomes singular if $u_1 = 0$. In summary, Kinematic Model 3.13 is dynamically feedback linearizable provided that the vehicle keeps moving. ◇

Example 3.4. Consider Model 1.35 of a car-like vehicle. It follows that

$$g_1 = \begin{bmatrix} \cos\theta & \sin\theta & \frac{1}{l}\tan\phi & 0 \end{bmatrix}^T, \quad g_2 = \begin{bmatrix} 0 & 0 & 0 & 1 \end{bmatrix}^T$$

and

$$ad_{g_1}g_2 = \begin{bmatrix} 0 & 0 & -\frac{1}{l\cos^2\phi} & 0 \end{bmatrix}^T \notin \text{span}\{g_1, g_2\}.$$

Thus, System 1.35 is not static feedback linearizable. On the other hand, choosing (x, y) as the output variables and differentiating them three times yield

$$\begin{bmatrix} x^{(3)} \\ y^{(3)} \end{bmatrix} = \begin{bmatrix} -\frac{3}{l}u_1\dot{u}_1 \sin\theta \tan\phi - \frac{1}{l^2}u_1^3 \tan^2\phi \cos\theta \\ \frac{3}{l}u_1\dot{u}_1 \cos\theta \tan\phi - \frac{1}{l^2}u_1^3 \tan^2\phi \sin\theta \end{bmatrix} + \begin{bmatrix} \cos\theta & -\frac{\sin\theta}{l\cos^2\phi}u_1^2 \\ \sin\theta & \frac{\cos\theta}{l\cos^2\phi}u_1^2 \end{bmatrix} \begin{bmatrix} \ddot{u}_1 \\ u_2 \end{bmatrix}$$

$$\triangleq \begin{bmatrix} v_1 \\ v_2 \end{bmatrix},$$

which is in the linear canonical form. Consequently, Model 1.35 is dynamic feedback linearizable under the dynamic feedback controller

$$u_1 = \xi_1,$$
$$\dot{\xi}_1 = \xi_2,$$
$$\begin{bmatrix} \dot{\xi}_2 \\ u_2 \end{bmatrix} = \begin{bmatrix} \cos\theta & -\frac{\sin\theta}{l\cos^2\phi}\xi_1^2 \\ \sin\theta & \frac{\cos\theta}{l\cos^2\phi}\xi_1^2 \end{bmatrix}^{-1} \left\{ \begin{bmatrix} v_1 \\ v_2 \end{bmatrix} \right.$$
$$\left. - \begin{bmatrix} -\frac{3}{l}\xi_1\xi_2\sin\theta\tan\phi - \frac{1}{l^2}\xi_1^3\tan^2\phi\cos\theta \\ \frac{3}{l}\xi_1\xi_2\cos\theta\tan\phi - \frac{1}{l^2}\xi_1^3\tan^2\phi\sin\theta \end{bmatrix} \right\},$$

provided that $u_1 \neq 0$.   ◇

Control design can easily be done for non-linear systems that are feedback linearizable or dynamic feedback linearizable. It is worth mentioning that *cascaded feedback linearization* [267] further enlarges the class of non-linear systems that can be handled using the differential geometric approach.

3.1.4 Options of Control Design

Control designs can be classified into two categories: an open-loop control to navigate the state to an equilibrium point or to make the system output track a desired trajectory, and a closed-loop feedback control to ensure stability and compensate for disturbances. For a linear system, the open-loop control design is basically trivial (since an asymptotically stable linear system has a unique equilibrium) and, with controllability, a closed-loop stabilizing control can be chosen to be continuous and of either static state feedback or dynamic output feedback. For a non-holonomic system, trajectory planning is usually required for the system to comply with non-holonomic constraints and, due to controllability, the open-loop control problem can always be solved as will be shown in Section 3.2. However, controllability does not necessarily imply that a continuous static state feedback control always exists for non-linear systems. The following theorem provides a necessary condition on existence of such a continuous static state feedback control, and existence of a solution to Algebraic Equation 3.24 is referred to as the *Brockett condition* [12, 30].

Theorem 3.5. *Suppose that non-linear system*

$$\dot{x} = f(x, u)$$

is asymptotically stabilized under a continuous static state feedback law $u = u(x)$ with $u(0) = 0$, where $f(x, u)$ is a smooth function. Then, for any given $\epsilon > 0$, there is a constant $\delta > 0$ such that, for every point $\xi \in \Re^n$ satisfying $\|\xi\| < \delta$, algebraic equation

$$\xi = f(x, u) \tag{3.24}$$

can be satisfied for some pair of $x \in \Re^n$ and $u \in \Re^m$ in the set of $\|x\| < \epsilon$ and $\|u\| < \epsilon$.

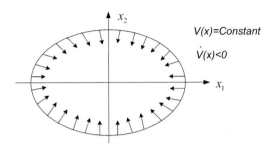

Fig. 3.1. Two-dimensional explanation of the Brockett condition

Proof of Theorem 3.5 is based on the fact that, under a continuous and stabilizing control $u(x)$, the system has a control Lyapunov function $V(x)$ whose time derivative is negative definite as $\dot{V} < 0$. Since $\dot{x}$ is continuous, since the level curves of $V(x) = c$ in a neighborhood of the origin are all closed, and since $\dot{V} < 0$ holds everywhere along a level curve of V, $\dot{x}$ must assume all the directions along the level curves and hence Algebraic Equation 3.24 must be solvable for all $\xi \in \Re^n$ of small magnitude. Figure 3.1 provides the graphical illustration that, given $\dot{V} < 0$, the corresponding trajectories passing through a closed level curve all move inwards and that, since $\dot{x}$ is continuous, $\dot{x}$ assumes all the possible directions in 2-D along the level curve. The following example illustrates an application of Theorem 3.5.

Example 3.6. Consider the system

$$\dot{x} = f(x, u) = \begin{bmatrix} u_1 \\ u_2 \\ x_2 u_1 - x_1 u_2 \end{bmatrix},$$

whose vector fields are

$$g_1(x) = \begin{bmatrix} 1 & 0 & x_2 \end{bmatrix}^T, \quad g_2(x) = \begin{bmatrix} 0 & 1 & -x_1 \end{bmatrix}^T.$$

It follows that

$$ad_{g_1} g_2 = \begin{bmatrix} 0 & 0 & -1 \end{bmatrix}^T$$

and that $\text{rank}\{g_1, g_2, ad_{g_1} g_2\} = 3$. Thus, the system is small-time controllable. On the other hand, algebraic equation

$$f(x, u) = \begin{bmatrix} 0 & 0 & \epsilon \end{bmatrix}^T$$

does not have a solution for any $\epsilon \neq 0$. By Theorem 3.5, the system cannot be stabilized under a continuous and static control of $u = u(x)$.   ◇

In general, consider the class of non-holonomic driftless systems:

$$\dot{x} = \sum_{i=1}^{m} g_i(x)u_i \triangleq G(x)u,$$

where $x \in \Re^n$, $m < n$, and vector fields $g_i(x)$ are linearly independent. Then, without loss of any generality, assume that the top m-by-m block of $G(x)$ is of full rank. Then, the algebraic equation

$$G(x)u = \begin{bmatrix} 0_{1 \times m} & \epsilon_1 & \cdots & \epsilon_{n-m} \end{bmatrix}^T$$

has no solution unless $\epsilon_1 = \cdots = \epsilon_{n-m} = 0$. By Theorem 3.5, the class of driftless systems including Chained Form 3.2 cannot be stabilized under any continuous static feedback control. Similarly, a continuous time-independent dynamic feedback control would not be stabilizing either. This means that, for driftless non-holonomic systems, a stabilizing feedback control must be either discontinuous or time-varying (as $u = u(x, t)$) or both. Such designs will be pursued in Section 3.3.

3.1.5 Uniform Complete Controllability

Given the fact that non-holonomic systems are non-linearly controllable but not stabilizable under any continuous static feedback control and that their partial or dynamic feedback linearizations have singularity, we need to search for a way to design systematically an appropriate feedback control. As revealed in Section 2.3.3, a successful control design often boils down to finding a control Lyapunov function. And, it is shown in Section 2.5 that, while there is no direct connection between non-linear controllability and existence of Lyapunov function, uniform complete controllability naturally renders a Lyapunov function. Although driftless non-holonomic systems do not have a controllable linearized system around the origin, their uniform complete controllability can be determined as follows.

Consider a non-holonomic system in the chained form of (3.2). It follows that the dynamics can be expressed as

$$\dot{x} = A(u_1(t))x + Bu,$$

or equivalently,

$$\dot{x}_1 = u_1, \quad \dot{z} = A_2(u_1(t))z + B_2 u_2, \tag{3.25}$$

where

$$z = \begin{bmatrix} x_2 & x_3 & \cdots & x_n \end{bmatrix}^T, \quad A = \text{diag}\{A_1, A_2\}, \quad B = \text{diag}\{B_1, B_2\},$$

$A_c \in \Re^{(n-1)\times(n-1)}$ and $B_c \in \Re^{(n-1)\times 1}$ are those defined in (2.43),

$$A_1 = 0, \quad B_1 = 1, \quad A_2(u_1(t)) = u_1(t)A_c, \quad B_2 = B_c. \tag{3.26}$$

Clearly, System 3.25 consists of two cascaded sub-systems. The first sub-system of x_1 is linear time-invariant and uniformly completely controllable. Hence, u_1 can always be designed (and, as will be shown in Section 3.3.2, u_1 should be chosen as a feedback control in terms of x rather than just x_1). By analyzing and then utilizing the properties of $u_1(t)$, the second non-linear sub-system of z can be treated as a linear time-varying system. This two-step process allows us to conclude the following result on uniform complete controllability of non-linear and non-holonomic systems. Once uniform complete controllability is established, the corresponding control Lyapunov function becomes known, and a feedback control design can be carried out for u_2.

Definition 3.7. *A time function $w(t) : [t_0, \infty) \to R$ is said to be* uniformly right continuous *if, for every $\epsilon > 0$, there exists $\eta > 0$ such that $t \le s \le t + \eta$ implies $|w(s) - w(t)| < \epsilon$ for all $t \in [t_0, \infty)$.*

Definition 3.8. *A time function $w(t) : [t_0, \infty) \to R$ is said to be* uniformly non-vanishing *if there exist constants $\delta > 0$ and $\underline{w} > 0$ such that, for any value of t, $|w(s)| \ge \underline{w}$ holds somewhere within the interval $[t, t + \delta]$. Function $w(t)$ is called* vanishing *if $\lim_{t \to \infty} w(t) = 0$.*

Lemma 3.9. *Suppose that scalar function $u_1(t)$ is uniformly right continuous, uniformly bounded, and uniformly non-vanishing. Then, the sub-system of z in (3.25) (that is, the pair of $\{u_1(t)A_c, B_c\}$) is uniformly completely controllable.*

Proof: Since $A_c^{(n-1)} = 0$, the state transition matrix of system $\dot{z} = u_1(t)A_c z$ can be calculated according to

$$\Phi(t, \tau) = e^{A_c \beta(t, \tau)} = \sum_{k=0}^{n-2} \frac{1}{k!} A_c^k \beta^k(t, \tau), \qquad (3.27)$$

where $\beta(t, \tau) = \int_\tau^t u_1(s)ds$. The proof is done by developing appropriate bounds on Controllability Grammian 2.45 in terms of $\Phi(t, \tau)$ in (3.27) and applying Definition 2.32.

Since $u_1(t)$ is uniformly bounded as $|u_{1d}(t)| \le \overline{u}_1$ for some constant $\overline{u}_1 > 0$, we know from (3.27) that

$$\|\Phi(t, t + \delta)\| \le e^{\|A_c\| \cdot |\beta(t, t+\delta)|} \le e^{\overline{u}_1 \|A_c\| \delta} \overset{\triangle}{=} \alpha_3(\delta).$$

Consequently, we have that, for any unit vector ξ,

$$\begin{aligned}
\xi^T W_c(t, t + \delta)\xi &\le \int_t^{t+\delta} \|\Phi(t, \tau)\|^2 d\tau \\
&\le \int_t^{t+\delta} e^{2\overline{u}_1 \|A_c\|(\tau - t)} d\tau \\
&= \int_0^\delta e^{2\overline{u}_1 \|A_c\| s} ds \overset{\triangle}{=} \alpha_2(\delta). \qquad (3.28)
\end{aligned}$$

For every $\delta > 0$, there exists constant $\underline{u}_1 > 0$ such that $|u_1(s)| \geq \underline{u}_1$ holds for some $s(t) \in [t, t + \delta]$ and for all t. It follows from uniform right continuity and uniform boundedness that, for some sub-intervals $[s(t), s(t) + \sigma(\delta, \underline{u}_1)] \subset [t, t + \delta]$ where value $\sigma(\cdot)$ is independent of t, $u_1(\tau)$ has a fixed sign and is uniformly bounded away from zero for all $\tau \in [s(t), s(t) + \sigma(\delta, \underline{u}_1)]$. Thus, it follows from (3.27) that, for any unit vector ξ,

$$\xi^T W_c(t, t + \delta)\xi \geq \xi^T W_c(s(t), s(t) + \sigma(\delta, \underline{u}_1))\xi$$
$$= \int_0^{\sigma(\delta, \underline{u}_1)} \left| \xi^T e^{A_c \int_0^\phi u_1(s(t)+\tau)d\tau} B_c \right|^2 d\phi. \qquad (3.29)$$

Defining variable substitution $\theta(\phi) = \int_0^\phi |u_1(s(t)+\tau)|d\tau$, we know from $u_1(\tau)$ being of fixed sign in interval $[s(t), s(t)+\sigma(\delta, \underline{u}_1)]$ that function $\theta(\phi)$ is strictly monotonically increasing over $[0, \sigma(\delta, \underline{u}_1)]$ and uniformly for all t, that

$$\theta(\phi) = \begin{cases} \int_0^\phi u_1(s(t) + \tau)d\tau & \text{if } u_1(s(t)) > 0 \\ -\int_0^\phi u_1(s(t) + \tau)d\tau & \text{if } u_1(s(t)) < 0 \end{cases},$$

and that, since $d\theta/d\phi \neq 0$, function $\theta(\phi)$ has a well defined inverse with

$$d\phi = \frac{d\theta}{|u_1(s(t) + \phi)|} \geq \frac{d\theta}{\overline{u}_1} > 0.$$

Applying the change of variable to (3.29) yields

$$\int_0^{\sigma(\delta, \underline{u}_1)} \left| \xi^T e^{A_c \int_0^\phi u_1(s(t)+\tau)d\tau} B_c \right|^2 d\phi$$

$$\geq \begin{cases} \dfrac{1}{\overline{u}_1} \int_0^{\sigma(\delta, \underline{u}_1)} \left| \xi^T e^{A_c \theta} B_c \right|^2 d\theta, & \text{if } u_1(s(t)) > 0 \\ \dfrac{1}{\overline{u}_1} \int_0^{\sigma(\delta, \underline{u}_1)} \left| \xi^T e^{-A_c \theta} B_c \right|^2 d\theta, & \text{if } u_1(s(t)) < 0 \end{cases}$$

$$\geq \frac{1}{\overline{u}_1} \min \left\{ \int_0^{\sigma(\delta, \underline{u}_1)} \left| \xi^T e^{A_c \theta} B_c \right|^2 d\theta, \int_0^{\sigma(\delta, \underline{u}_1)} \left| \xi^T e^{-A_c \theta} B_c \right|^2 d\theta \right\}$$

$$\overset{\triangle}{=} \alpha_1(\delta, \overline{u}_1, \underline{u}_1). \qquad (3.30)$$

In (3.30), the value of $\alpha_1(\cdot)$ is positive because both time-invariant pairs of $\{\pm A_c, B_c\}$ are controllable. The proof is completed by combining (3.29) and (3.30).   □

Since $u_1(t)$ is the control for the first sub-system in (3.25), its property depends upon the control objective for x_1. Should x_1 be commanded to track a non-vanishing time function, $u_1(t)$ is non-vanishing and hence uniform complete controllability is ensured for the design of u_2. If stabilization of x_1 is desired, $u_1(t)$ must be vanishing in which case the second sub-system in (3.25) is no longer uniformly completely controllable. Nonetheless, an appropriate transformation can be applied to the sub-system of x_2 so that uniform complete controllability is recovered for the purpose of finding Lyapunov function and carrying out the design of u_2. In what follows, two examples adopted from [208] are included to illustrate transformations used to recover uniform complete controllability in two different cases. In Example 3.10, $u_1(t)$ is vanishing but does not belong to L_1 space, and a time-unfolding transformation is applied. It also is shown in [208] that, if $u_1(t)$ is non-vanishing but not uniformly non-vanishing, a time-folding transformation can be applied. In Example 3.11, $u_1(t)$ is vanishing and belongs to L_1 space, and a time-varying state transformation is prescribed.

Example 3.10. Consider

$$u_1(t) = \frac{1}{\kappa(t)} w(t),$$

where $w(t)$ is continuous and uniformly bounded, $\kappa(t) > 0$ for all $t \geq 0$, $\lim_{t \to \infty} \kappa(t) = +\infty$, but $1/\kappa(t) \notin L_1$. Let us introduce the following transformation of time and control:

$$\tau = \int_0^t \frac{1}{\kappa(s)} ds, \quad \text{and} \quad u_2(t) = \frac{1}{\kappa(t)} u_2'(\tau).$$

The transformation essentially unfolds the time, and it is diffeomorphic. Applying the transformation to the sub-system of z in (3.25) yields

$$\frac{dz(\tau)}{d\tau} = w'(\tau) A_c z(\tau) + B_c u_2', \tag{3.31}$$

where $w'(\tau) = w(t)$ with t being replaced by the inverse of the above time transformation (which can be found once $\kappa(t)$ is specified). As long as $w'(\tau)$ in (3.31) is uniformly non-vanishing, design of control u_2 can be done through the design of u_2' for System 3.31. $\diamond$

Example 3.11. Suppose that

$$u_1(t) = e^{-t} w(t),$$

where $w(t)$ is continuous, uniformly non-vanishing, and uniformly bounded. Consider the time-dependent state transformation

$$z' = \text{diag}\{e^{(n-2)t}, \cdots, e^t, 1\} z. \tag{3.32}$$

Applying the transformation to the second sub-system in (3.25) yields

$$\begin{aligned} \dot{z}' &= \text{diag}\{(n-2), \cdots, 1, 0\} z' + \text{diag}\{e^{(n-2)t}, \cdots, e^t, 1\}[u_1(t) A_c z + B_c u_2] \\ &= \text{diag}\{(n-2), \cdots, 1, 0\} z' + w(t) A_c z' + B_c u_2 \\ &= [\text{diag}\{(n-2), \cdots, 1, 0\} + w(t) A_c] z' + B_c u_2. \end{aligned} \tag{3.33}$$

If $w(t) = 1$, Transformed System 3.33 is time-invariant and controllable (hence uniformly completely controllable). Given any uniform non-vanishing function $w(t)$, the state transition matrix of System 3.33 can be found to check uniform complete controllability. Stability analysis is needed (and will be shown in Section 3.3.2) to guarantee that Transformation 3.32 is well defined.   ◇

3.1.6 Equivalence and Extension of Chained Form

Under the global state transformation

$$z_1 = x_1,$$
$$z_{j1} = x_{jn_j}, \ 2 \le j \le m,$$
$$z_{jk} = (-1)^k x_{j(n_j-k+1)} + \sum_{l=1}^{k} (-1)^l \frac{1}{(k-l)!} (x_{11})^{k-l} x_{j(n_j-l+1)},$$
$$2 \le k \le n_j, 2 \le j \le m,$$

Chained System 3.1 becomes the so-called power form defined by

$$
\begin{cases}
\dot{z}_1 = u_1, \\
\dot{z}_{21} = u_2, \ \dot{z}_{22} = z_1 u_2, \ \dot{z}_{23} = \dfrac{z_1^2}{2!} u_2, \ \cdots \ \dot{z}_{2n_2} = \dfrac{z_1^{n_2-1}}{(n_2-1)!} u_2, \\
\vdots \\
\dot{z}_{m1} = u_m, \ \dot{z}_{m2} = z_1 u_m, \ \dot{z}_{m3} = \dfrac{z_1^2}{2!} u_m, \ \cdots \ \dot{z}_{mn_m} = \dfrac{z_1^{n_m-1}}{(n_m-1)!} u_m.
\end{cases}
\tag{3.34}
$$

Another equivalent model for non-holonomic systems is the so-called skew-symmetric chained form [222]. Specifically, consider a system in Chained Form 3.3, that is,

$$\dot{x}_1 = u_1; \ \dot{x}_i = x_{i+1} u_1, \ i = 2, \cdots, n-1; \ \dot{x}_n = u_2. \tag{3.35}$$

Define the following coordinate transformation:

$$z_1 = x_1, \ z_2 = x_2, \ z_3 = x_3; \ z_{j+3} = k_j z_{j+1} + L_{g_1} z_{j+2}, \ 1 \le j \le n-3;$$

where $k_j > 0$ are positive constants (for $1 \le j \le n-3$), and $g_1 = [1, x_3, x_4, \cdots, x_n, 0]^T$. Then, under the transformation, Chained System 3.35 is converted into the skew-symmetric chained form:

$$
\begin{cases}
\dot{z}_1 = u_1, \\
\dot{z}_2 = u_1 z_3, \\
\dot{z}_{j+3} = -k_{j+1} u_1 z_{j+2} + u_1 z_{j+4}, \ 0 \le j \le n-4, \\
\dot{z}_n = -k_{n-2} u_1 z_{n-1} + w_2,
\end{cases}
\tag{3.36}
$$

where $w_2 = (k_{n-2} z_{n-1} + L_{g_1} z_n) u_1 + u_2$.

On the other hand, Chained Form 3.1 can be generalized by admitting both draft terms and higher-order derivatives. Such a generalization of (3.2) leads to the so-called extended chained form:

$$
\begin{cases}
x_1^{(k_1)} = u_1, \\
x_2^{(k_2)} = \alpha_2\left(x_1,\cdots,\dot{x}_1^{(k_1-1)}, u_1\right)x_3 + \beta_2\left(x_2,\cdots,\dot{x}_2^{(k_2-1)}\right), \\
\vdots \\
x_{n-1}^{(k_{n-1})} = \alpha_{n-1}\left(x_1,\cdots,\dot{x}_1^{(k_{n-1}-1)}, u_1\right)x_n + \beta_{n-1}\left(x_{n-1},\cdots,\dot{x}_{n-1}^{(k_{n-1}-1)}\right), \\
x_n^{(k_n)} = u_2 + \beta_n\left(x_n,\cdots,\dot{x}_n^{(k_n-1)}\right),
\end{cases}
$$

$$(3.37)$$

where k_i are positive integers, $\alpha_i(\cdot)$ are smooth functions (some or all of which are vanishing with respect to their arguments), and $\beta_i(\cdot)$ are smooth drift functions. If $\alpha_i(\cdot) = u_1$ and $\beta_i(\cdot) = 0$, Model 3.37 reduces to that in [117]. An extended power form can be similarly defined [112]. As illustrated by the following example, the extended chained model should be used as the canonical model for certain non-holonomic systems.

Example 3.12. Consider Model 1.49 and 1.50 of a surface vessel. By direction computation, the model is transformed into the following extended chained form:

$$
\begin{cases}
\ddot{x} = v_1, \\
\ddot{y} = v_1\xi + \beta(\dot{x}, \dot{y}, \dot{\psi}, \psi), \\
\dot{\xi} = v_2,
\end{cases}
$$

where $\xi = \tan\psi$ is the transformed state variable, hence $\psi = \arctan\xi$ and $\dot{\psi} = \dot{\xi}/(1+\xi^2)$,

$$
\beta(\dot{x}, \dot{y}, \dot{\psi}, \psi) = \left[\left(1 - \frac{m_{11}}{m_{22}}\right)\dot{x}\dot{\psi} - \frac{d_{22}}{m_{22}}\dot{y}\right](\sin\psi\tan\psi + \cos\psi),
$$

and transformed control inputs v_i are defined by

$$
v_1 = \left(\frac{m_{22}}{m_{11}}\dot{y}\dot{\psi} - \frac{d_{11}}{m_{11}}\dot{x} + \frac{1}{m_{11}}\tau_1\right)\cos\psi + \left(\frac{m_{11}}{m_{22}}\dot{x}\dot{\psi} + \frac{d_{22}}{m_{22}}\dot{y}\right)\sin\psi
$$
$$
-(\dot{x}\sin\psi + \dot{y}\cos\psi)\dot{\psi},
$$

and

$$
v_2 = \sec^2\psi\left(\frac{m_{11} - m_{22}}{m_{33}}\dot{x}\dot{y} - \frac{d_{33}}{m_{33}}\dot{\psi} + \frac{1}{m_{33}}\tau_2\right) + 2(\dot{\psi})^2\sec^2\psi\tan\psi,
$$

respectively.   ◇

The dynamic augmentation from (3.2) to (3.37) captures a larger class of non-holonomic kinematic constraints, and it can be further extended to include a torque-level dynamic model such as the second equation in (1.24).

From the control design point of view, these augmentations of introducing integrators can be handled in principle by employing the backstepping approach discussed in Section 2.6.1. If drift functions $\beta_i(\cdot)$ have similar vanishing properties as $\alpha_i(\cdot)$, the aforementioned design methods such as dynamic feedback linearization and state transformation can be applied to Extended Chained Model 3.37 in a similar fashion as Standard Chained Model 3.2.

Introduction of the chained forms enables us to have one canonical form for different non-holonomic systems, to study their common properties, and to develop systematic procedures of analysis and control design. Since there is a well defined transformation between the original system and its chained form, all the results such as properties and control designs can be mapped back to the original systems. Indeed, the transformation typically retains some (output) variables of the original model. For instance, in Chained Form 3.10 of a front-steer back-drive vehicle, Cartesian coordinators of its guidepoint remain as the output (and state) variables, but steering angle and body angle no longer appear explicitly. Nonetheless, it is always possible and often better to conduct analysis and control design on the original model in a specific application such that physical meanings of the original variables can be exploited and that additional constraints (such as maximum steering rate, maximum steering angle, *etc.*) can be taken into consideration.

3.2 Steering Control and Real-time Trajectory Planning

The basic issue of motion planning is to find a feasible trajectory (or the desired trajectory) which satisfies such motion requirements as boundary conditions, non-holonomic constraints, and other geometrical constraints imposed by the environment. Non-holonomic systems are small-time non-linearly controllable, and hence there is a solution to their navigation problem of planning a feasible motion and determining the corresponding steering inputs. Specifically, we first present several simple steering control laws that generate feasible trajectories for a non-holonomic chained system in a free configuration space. Then, we formulate the real-time trajectory planning problem by considering a dynamic environment in which both static and moving obstacles are present. As an illustration, a real-time algorithm is presented to generate an optimized collision-free and feasible trajectory for a car-like vehicle with a limited sensing range. Based on the trajectory planning and replanning in the dynamic environment, a feedback control can be designed (in Section 3.3) so that the motion of a non-holonomic vehicle becomes pliable.

3.2.1 Navigation of Chained Systems

Consider a non-holonomic system in the chained form of (3.3), that is,

$$\dot{z}_1 = u_1, \quad \dot{z}_2 = u_2, \quad \dot{z}_3 = z_2 u_1, \quad \cdots, \quad \dot{z}_n = z_{n-1} u_1. \tag{3.38}$$

The basic problem is to choose steering control u such that $z(t_0) = z_0$ and $z(t_f) = z_f$, where $z = [z_1, \cdots, z_n]^T$ is the solution to (3.38), z_0 is the given initial condition, $z_f = [z_{f,1}, \cdots, z_{f,n}]^T$ is the given final condition, and t_f may be prescribed. In what follows, three typical laws of steering control are presented. Other approaches such as differential geometry [251], differential flatness [67], input parameterization [170, 259], and optimal control theory [64] can also be applied to synthesize steering controls.

Sinusoidal Steering Inputs

The method of synthesizing a sinusoidal steering control is based on the simple mathematical facts that the indefinite integral of product $\sin \omega t \sin k\omega t$ produces the higher-frequency component $\sin(k+1)\omega$, that

$$\int_0^{T_s} \sin \omega t\, dt = \int_0^{T_s} \cos k\omega t\, dt = 0,$$

and that

$$\int_0^{T_s} \sin j\omega t \sin k\omega t\, dt = \begin{cases} 0 & \text{if } j \neq k \\ \frac{\pi}{\omega} & \text{if } j = k \end{cases},$$

where $T_s = 2\pi/\omega$. Based on these facts, the basic step-by-step process of generating a sinusoidal steering control is to steer state variables z_i one-by-one [169] as follows:

Step 1: In the first interval of $t \in [t_0, t_0 + T_s]$, choose inputs $u_1 = \alpha_1$ and $u_2 = \beta_1$ for constants α_1, β_1 such that $z_1(t_0 + T_s) = z_{f,1}$ and $z_2(t_0 + T_s) = z_{f,2}$, and calculate $z_i(t_0 + T_s)$ for $3 \leq i \leq n$ by integrating (3.38).

Step 2: In the second interval of $t \in [t_0 + T_s, t_0 + 2T_s]$, let the inputs be

$$u_1(t) = \alpha_2 \sin \omega t, \quad u_2(t) = \beta_2 \cos \omega t. \tag{3.39}$$

Direct integration of (3.38) under (3.39) yields

$$z_1(t_0 + 2T_s) = z_{f,1}, \ z_2(t_0 + 2T_s) = z_{f,2}, \ z_3(t_0 + 2T_s) = z_3(t_0 + T_s) + \frac{\alpha_2 \beta_2}{2\omega} T_s,$$

and $z_i(t_0 + 2T_s)$ for $i > 3$. Hence, β_2 can be chosen such that $z_3(t_0 + 2T_s) = z_{f,3}$.

Step k (3 ≤ k ≤ (n − 1)): In the kth interval of $t \in [t_0 + (k-1)T_s, t_0 + kT_s]$, let the steering inputs be

$$u_1(t) = \alpha_2 \sin \omega t, \quad u_2(t) = \beta_k \cos(k-1)\omega t. \tag{3.40}$$

It follows from (3.38) under (3.40) that

$$\begin{cases} z_1(t_0 + 2T_s) = z_{f,1}, \\ \vdots \\ z_k(t_0 + kT_s) = z_{f,k}, \\ z_{k+1}(t_0 + kT_s) = z_{k+1}(t_0 + (k-1)T_s) + \dfrac{\alpha_2^k \beta_k}{k!(2\omega)^k} T_s, \end{cases}$$

and $z_i(t_0 + kT_s)$ for $k + 1 < i \leq n$. Hence, β_k can be chosen such that $z_{k+1}(t_0 + kT_s) = z_{f,(k+1)}$. The step repeats itself till $k = n - 1$.

Intuitively, the above steering controls can be designed in terms of simple properties of sinusoidal functions because of the mathematical properties of Lie brackets of the chained system. Recall from Section 3.1.2 that $ad_{g_1}^k g_2 = [0 \cdots 0\, 1\, 0 \cdots 0]^T$ which is precisely the motion direction of z_{k+2}, and hence the motion of z_{k+2} can be achieved by letting the frequency of u_2 be k times that of u_1.

The above steering control algorithm based on piecewise-sinusoidal functions in (3.40) has a completion time of $t_f = t_0 + (n - 1)T_s$. Combining all the piecewise-sinusoidal functions together yields the so-called all-at-once sinusoidal steering method [259] in which the inputs are defined to be

$$u_1 = \alpha_1 + \alpha_2 \sin \omega t, \quad u_2 = \beta_1 + \beta_2 \cos \omega t + \cdots + \beta_{n-1} \cos(n - 2)\omega t. \quad (3.41)$$

Integrating (3.38) under (3.41) yields n algebraic equations in terms of $(n+1)$ design parameters. It is straightforward to show that, if $z_{1f} \neq z_{10}$, these equations can be solved to meet terminal condition $z(t_f) = z_f$ with completion time $t_f = t_0 + T_s$.

Example 3.13. Consider a fourth-order chained system in the form of (3.38). Sinusoidal inputs in (3.41) can be used to steer the system from initial value $z_0 \overset{\triangle}{=} [z_{10}\ z_{20}\ z_{30}\ z_{40}]^T$ to final value $z_f \overset{\triangle}{=} [z_{1f}\ z_{2f}\ z_{3f}\ z_{4f}]^T$. It is elementary to solve for control parameters coefficients and obtain

$$\begin{cases} \alpha_1 = \dfrac{z_{1f} - z_{10}}{T_s}, \quad \alpha_2 \neq 0, \\[2mm] \beta_1 = \dfrac{z_{2f} - z_{20}}{T_s}, \quad \beta_2 = \dfrac{2\omega}{\alpha_2 T_s}\left(z_{3f} - z_{30} - \alpha_1 z_{20} T_s - \dfrac{\alpha_1 \beta_1 T_s^2}{2} + \dfrac{\alpha_2 \beta_1 T_s}{\omega}\right), \\[3mm] \beta_3 = \dfrac{8\omega^2}{(2\alpha_1^2 + \alpha_2^2)T_s}\left(z_{4f} - z_{40} - \alpha_1 z_{30} T_s - \dfrac{\alpha_1^2 T_s^2 z_{20}}{2} - \dfrac{\alpha_1^2 T_s^2 \beta_1 T_s^3}{6}\right. \\[3mm] \qquad \left. - \dfrac{\alpha_1^2 \beta_2 T_s}{\omega^2} - \dfrac{\alpha_1 \alpha_2 z_{20} T_s}{\omega} - \dfrac{\alpha_1 \alpha_2 \beta_2 T_s^2}{4\omega} - \dfrac{\alpha_2^2 \beta_1 T_s}{2\omega^2}\right). \end{cases}$$

Given boundary conditions $z_0 = [0\ 0\ 0\ 0]^T$ and $z_f = [1\ 0\ 0\ 2]^T$, we choose $\omega = 2\pi/T_s$, $T_s = 3$, and $\alpha_2 = 0.2$. The corresponding values of design parameters $\alpha_1, \beta_1, \beta_2$ and β_3 are $0.3333, 0, 0, 89.2168$, respectively. Figure 3.2 shows the trajectories under the sinusoidal steering inputs.   ◇

Piecewise-constant Steering Inputs

The piecewise-constant steering method [159] is to choose the inputs as: for $t \in [t_0 + (k - 1)T_s, t_0 + kT_s)$,

$$u_1(t) = \alpha_k, \quad u_2(t) = \beta_k, \quad (3.42)$$

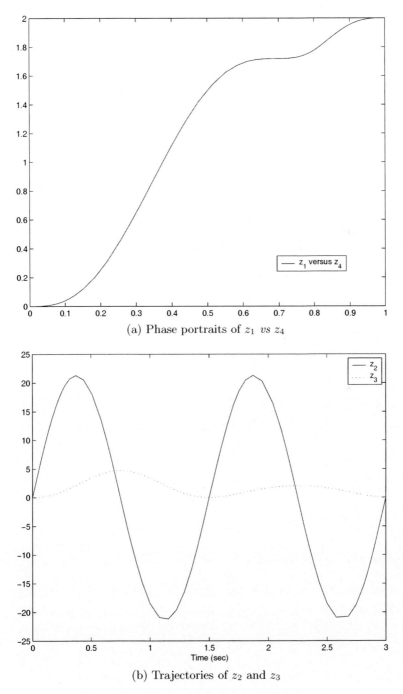

(a) Phase portraits of z_1 vs z_4

(b) Trajectories of z_2 and z_3

Fig. 3.2. Trajectories under sinusoidal inputs

where $k = 1, 2, \cdots, (n-1)$, α_k and β_k are design constants, and T_s is the sampling period. Integrating Chained System 3.38 over the interval yields

$$
\begin{cases}
z_1(t_0 + kT_s) = z_1(t_0 + (k-1)T_s) + \alpha_k T_s, \\
z_2(t_0 + kT_s) = z_2(t_0 + (k-1)T_s) + \beta_k T_s, \\
z_3(t_0 + kT_s) = z_3(t_0 + (k-1)T_s) + z_2(t_0 + (k-1)T_s)\alpha_k T_s + \alpha_k \beta_k \dfrac{T_s^2}{2}, \\
\vdots \\
z_n(t_0 + kT_s) = z_n(t_0 + (k-1)T_s) + z_{n-1}(t_0 + (k-1)T_s)\alpha_k T_s \\
\qquad\qquad\quad + \cdots + \beta_k \alpha_k^{n-2} \dfrac{T_s^{n-1}}{(n-1)!}.
\end{cases}
$$

By setting $z(t_f) = z_f$ with $t_f = t_0 + (n-1)T_s$ and by repeatedly applying the above equations, we obtain a set of algebraic equations in terms of z_0, z_f, α_k, and β_k. To simply the calculations, $\alpha_k = \alpha$ can be set, and $z(t_f) = z_f$ becomes n linear algebraic equations in terms of n design parameters α and β_1 up to β_{n-1}. Solution to the resulting equations provides the piecewise-constant steering inputs.

Example 3.14. Consider the same steering control problem stated in Example 3.13. If the piecewise-constant inputs in (3.42) are used, the control parameters are determined as

$$
\begin{cases}
\alpha = \dfrac{z_{1f} - z_{10}}{3T_s}, \\[2mm]
\begin{bmatrix} \beta_1 \\ \beta_2 \\ \beta_3 \end{bmatrix} =
\begin{bmatrix}
T_s & T_s & T_s \\
\dfrac{5\alpha T_s^2}{2} & \dfrac{3\alpha T_s^2}{2} & \dfrac{\alpha T_s^2}{2} \\
\dfrac{19\alpha^2 T_s^3}{6} & \dfrac{7\alpha^2 T_s^3}{6} & \dfrac{\alpha^2 T_s^3}{6}
\end{bmatrix}^{-1}
\begin{bmatrix}
z_{2f} - z_{20} \\
z_{3f} - z_{30} - 3z_{20}\alpha T_s \\
z_{4f} - z_{40} - 3z_{30}\alpha T_s - 4.5z_{20}\alpha^2 T_s^2
\end{bmatrix}.
\end{cases}
$$

Under the choice of $T_s = 1$, the total steering time is identical to that in Example 3.13, the trajectories are shown in Fig. 3.3, and the values of β_1, β_2 and β_3 are $18, -36, 18$, respectively. $\diamondsuit$

Polynomial Steering Inputs

The steering inputs can also be chosen to be polynomial time functions as [169], if $z_{1f} \neq z_{10}$,

$$
u_1(t) = c_{10}, \quad u_2(t) = c_{20} + c_{21}(t - t_0) + \cdots + c_{2(n-2)}(t - t_0)^{n-2}, \tag{3.43}
$$

where $c_{10}, c_{20}, c_{21}, \cdots$, and $c_{2(n-2)}$ are design constants. Integrating Chained System 3.38 under Input 3.43 yields

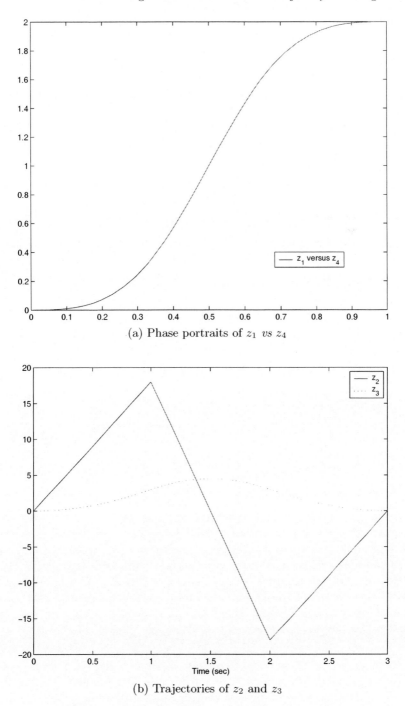

(a) Phase portraits of z_1 vs z_4

(b) Trajectories of z_2 and z_3

Fig. 3.3. Trajectories under piecewise-constant inputs

$$
\begin{cases}
z_1(t) = z_1(t_0) + c_{10}(t - t_0), \\
z_2(t) = z_2(t_0) + c_{20}(t - t_0) + \dfrac{c_{21}(t - t_0)^2}{2} + \cdots + \dfrac{c_{2(n-2)}(t - t_0)^{n-1}}{n - 1}, \\
\vdots \\
z_n(t) = z_n(t_0) + \displaystyle\sum_{k=0}^{n-2} \dfrac{k! c_{10}^{n-2} c_{2k}(t - t_0)^{n+k-1}}{(n + k - 1)!} + \sum_{k=2}^{n-1} \dfrac{c_{10}^{n-k}(t - t_0)^{n-k}}{(n - k)!} z_k(t_0).
\end{cases}
$$

$$(3.44)$$

If $z_{f,1} \neq z_1(t_0)$, c_{10} can be solved for any $t_f > t_0$ as

$$
c_{10} = \frac{z_{f,1} - z_1(t_0)}{t_f - t_0}.
$$

Then, the rest of the $(n-1)$ equations in (3.44) are linear in terms of c_{2j}, and they can be solved upon setting $z(t_f) = z_f$. If $z_{f,1} = z_1(t_0)$, an intermediate point $z(t_1)$ with $z_1(t_1) \neq z_1(t_0)$ and $t_0 < t_1 < t_f$ can be chosen to generate two segments of the trajectory: one from $z(t_0)$ to $z_1(t_1)$, and the other from $z_1(t_1)$ to z_f.

Example 3.15. For the steering problem studied in Example 3.13, polynomial inputs in (3.43) can be used, and their corresponding parameters are: letting $T_s = t_f - t_0$,

$$
\begin{cases}
c_{10} = \dfrac{z_{1f} - z_{10}}{T_s}, \\
\begin{bmatrix} c_{20} \\ c_{21} \\ c_{22} \end{bmatrix}
=
\begin{bmatrix}
T_s & \dfrac{T_s^2}{2} & \dfrac{T_s^3}{3} \\
\dfrac{c_{10} T_s^2}{2} & \dfrac{c_{10} T_s^3}{6} & \dfrac{c_{10} T_s^4}{12} \\
\dfrac{c_{10}^2 T_s^3}{6} & \dfrac{c_{10}^2 T_s^4}{24} & \dfrac{c_{10}^2 T_s^5}{60}
\end{bmatrix}^{-1}
\begin{bmatrix}
z_{2f} - z_{20} \\
z_{3f} - z_{30} - c_{10} z_{20} T_s \\
z_{4f} - z_{40} - c_{10} z_{30} T_s - 0.5 c_{10}^2 z_{20} T_s^2
\end{bmatrix}.
\end{cases}
$$

If $T_s = 3$ is set, the values of $c_{10}, c_{20}, c_{21}, c_{22}$ are $0.3333, 40, -80, 26.6667$, respectively. The corresponding trajectories are shown in Fig. 3.4. $\diamond$

Although the above three steering control designs are all comparable, we see that the polynomial control is arguably the most user-friendly since both the steering controls and the resulting trajectories are smooth. It should be noted that, except for more computations, $u_1(t)$ in (3.43) can also be chosen to be a polynomial function of time. Based on (3.44), t_f can be adjusted to satisfy any additional constraints on u_i such as kinematic control saturation.

3.2.2 Path Planning in a Dynamic Environment

Autonomous vehicles are likely to operate in an environment where there are static and moving obstacles. In order for a vehicle to maneuver successfully in such a dynamic environment, a feasible and collision-free trajectory needs to be planned in the physical configuration space. In order to illustrate the process of synthesizing the corresponding polynomial steering control in the

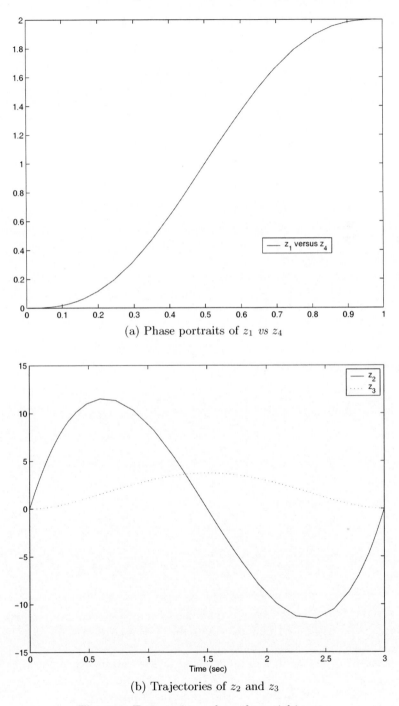

(a) Phase portraits of z_1 vs z_4

(b) Trajectories of z_2 and z_3

Fig. 3.4. Trajectories under polynomial inputs

form of (3.43), a specific vehicle model should be considered, and its boundary conditions should be specified in the physical configuration space. Accordingly, we adopt the front-steering back-driving vehicle in Fig. 3.5 as the illustrative example. Unless mentioned otherwise, a circle of radius r_0 is used to represent the vehicle's physical envelope, and the guidepoint is placed at the vehicle's center in order to minimize the envelope. Parallel to the modeling in Section 1.3.2, the kinematic model of this vehicle is given by

$$
\begin{bmatrix} \dot{x} \\ \dot{y} \\ \dot{\theta} \\ \dot{\phi} \end{bmatrix} = \begin{bmatrix} \rho\cos\theta - \frac{\rho}{2}\tan\phi\sin\theta & 0 \\ \rho\sin\theta + \frac{\rho}{2}\tan\phi\cos\theta & 0 \\ \frac{\rho}{l}\tan\phi & 0 \\ 0 & 1 \end{bmatrix} \begin{bmatrix} w_1 \\ w_2 \end{bmatrix},
\tag{3.45}
$$

where ρ is the radius of the driving wheels, l is the distance between centers of two wheel axles, w_1 is the angular velocity of the driving wheels, w_2 is the steering rate of the guiding wheels, $q = [x, y, \theta, \phi]^T$ is the state, and their initial and final configurations are $q_0 = [x_0, y_0, \theta_0, \phi_0]^T$ and $q_f = [x_f, y_f, \theta_f, \phi_f]^T$, respectively. In order to apply the design of polynomial steering control, Non-holonomic Model 3.45 is mapped into the chained form

$$
\dot{z}_1 = u_1, \quad \dot{z}_2 = u_2, \quad \dot{z}_3 = z_2 u_1, \quad \dot{z}_4 = z_3 u_1.
\tag{3.46}
$$

under coordinate transformation

$$
\begin{cases}
z_1 = x - \dfrac{l}{2}\cos\theta, \\
z_2 = \dfrac{\tan\phi}{l\cos^3\theta}, \\
z_3 = \tan\theta, \\
z_4 = y - \dfrac{l}{2}\sin\theta,
\end{cases}
\tag{3.47}
$$

and control transformation

$$
\begin{cases}
w_1 = \dfrac{u_1}{\rho\cos\theta}, \\
w_2 = \dfrac{3\sin\theta\sin^2\phi}{l\cos^2\theta}u_1 + \left(l\cos^3\theta\cos^2\phi\right)u_2.
\end{cases}
\tag{3.48}
$$

A typical scenario of motion planning is that, for any given vehicle, its sensing range is limited and its environmental changes are due to appearance/disappearance and/or motion of objects in the vicinity. For the illustrative example, the 2-D version is depicted in Fig. 3.6 in which the vehicle is represented by the circle centered at $O(t) = (x, y)$ and of radius r_0, its sensor range is also circular and of radius R_s, and the ith obstacle ($i = 1, \cdots, n_o$) are represented by the circle centered at point $O_i(t)$ and of radius r_i. For moving objects, origin $O_i(t)$ is moving with linear velocity vector $v_i(t)$.

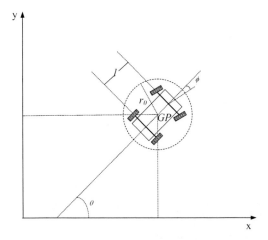

Fig. 3.5. Physical envelope of a car-like robot

The real-time trajectory planning problem is to find trajectory $q(t)$ to satisfy Kinematic Model 3.45 or 3.46, to meet the boundary conditions of $q(t_0) = q_0$ and $q(t_f) = q_f$, and to avoid all the obstacles in the environment during the motion. To ensure solvability and to simplify the technical development, the following conditions are introduced:

(a) Physical envelopes of the vehicle and all the obstacles are known. Unless stated otherwise, the envelopes are assumed to be circular.

(b) Sampling period T_s used by onboard sensors and steering controls is chosen such that T_s is small, that $\overline{k} = (t_f - t_0)/T_s$ is an integer, that position O_i (i.e., $O_i = (x_i^k, y_i^k)$ at $t = t_0 + kT_s$) of all the obstacles within the sensing range are detected at the beginning of each sampling period, and that their velocities $v_i^k \triangleq [\ v_{i,x}^k\ \ v_{i,y}^k\ ]^T$ are known and (approximately) constant for $t \in [t_0 + kT_s, t_0 + (k+1)T_s)$.

(c) All obstacles must be avoided. The obstacles do not form an inescapable trap, nor does any of the obstacles prevent the vehicle arriving at q_f indefinitely, and the vehicle can avoid any of the obstacles by moving faster than them. If needed, intermediate waypoints (and their configurations) can be found such that the feasible trajectory can be expressed by segments parameterized by a class of polynomial functions (of sixth-order). Furthermore, the feasible trajectory is updated with respect to sampling period T_s in order to accommodate the environmental changes.

(d) Boundary configurations q_0 and q_f have the properties that $x_0 - \frac{l}{2}\sin\theta_0 \neq x_f - \frac{l}{2}\sin\theta_f$ and $|\theta_0 - \theta_f| < \pi$.

(e) For simplicity, no consideration is given to any of additional constraints such as maximum speed, minimum turning radius, etc.

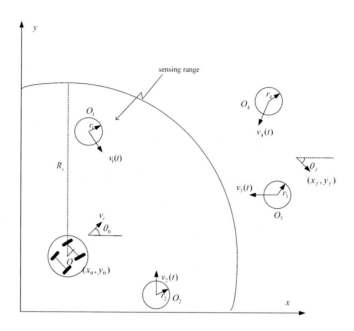

Fig. 3.6. Trajectory planning in a dynamic environment

Under these conditions, an optimized trajectory planning algorithm can be developed as will be shown in Section 3.2.3.

In applications, Conditions (a) - (e) can be relaxed in the following ways. Collision avoidance is achieved by imposing an appropriate minimum distance between any two objects of certain physical envelopes. If the envelopes are all circular (spherical), the minimum distance is in terms of distance between two centers, and hence the corresponding condition is a second-order inequality. It is straightforward to relax Condition (a) by including polygonal envelopes, in which case some of the collision-free conditions are of first-order inequality. Though small sampling period T_s is required to detect the changes in obstacle's position and velocity, the velocities in Condition (b) may be estimated from position measurements. If some of the obstacles are allowed to be overrun, whether to avoid these obstacles can also be optimized [280]. The intermediate waypoints required in Conditions (c) and (d) can be determined by applying the heuristic approach of either A^* or D^* search [243, 244]. Should the vehicle velocity be limited, it may not successfully avoid a fast-moving constant-speed obstacle unless the sensing range is properly increased. Should the vehicle be subject to certain minimum turn radius, Condition (e) can be relaxed by combining the Dubin's algorithm [25, 56, 251].

3.2.3 A Real-time and Optimized Path Planning Algorithm

It follows from State Transformation 3.47, Chained Form 3.46 and Solution 3.44 that polynomial trajectories can be parameterized as

$$z_4(z_1) = ab(z_1),$$

where $a = [a_0, a_1, \cdots, a_p]$ is the vector of design parameters, and $b(z_1) = [1, z_1(t), (z_1(t))^2, \cdots, (z_1(t))^p]^T$ is the vector of basis functions of $z_1(t)$. Within time interval $[t_0 + kT_s, t_0 + (k + 1)T_s)$, the updated class of trajectories is denoted by

$$z_4(z_1) = a^k b(z_1). \tag{3.49}$$

By continually updating parameter vector $a^k = [a_0^k, a_1^k, \cdots, a_p^k]$, an optimized feasible trajectory is planned based on vehicle's current initial configuration $(q(t_0+kT_s)$ or $z(t_0+kT_s))$, final configuration q_f, initial position $O_i = (x_i^k, y_i^k)$ and constant velocity v_i^k of neighboring obstacles. The planning is done in three steps: define a class of feasible trajectories under polynomial steering controls, determine the sub-set of collision-free trajectories, and find the optimized trajectory.

Parameterized Polynomial Trajectories

A trajectory in (3.49) is *feasible* if it corresponds to certain polynomial steering inputs and satisfies the following boundary conditions on Chained Form 3.46:

$$a^k b(z_1^k) = z_4^k, \quad a^k \left.\frac{db(z_1)}{dz_1}\right|_{z_1=z_1^k} = \tan\theta^k, \quad a^k \left.\frac{d^2 b(z_1)}{d(z_1)^2}\right|_{z_1=z_1^k} = \frac{\tan\phi^k}{l\cos^3\theta^k}, \tag{3.50}$$

$$a^k b(z_1^f) = z_4^f, \quad a^k \left.\frac{db(z_1)}{dz_1}\right|_{z_1=z_1^f} = \tan\theta_f, \quad a^k \left.\frac{d^2 b(z_1)}{d(z_1)^2}\right|_{z_1=z_1^f} = \frac{\tan\phi_f}{l\cos^3\theta_f}, \tag{3.51}$$

where $z_i(t)$ are defined by (3.47):

$$z_i^k \triangleq z_i(t_0 + kT_s), \quad \text{and} \quad z_i^f \triangleq z_i(t_f).$$

It follows from (3.50) and (3.51) that, if $p = 5$, there is a unique solution to the parameterized feasible trajectory of (3.49) for any pair $q(t_0 + kT_s)$ and $q(t_f)$. By choosing $p \geq 6$, we obtain a family of feasible trajectories from which a collision-free and optimized trajectory can be found. To simplify computation, we set $p = 6$ in the subsequent development. In other words, we choose to parameterize the class of feasible trajectories in (3.49) by one free parameter a_6^k. The rest of design parameters are chosen to satisfy the "current" boundary conditions in (3.50) and (3.51), that is, they are the solution to the following algebraic equation:

$$[a_0^k, a_1^k, a_2^k, a_3^k, a_4^k, a_5^k]^T = (B^k)^{-1}[Z^k - A^k a_6^k], \tag{3.52}$$

where

$$Z^k = \begin{bmatrix} z_4^k \\ z_3^k \\ z_2^k \\ z_4^f \\ z_3^f \\ z_2^f \end{bmatrix}, \quad A^k = \begin{bmatrix} (z_1^k)^6 \\ 6(z_1^k)^5 \\ 30(z_1^k)^4 \\ (z_1^f)^6 \\ 6(z_1^f)^5 \\ 30(z_1^f)^4 \end{bmatrix}, \tag{3.53}$$

$$B^k = \begin{bmatrix} 1 & z_1^k & (z_1^k)^2 & (z_1^k)^3 & (z_1^k)^4 & (z_1^k)^5 \\ 0 & 1 & 2z_1^k & 3(z_1^k)^2 & 4(z_1^k)^3 & 5(z_1^k)^4 \\ 0 & 0 & 2 & 6z_1^k & 12(z_1^k)^2 & 20(z_1^k)^3 \\ 1 & z_1^f & (z_1^f)^2 & (z_1^f)^3 & (z_1^f)^4 & (z_1^f)^5 \\ 0 & 1 & 2z_1^f & 3(z_1^f)^2 & 4(z_1^f)^3 & 5(z_1^f)^4 \\ 0 & 0 & 2 & 6z_1^f & 12(z_1^f)^2 & 20(z_1^f)^3 \end{bmatrix}. \tag{3.54}$$

To determine explicitly the feasible class of parameterized trajectories in (3.49), we need to find their corresponding steering controls u_i in Chained Form 3.46 and also to calculate z_i^k required in (3.53) and (3.54). To this end, we choose steering inputs u_i to be the polynomials in (3.43). That is, after the "current" kth sampling and for the rest of time $t \in (t_0 + kT_s, t_f]$, the updated steering inputs are of the form

$$u_1^k(t) = c_{10},$$
$$u_2^k(t) = c_{20}^k + c_{21}^k(t - t_0 - kT_s) + c_{22}^k(t - t_0 - kT_s)^2 + c_{23}^k(t - t_0 - kT_s)^3.$$

Substituting the above steering inputs into Chained Form 3.46, we can integrate the equations, find the expressions for $z_1(t)$ up to $z_4(t)$, and conclude that parameterized trajectories in (3.49) are generated by the following steering inputs:

$$u_1^k(t) = \frac{z_1^f - z_1(t_0)}{t_f - t_0} \triangleq c_{10}, \tag{3.55}$$

$$u_2^k(t) = 6[a_3^k + 4a_4^k z_1^k + 10a_5^k(z_1^k)^2 + 20a_6^k(z_1^k)^3]c_{10} + 24[a_4^k + 5a_5^k z_1^k$$
$$+15a_6^k(z_1^k)^2](t - t_0 - kT_s)(c_{10})^2 + 60(a_5^k + 6a_6^k z_1^k)(t - t_0 - kT_s)^2(c_{10})^3$$
$$+120a_6^k(t - t_0 - kT_s)^3(c_{10})^4. \tag{3.56}$$

Under Steering Controls 3.55 and 3.56, z_i^{k+1} is updated from z_i^k according to the following expressions:

$$z_1^{k+1} = z_1^k + c_{10}T_s,$$

$$z_2^{k+1} = z_2^k + \int_{t_0+kT_s}^{t_0+(k+1)T_s} u_2^k(t)dt,$$

$$z_3^{k+1} = z_3^k + c_{10}T_s z_2^k + c_{10} \int_{t_0+kT_s}^{t_0+(k+1)T_s} \int_{t_0+kT_s}^{s} u_2^k(t)dtds,$$

$$z_4^{k+1} = z_4^k + c_{10}T_s z_3^k + \frac{T_s^2}{2}c_{10}z_2^k$$

$$+ c_{10} \int_{t_0+kT_s}^{t_0+(k+1)T_s} \int_{t_0+kT_s}^{\tau} \int_{t_0+kT_s}^{s} u_2^k(t)\,dt\,ds\,d\tau. \qquad (3.57)$$

Initial transformed state z_i^0 is determined using (3.47) and initial configuration q_0. Then, the process of updating the feasible trajectory is done recursively as follows. Given the "current" initial condition z_i^k and based on an adequate choice of a_6^k, the updated trajectory is given by (3.49) after solving Linear Equation 3.52, then the updated steering controls are provided by (3.55) and (3.56), and the trajectory and the steering control laws are maintained for time interval $[t_0 + kT_s, t_0 + (k+1)T_s)$. At the $(k+1)$th sampling, the current initial condition is updated using the equations in (3.57), and the updating cycle repeats. For implementation, Steering Controls 3.55 and 3.56 defined over each of the sampling periods are mapped back to physical control variables w_1 and w_2 through Control Mapping 3.48.

Collision Avoidance Criterion

The class of polynomial trajectories in (3.49) and with $p = 6$ contains the free parameter a_6^k. To achieve collision-free motion in a dynamically changing environment, a collision avoidance criterion should be developed, and an analytical solution is needed to update a_6^k real-time according to sampling period T_s.

Consider the typical setting in Fig. 3.6. Within time interval $[t_0 + kT_s, t_0 + (k+1)T_s)$, the vehicle has coordinates $(x(t), y(t))$, and the ith obstacle has initial coordinates (x_i^k, y_i^k) and constant velocity $[\,v_{i,x}^k\ v_{i,y}^k\,]^T$. Therefore, there is no collision between the vehicle and the obstacle if

$$[y - (y_i^k + v_{i,y}^k\tau)]^2 + [x - (x_i^k + v_{i,x}^k\tau)]^2 \geq (r_i + r_0)^2, \qquad (3.58)$$

where $\tau = t - (t_0 + kT_s)$ and $t \in [t_0 + kT_s, t_f]$. The relationships between $(x(t), y(t))$ and $(z_1(t), z_4(t))$ are defined by Transformation 3.47, and hence Inequality 3.58 can be rewritten in the transformed space of z_4 vs z_1 as:

$$\left(z_{4,i}' + \frac{l}{2}\sin(\theta) - y_i^k\right)^2 + \left(z_{1,i}' + \frac{l}{2}\cos(\theta) - x_i^k\right)^2 \geq (r_i + r_0)^2, \qquad (3.59)$$

where $z_{1,i}' = z_1 - v_{i,x}^k\tau$ and $z_{4,i}' = z_4 - v_{i,y}^k\tau$. It is apparent that Inequality 3.59 should be checked only for those time instants at which

$$x_i^k \in [z_{1,i}' + 0.5l\cos(\theta) - r_i - r_0,\ z_{1,i}' + 0.5l\cos(\theta) + r_i + r_0]. \qquad (3.60)$$

Since there is no closed-form expression for θ only in terms of z_1 and z_4, Inequality 3.59 needs to be improved to yield a useful criterion. To this end, let us introduce intermediate variables

$$(x_i', y_i') = \left(z_{4,i}' + \frac{l}{2}\sin(\theta), \; z_{1,i}' + \frac{l}{2}\cos(\theta) \right).$$

It is shown in Fig. 3.7 that, for $\theta \in [-\pi/2, \pi/2]$, all possible locations of point (x_i', y_i') are on the right semi circle centered at $(z_{1,i}', z_{4,i}')$ and of radius $l/2$. Therefore, without the knowledge of θ *a priori*, the ith obstacle must stay clear from all the circles of radius $(r_i + r_0)$ centered and along the previous right semi circle. In other words, the region of possible collision is completely covered by the unshaded portion of the circle centered at $(z_{1,i}', z_{1,i}')$ and of radius $(r_i + r_0 + l/2)$. That is, the collision avoidance criterion in the (z_1, z_4) plane is mathematically given by

$$[z_4 - (y_i^k + v_{i,y}^k \tau)]^2 + [z_1 - (x_i^k + v_{i,x}^k \tau)]^2 \geq \left(r_i + r_0 + \frac{l}{2} \right)^2, \qquad (3.61)$$

which needs to be checked only if

$$x_i^k \in [z_{1,i}' - r_i - r_0, \; z_{1,i}' + 0.5l + r_i + r_0]. \qquad (3.62)$$

Recalling from (3.55) that z_1 and hence $z_{1,i}'$ change linearly with respect to time, we can rewrite Time Interval 3.62 as

$$t \in [\underline{t}_i^k, \; \bar{t}_i^k], \qquad (3.63)$$

where

$$\underline{t}_i^k = \begin{cases} t_0 + kT_s & \text{if } x_i^k \in [z_1^k - r_i - r_0, \; z_1^k + 0.5l + r_i + r_0] \\ t_f & \text{if } z_1^k - x_i^k - r_i - r_0 > 0 \text{ and if } c_{10} - v_{i,x}^k \geq 0 \\ \min\left\{ t_f, \; t_0 + kT_s + \frac{z_1^k - x_i^k - r_i - r_0}{|c_{10} - v_{i,x}^k|} \right\} & \\ & \text{if } z_1^k - x_i^k - r_i - r_0 > 0 \text{ and if } c_{10} - v_{i,x}^k \leq 0 \\ \min\left\{ t_f, \; t_0 + kT_s + \frac{|z_1^k - x_i^k + 0.5l + r_i + r_0|}{c_{10} - v_{i,x}^k} \right\} & \\ & \text{if } z_1^k - x_i^k + 0.5l + r_i + r_0 < 0 \text{ and if } c_{10} - v_{i,x}^k \geq 0 \\ t_f & \text{if } z_1^k - x_i^k + 0.5l + r_i + r_0 < 0 \text{ and if } c_{10} - v_{i,x}^k \leq 0 \end{cases},$$

and

$$\bar{t}_i^k = \begin{cases} \min\left\{ t_f, \; t_0 + kT_s + \frac{x_i^k - z_1^k + r_i + r_0}{c_{10} - v_{i,x}^k} \right\} & \\ & \text{if } x_i^k \in [z_1^k - r_i - r_0, \; z_1^k + 0.5l + r_i + r_0] \text{ and if } c_{10} - v_{i,x}^k \geq 0 \\ \min\left\{ t_f, \; t_0 + kT_s + \frac{z_1^k - x_i^k + 0.5l + r_i + r_0}{|c_{10} - v_{i,x}^k|} \right\} & \\ & \text{if } x_i^k \in [z_1^k - r_i - r_0, \; z_1^k + 0.5l + r_i + r_0] \text{ and if } c_{10} - v_{i,x}^k \leq 0 \\ t_f & \text{if } z_1^k - x_i^k - r_i - r_0 > 0 \text{ and if } c_{10} - v_{i,x}^k \geq 0 \\ \min\left\{ t_f, \; t_0 + kT_s + \frac{z_1^k - x_i^k + 0.5l + r_i + r_0}{|c_{10} - v_{i,x}^k|} \right\} & \\ & \text{if } z_1^k - x_i^k - r_i - r_0 > 0 \text{ and if } c_{10} - v_{i,x}^k \leq 0 \\ \min\left\{ t_f, \; t_0 + kT_s + \frac{|z_1^k - x_i^k - r_i - r_0|}{c_{10} - v_{i,x}^k} \right\} & \\ & \text{if } z_1^k - x_i^k + 0.5l + r_i + r_0 < 0 \text{ and if } c_{10} - v_{i,x}^k \geq 0 \\ t_f & \text{if } z_1^k - x_i^k + 0.5l + r_i + r_0 < 0 \text{ and if } c_{10} - v_{i,x}^k \leq 0 \end{cases}.$$

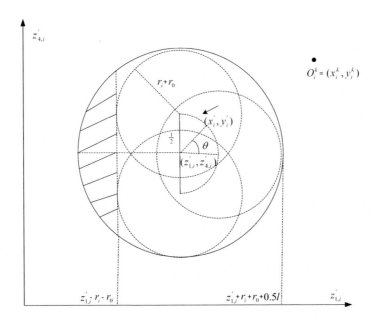

Fig. 3.7. Illustration of collision avoidance in the transformed plane

In summary, we have two *collision avoidance criteria*: Time Criterion 3.63 and Geometric Criterion 3.61.

To determine a class of feasible and collision-free trajectories, substituting (3.49) into (3.61) yields the following second-order inequality:

$$\min_{t \in [\underline{t}_i^k, \overline{t}_i^k]} \left[\zeta_2(z_1(t), t)(a_6^k)^2 + \zeta_{1,i}(z_1(t), t)a_3^k + \zeta_{0,i}(z_1(t), t) \right] \geq 0, \quad (3.64)$$

where $b'(z_1) = [1, z_1(t), (z_1(t))^2, \cdots, (z_1(t))^5]^T$,

$$\zeta_2(z_1(t), t) = \left[(z_1(t))^6 - b'(z_1(t))(B^k)^{-1}A^k \right]^2,$$

$$\zeta_{1,i}(z_1(t), t) = 2 \left[(z_1(t))^6 - b'(z_1(t))(B^k)^{-1}A^k \right] \cdot \left[b'(z_1(t))(B^k)^{-1}Y^k - y_i^k \right.$$
$$\left. - v_{i,y}^k \tau \right],$$

$$\zeta_{0,i}(z_1(t), t) = \left[b'(z_1(t))(B^k)^{-1}Y^k - y_i^k - v_{i,y}^k \tau \right]^2 + (z_1(t) - x_i^k - v_{i,x}^k \tau)^2$$
$$- (r_i + r_0 + 0.5l)^2.$$

It is shown in [207] that, unless a collision already occurred at initial configuration q_0 or will occur at final configuration q_f when $t = t_f$ (under Condition (c), the latter case can be avoided by adjusting t_f), $\zeta_2(z_1(t), t) > 0$ and Inequality 3.64 is always solvable. That is, to ensure that the trajectories in

(3.49) are feasible and collision-free, we can choose a_6^k in set Ω_o^k where Ω_o^k is the interval defined by

$$\Omega_o^k \triangleq \bigcap_{i \in \{1, \cdots, n_o\}} \left[(-\infty, \underline{a}_{6,i}^k] \cup [\overline{a}_{6,i}^k, +\infty) \right], \tag{3.65}$$

where

$$\underline{a}_{6,i}^k = \min_{t \in [\underline{t}_i^k, \overline{t}_i^k]} \frac{-\zeta_{1,i} - \sqrt{(\zeta_{1,i})^2 - 4\zeta_2 \zeta_{0,i}}}{2\zeta_2},$$

$$\overline{a}_{6,i}^k = \max_{t \in [\underline{t}_i^k, \overline{t}_i^k]} \frac{-\zeta_{1,i} + \sqrt{(\zeta_{1,i})^2 - 4\zeta_2 \zeta_{0,i}}}{2\zeta_2}.$$

Note that, in (3.65), the union operation is done for all the obstacles within the sensing range and hence the knowledge of n_o is not necessarily required.

Optimized Trajectory Planning

Given the class of feasible trajectories in (3.49), an optimized trajectory can be selected according to certain performance index. In what follows, the following L_2 performance index is utilized:

$$J_k = \int_{z_1^k}^{z_1^f} \left[z_4 - \frac{z_4^f - z_4^0}{z_1^f - z_1^0}(z_1 - z_1^0) - z_4^0 \right]^2 dz_1. \tag{3.66}$$

Intuitively, the above performance index is suitable for three reasons. First, between a pair of initial and final points in the 2-D (or 3-D) space, the minimum distance is the straight line segment connecting them. Second, in the presence of non-holonomic constraints, the straight line segment is usually not feasible (because of other configuration variables), and the problem of minimizing the distance is generally too difficult to solve analytically for non-holonomic systems. For real-time implementation, one would like to choose a performance index under which the optimization problem is solvable in closed-form. Third, as shown in Section 1.3.2 and in Transformation 3.47, $(z_1(t), z_4(t))$ are the coordinates of the center along the back axle. Figure 3.8 provides a graphical interpretation of Performance Index 3.66. Thus, through minimizing Performance Index 3.66, the vehicle motion should not deviate much from the straight line segment between the initial and final coordinates.[1]

To determine the optimal value of a_6^k, we substitute (3.49) and (3.52) into (3.66) and obtain

---

[1] There are cases that the system may have to deviate far away from the straight line segment in order to avoid obstacles. In these cases, Performance Index 3.66 could be modified by inserting z_1^k and z_4^k for z_1^0 and z_4^0, respectively.

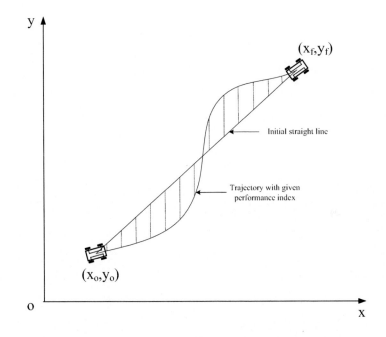

Fig. 3.8. The geometrical meaning of performance index

$$J_k(a_6^k) = \int_{z_1^k}^{z_1^f} \left[b'(z_1)(B^k)^{-1}(Y^k - A^k a_6^k) + a_6^k(z_1)^6 - \frac{z_4^f - z_4^0}{z_1^f - z_1^0}(z_1 - z_1^0) \right.$$
$$\left. - z_4^0 \right]^2 dz_1.$$

Direct computation yields

$$J_k(a_6^k) = \alpha_1(a_6^k)^2 + \alpha_2 a_6^k + \alpha_3, \qquad (3.67)$$

where

$$\alpha_1 \triangleq \int_{z_1^k}^{z_1^f} \left[(z_1(t))^6 - b'(z_1(t))(B^k)^{-1}A^k \right]^2 dz_1,$$

$$\alpha_2 \triangleq 2 \int_{z_1^k}^{z_1^f} \left[(z_1(t))^6 - b'(z_1(t))(B^k)^{-1}A^k \right] \left[b'(z_1(t))(B^k)^{-1}Y^k \right.$$
$$\left. - \frac{z_4^f - z_4^0}{z_1^f - z_1^0}(z_1(t) - z_1^0) - z_4^0 \right] dz_1,$$

and

$$\alpha_3 \triangleq \int_{z_1^k}^{z_1^f} \left[b'(z_1(t))(B^k)^{-1} Y^k - \frac{z_4^f - z_4^0}{z_1^f - z_1^0}(z_1(t) - z_1^0) - z_4^0 \right]^2 dz_1.$$

Clearly, $\alpha_1 > 0$ unless $z_1^k = z_f$. Hence, the unconstrained optimal value of a_6^k can easily be found by differentiating (3.67), that is,

$$a_6^k = -\frac{\alpha_2}{2\alpha_1}.$$

Using the boundary conditions of $\partial^l z_4 / \partial z_1^l$ as defined in (3.50) and (3.51) for $l = 0, 1, 2$, one can show through direct integration that

$$\frac{\alpha_2}{2\alpha_1} = \frac{13 \left(\frac{\partial^2 z_4}{\partial z_1^2} \Big|_{z_1 = z_1^k} + \frac{\partial^2 z_4}{\partial z_1^2} \Big|_{z_1 = z_1^f} \right)}{12(z_1^k - z_1^f)^4} + \frac{117 \left(\frac{\partial z_4}{\partial z_1} \Big|_{z_1 = z_1^k} - \frac{\partial z_4}{\partial z_1} \Big|_{z_1 = z_1^k} \right)}{10(z_1^k - z_1^f)^5}$$
$$+ \frac{429 \left[\frac{z_4^f - z_4^0}{z_1^f - z_1^0}(z_1^k - z_1^f) - (z_4^k - z_4^f) \right]}{10(z_1^k - z_1^f)^6}. \tag{3.68}$$

In the presence of obstacles, a_6^k has to be chosen from set Ω_o^k in (3.65). Hence, the constrained optimal value of a_6^k is the projection of the unconstrained optimal solution onto Ω_o^k, that is,

$$a_6^{k*} \in \Omega_o^k : \left\| a_6^{k*} + \frac{\alpha_2}{2\alpha_1} \right\| = \min_{a_6^k \in \Omega_o^k} \left\| a_6^k + \frac{\alpha_2}{2\alpha_1} \right\|. \tag{3.69}$$

In summary, the optimized collision-free and feasible trajectory is given by (3.49) with Ω_o^k in (3.65), with a_6^k from (3.68) and (3.69), and with the rest of a^k from (3.52).

As an illustration, consider the following simulation setup:

(a) Vehicle parameters: $r_0 = 1$, $l = 0.8$ and $\rho = 0.1$.
(b) Boundary conditions: $(x_0, y_0, \theta_0, \phi_0) = (0, 12, -\frac{\pi}{4}, 0)$ and $(x_f, y_f, \theta_f, \phi_f) = (17, 0, \frac{\pi}{4}, 0)$.
(c) Time interval and sampling period: $t_0 = 0$, $t_f = 40$, and $T_s = 1$.
(d) Sensor range: $R_s = 8$.
(e) Obstacles: three of size $r_i = 0.5$.

Based on motion profiles of the obstacles, the aforementioned algorithm can be used to determine the corresponding optimized trajectory. For the choices of obstacles in Table 3.1, the optimized collision-free trajectory is shown in Fig. 3.9. In the figure, all the four objects are drawn using the same scale and with display period of $T_d = 3$.

3.3 Feedback Control of Non-holonomic Systems

In this section, feedback controls are designed for non-holonomic systems mapped into Chained Form 3.2. The so-called tracking control is the one that

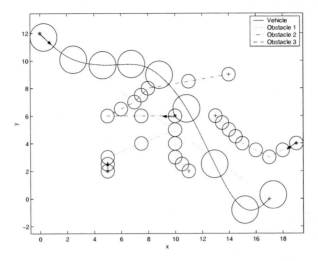

Fig. 3.9. The optimized trajectory under the choices in Table 3.1

Table 3.1. A setting of three moving obstacles

	Initial locations	$t \in [0, 10]$	$t \in [10, 20]$	$t \in [20, 30]$	$t \in [30, 40]$
Obst. 1	(5,2)	$v_{11} = \begin{bmatrix} 0 \\ 0.1 \end{bmatrix}$	$v_{12} = \begin{bmatrix} 0.5 \\ 0.2 \end{bmatrix}$	$v_{13} = \begin{bmatrix} 0 \\ -0.2 \end{bmatrix}$	$v_{14} = \begin{bmatrix} 0.1 \\ -0.1 \end{bmatrix}$
Obst. 2	(10,6)	$v_{21} = \begin{bmatrix} -0.5 \\ 0 \end{bmatrix}$	$v_{22} = \begin{bmatrix} 0.2 \\ 0.1 \end{bmatrix}$	$v_{23} = \begin{bmatrix} 0.1 \\ 0.1 \end{bmatrix}$	$v_{24} = \begin{bmatrix} 0.6 \\ 0.1 \end{bmatrix}$
Obst. 3	(19,4)	$v_{31} = \begin{bmatrix} -0.2 \\ -0.1 \end{bmatrix}$	$v_{32} = \begin{bmatrix} -0.2 \\ 0.1 \end{bmatrix}$	$v_{33} = \begin{bmatrix} -0.1 \\ 0.1 \end{bmatrix}$	$v_{34} = \begin{bmatrix} -0.1 \\ 0.1 \end{bmatrix}$

makes the system asymptotically follow a desired trajectory. The desired trajectory must be feasible, that is, it has been planned to satisfy non-holonomic constraints and hence is given by

$$\dot{x}_{1d} = u_{1d}, \;\; \dot{x}_{2d} = x_{3d}u_{1d}, \;\; \cdots, \;\; \dot{x}_{(n-1)d} = x_{nd}u_{1d}, \;\; \dot{x}_{nd} = u_{2d}, \qquad (3.70)$$

where $x_d = [x_{1d}, \cdots, x_{nd}]^T \in \Re^n$ is the desired state trajectory, $y_d = [x_{1d}, x_{2d}]^T \in \Re^2$ is the desired output trajectory, and $u_d(t) = [u_{1d}(t), u_{2d}(t)]^T \in \Re^2$ is the corresponding open-loop steering control. If x_d is a constant and hence $u_d = 0$, the tracking control problem reduces to the stabilization problem, also called the regulation problem. Should x_d converge to a constant, u_d would have to be vanishing, and the control problem would be treated as the stabilization problem. As discussed in Section 3.1, the fact of $u_1 \to 0$ requires specific remedies in the control design, and stabilizing feedback controls also need to be either continuous but time-varying or time-independent but discontinuous. Consequently, tracking control and stabilizing controls are designed separately in the subsequent subsections.

3.3.1 Tracking Control Design

It is assumed in the tracking control design that desired trajectory x_d is uniformly bounded and that steering control $u_{1d}(t)$ is uniformly non-vanishing. Let

$$x_e = [x_{1e}, \cdots, x_{ne}]^T \triangleq x - x_d, \quad y_e = [y_{1e}, y_{2e}]^T \triangleq y - y_d, \quad v = [v_1, v_2]^T \triangleq u - u_d$$

be the state tracking error, the output tracking error, and the feedback control to be designed, respectively. Then, it follows from (3.2) and (3.70) that the tracking error system consists of the following two cascaded sub-systems:

$$\dot{x}_{1e} = v_1, \quad y_{1e} = x_{1e}, \tag{3.71}$$

$$\dot{z} = u_{1d}(t)A_c z + B_c v_2 + G(x_d, z)v_1, \quad y_{2e} = C_2 z, \tag{3.72}$$

where $z = [z_1, \cdots, z_{n-1}]^T \triangleq [x_{2e}, \cdots, x_{ne}]^T \in \Re^{n-1}$, $A_c \in \Re^{(n-1)\times(n-1)}$ and $B_c \in \Re^{n-1}$ are those in (2.43),

$$C_2 = \begin{bmatrix} 1 & 0 & \cdots & 0 \end{bmatrix}, \quad \text{and} \quad G(x_d, z) = \begin{bmatrix} z_2 + x_{3d} & z_3 + x_{4d} & \cdots & z_{n-1} + x_{nd} & 0 \end{bmatrix}.$$

The cascaded structure of Error System 3.71 and 3.72 enables us to apply the backstepping design. Specifically, Sub-system 3.71 is of first-order, and a stabilizing control v_1 can easily be designed; Sub-system 3.72 has a linear time-varying nominal system defined by

$$\dot{z} = u_{1d}(t)A_c z + B_c v_2, \tag{3.73}$$

and the non-linear coupling from the first sub-system to the second is through $G(x_d, z)v_1$ and does not have to be explicitly compensated for because v_1 is stabilizing and hence is vanishing. Given these observations, we can choose the tracking control to be

$$v_1 = -r_1^{-1} p_1 x_{1e}, \quad v_2 = -r_2^{-1} B_c^T P_2(t)z, \tag{3.74}$$

where $r_1, r_2, q_1, q_2 > 0$ are scalar constants, $p_1 = \sqrt{q_1 r_1}$, Q_2 is a positive definite matrix, and $P_2(t)$ is the solution to the differential Riccati equation

$$0 = \dot{P}_2 + P_2 A_c u_{1d}(t) + u_{1d}(t)A_c^T P_2 - \frac{1}{r_2} P_2 B_c B_c^T P_2 + Q_2 \tag{3.75}$$

under the terminal condition that $P_2(+\infty)$ is positive definite. It follows from $u_{1d}(t)$ being uniformly non-vanishing, Lemma 3.9 and Theorem 2.33 that solution P_2 to Riccati Equation 3.75 is positive definite and uniformly bounded. Hence, the Lyapunov function can be chosen to be

$$V(x_e, t) = \alpha p_1 x_{1e}^2 + z^T P_2(t) z \overset{\triangle}{=} V_1(x_{1e}) + V_2(z, t), \qquad (3.76)$$

where $\alpha > 0$ is a constant. Taking the time derivatives of $V_i(\cdot)$ along the trajectories of Error System 3.71 and 3.72 and under Control 3.74 yields

$$\begin{aligned}
\dot{V} &= -2\alpha q_1 x_{1e}^2 - z^T Q_2 z - r_2^{-1} z^T P_2 B_c B_c^T P_2 z - 2z^T P_2 G(x_d, z) r_1^{-1} p_1 x_{1e} \\
&\leq -2\alpha q_1 x_{1e}^2 - 2c_1 \|z\|^2 + 2c_2 \|z\|^2 |x_{1e}| + 2c_3 \|z\| |x_{1e}|, \qquad (3.77)
\end{aligned}$$

where

$$c_1 = 0.5 \lambda_{min}(Q_2 + r_2^{-1} P_2 B_c B_c^T P_2), \quad c_2 = \lambda_{max}(P_2) r_1^{-1} p_1, \quad c_3 = c_2 \|x_d\|.$$

On the other hand, the solution to the first sub-system is

$$x_{1e}(t) = x_{1e}(t_0) e^{-\frac{p_1}{r_1}(t - t_0)}. \qquad (3.78)$$

Choosing $\alpha > c_3^2/(c_1 q_1)$ in (3.77), applying inequality $2ab \leq a^2 + b^2$, and then substituting (3.78) into the expression yield

$$\begin{aligned}
\dot{V} &\leq -\alpha q_1 x_{1e}^2 - c_1 \|z\|^2 + 2c_2 \|z\|^2 |x_{1e}| \\
&\leq -\beta_2 V + \beta_3 V e^{-\beta_1 (t - t_0)}, \qquad (3.79)
\end{aligned}$$

where

$$\beta_1 = \frac{q_1}{p_1}, \quad \beta_2 = \min\left\{\beta_1, \frac{c_1}{\lambda_{max}(P_2)}\right\}, \quad \beta_3 = \frac{2c_2 |x_{1e}(t_0)|}{\lambda_{min}(P_2)}.$$

Based on Lyapunov Function 3.76, asymptotic stability and exponential convergence of the tracking error system under Control 3.74 can be concluded by invoking Comparison Theorem 2.8 to solve for V from Inequality 3.79.

It is worth noting that, as shown in [104] as well as Section 2.6.4, the two components v_1 and v_2 in (3.74) are optimal controls for Linear Sub-system 3.71 and Linear Nominal System 3.73 with respect to performance indices

$$J_1 = \int_{t_0}^{\infty} [q_1 x_{1e}^2 + r_1 v_1^2] dt, \quad \text{and} \quad J_2 = \int_{t_0}^{\infty} [z^T Q_2 z + r_2 v_2^2] dt, \qquad (3.80)$$

respectively. For the non-linear tracking error system of (3.71) and (3.72), Control 3.74 is not optimal, nor can the optimal control be found analytically. Nonetheless, given performance index $J = J_1 + J_2$, a near-optimal control can analytically be designed [208]. It is also shown in [208] that output feedback tracking control can also be designed and that, if a Lie group operation [163] is used to define error state x_e, the desired trajectory x_d does not explicitly appear in the resulting tracking error system and hence is not required to be uniformly bounded. Control 3.74 requires Lyapunov function matrix $P_2(t)$ which only depends on $u_{1d}(t)$ and hence can be pre-computed off line and stored with an adequate sampling period. If $u_{1d}(t)$ is periodic, so is solution $P_2(t)$.

To illustrate performance of Control 3.74, consider the fourth-order Chained System 3.2 and choose desired reference trajectory x_d to be that generated under the sinusoidal steering inputs in (3.41) with $\omega = 0.1$, $\alpha_1 = 0.3183$, $\alpha_2 = 0.2$, $\beta_1 = 0$, $\beta_2 = 0$ and $\beta_3 = 0.0525$. Feedback Control 3.74 is simulated with control parameters $r_1 = r_2 = 1$, $q_1 = 10$, and $q_2 = 20$, and its performance is shown in Fig. 3.10(a) for the chained system with "perturbed" initial condition $x(t_0) = [-1, 1, 0.2, 0]^T$. The phase portrait is given in Fig. 3.10(b), where the solid curve is the portion of the reference trajectory over time interval $[0, 20\pi]$.

3.3.2 Quadratic Lyapunov Designs of Feedback Control

In this subsection, feedback controls are designed to stabilize Chained System 3.2 using quadratic Lyapunov functions in a form similar to (3.76) based on uniform complete controllability. To stabilize Chained System 3.2 to any given equilibrium point, control u_1 must be vanishing. It follows from the discussions in Section 3.1.5 that uniform complete controllability could be recovered even if control u_1 is vanishing. This means that $u_1(t)$ should asymptotically converge to zero even if $x_1(t_0) = 0$. In other words, a simple backstepping of first designing u_1 for the first sub-system in (3.25) and then synthesizing u_2 for the second sub-system does not work. Rather, we need to meet two basic requirements in order to carry out quadratic Lyapunov designs:

(a) If $\|x(t_0)\| \neq 0$, $u_1(t) \neq 0$ for any finite time t even if $x_1(t_0) = 0$.
(b) Control $u_1(t)$ should be chosen such that, under an appropriate transformation, the second sub-system in (3.25) becomes uniformly completely controllable.

In what follows, a few different choices are made for $u_1(t)$ to meet the above requirements, and the corresponding controls of $u_2(t)$ are designed.

A Design Based on Exponential Time Function

Consider the following control:

$$u_1(t) = -(k_1 + \zeta)x_1 + w(x_\tau(t))e^{-\zeta(t-t_0)}, \tag{3.81}$$

where $k_1, \zeta > 0$ are design constants, $x_\tau(t) \triangleq \{x(\tau) : \; t_0 \leq \tau \leq t\}$ represents the history of $x(\cdot)$, and scalar function $w(\cdot)$ has the properties that $0 \leq w(x_\tau(t)) \leq c_0$ for some $c_0 > 0$, that $w(x_\tau(t)) = 0$ only if $x(t_0) = 0$, and that $w(x_\tau(t))$ is non-decreasing with respect to time and consequently has the limit of $\lim_{t\to\infty} w(x_\tau(t)) = c_1$ for some non-negative constant $0 < c_1 \leq c_0$ for all $x_\tau(t)$ with $x(t_0) \neq 0$. There are many choices for $w(x_\tau(t))$; for instance, the choice of

$$w(x_\tau(t)) = c_0 \frac{\|x(t_0)\|}{1 + \|x(t_0)\|} \triangleq c_1(x_0)$$

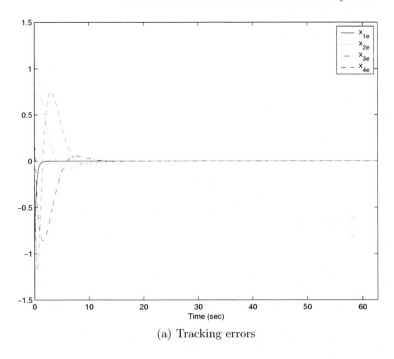

(a) Tracking errors

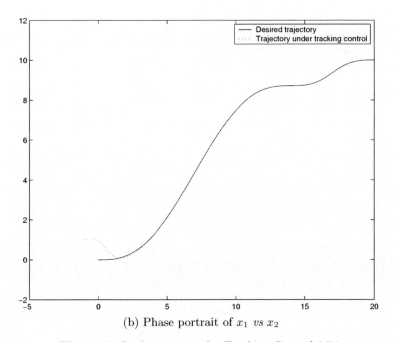

(b) Phase portrait of x_1 vs x_2

Fig. 3.10. Performance under Tracking Control 3.74

is constant, while the choice of

$$w(x_\tau(t)) = c_0 \frac{\max_{\tau \in [t_0,t]} \|x(\tau)\|^2}{\|x(t)\| + \max_{\tau \in [t_0,t]} \|x(\tau)\|^2}$$

can have the following values for its limit c_1: $c_1 = c_0$ for all convergent trajectories of $x(t)$, $c_1 \leq c_0$ for all bounded trajectories, and $c_1 = c_0$ for all divergent trajectories.

Under Control 3.81, the first sub-system in (3.25) has the property that

$$\frac{dx_1^2}{dt} = -2(k_1 + \zeta)x_1^2 + 2x_1 w(x_\tau(t))e^{-\zeta(t-t_0)} \leq -(2k_1 + \zeta)x_1^2 + \frac{c_0^2}{\zeta}e^{-2\zeta(t-t_0)},$$

from which stability and asymptotic convergence of $x_1(t)$ can be concluded. In addition, it follows from the system equation and Control 3.81 that

$$\frac{d[x_1(t)e^{\zeta(t-t_0)}]}{dt} = -k_1[x_1 e^{\zeta(t-t_0)}] + w(x_\tau(t))$$

and that, unless $\|x(t_0)\| = 0$,

$$\lim_{t \to \infty}[x_1(t)e^{\zeta(t-t_0)}] = c_1/k_1, \quad \lim_{t \to \infty}[u_1(t)e^{\zeta(t-t_0)}] = -c_1/k_1 \neq 0.$$

Since $[u_1(t)e^{\zeta(t-t_0)}]$ is uniformly bounded and uniformly non-vanishing, a state transformation similar to (3.32) in Example 3.11 can be applied to the second sub-system of (3.25). That is, we have

$$\dot{z}' = A'z' + B_c u_2 + \eta(x,t)A_c z', \tag{3.82}$$

where

$$z' = \text{diag}\{e^{(n-2)\zeta(t-t_0)}, \cdots, e^{\zeta(t-t_0)}, 1\}z \tag{3.83}$$

is the transformed state,

$$A' \triangleq \zeta \text{diag}\{(n-2), \cdots, 1, 0\} - \frac{c_1}{k_1}A_c,$$

and the residual function

$$\eta(x,t) \triangleq u_1(t)e^{\zeta(t-t_0)} + c_1/k_1$$

is uniformly bounded and vanishing. Since the pair $\{A', B_c\}$ is controllable, control u_2 can be chosen to be

$$u_2 = -B_c^T P_2 z' = -B_c^T P_2 \text{diag}\{e^{(n-2)\zeta(t-t_0)}, \cdots, e^{\zeta(t-t_0)}, 1\}z, \tag{3.84}$$

where P_2 is the solution to algebraic Riccati equation: for some positive definite matrix Q_2,

$$P_2 A' + (A')^T P_2 - P_2 B_c B_c^T P_2 + Q_2 = 0. \tag{3.85}$$

To demonstrate that transformed state z' is uniformly bounded and asymptotically convergent under Control 3.84, consider Lyapunov function $V_2 = (z')^T P_2 z'$ whose time derivative is

$$\dot{V}_2 = -(z')^T (Q_2 + P_2 B_c B_c^T P_2) z' + 2\eta(x,t)(z')^T A_c z'$$
$$\leq -\gamma_1 V_2 + \gamma_2 |\eta(x,t)| V_2,$$

where

$$\gamma_1 = \frac{\lambda_{min}(Q_2 + P_2 B_c B_c^T P_2)}{\lambda_{max}(P_2)}, \quad \gamma_2 = \frac{2\lambda_{max}(A_c)}{\lambda_{min}(P_2)}.$$

Applying Comparison Theorem 2.8 and solving V_2 from the above inequality yield

$$\lim_{t \to +\infty} V_2(t) \leq \lim_{t \to +\infty} e^{-\gamma_1 (t-t_0) \left[1 - \frac{\gamma_2}{\gamma_1(t-t_0)} \int_{t_0}^t |\eta(x,\tau)| d\tau \right]} V_2(t_0) = 0,$$

since $\eta(x,t)$ is uniformly bounded and vanishing. Thus, z' is asymptotically stable and exponentially convergent, and so is z. In summary, Control Components 3.81 and 3.84 together make Chained System 3.76 asymptotically stable and exponentially convergent, and they are smooth and time-varying in terms of exponential time function $e^{\pm \zeta t}$.

A Dynamic Control Design

Instead of using an exponential function, u_1 can be designed to be a simple dynamic controller as

$$\dot{u}_1 = -(k_1 + \zeta)u_1 - k_1 \zeta x_1, \tag{3.86}$$

where $u_1(t_0) = -k_1 x_1(t_0) + \|x(t_0)\|$, and $k_1 > 0$ and $0 < \zeta < k_1$ are design parameters. It follows that, under Control 3.86, the solution to the first subsystem of (3.25) is

$$x_1(t) = c_3 e^{-k_1(t-t_0)} + c_2 e^{-\zeta(t-t_0)}.$$

Hence, the solution to Control 3.86 is

$$u_1(t) = -k_1 c_3 e^{-k_1(t-t_0)} - \zeta c_2 e^{-\zeta(t-t_0)}, \tag{3.87}$$

where $c_2(x_0) \overset{\triangle}{=} \|x(t_0)\|/(k_1 - \zeta)$ and $c_3(x_0) \overset{\triangle}{=} x_1(t_0) - c_2(x_0)$. Through injection of $u_1(t_0)$, Control 3.86 is to excite state variable x_1 whenever $\|x(t_0)\| \neq 0$ while making both u_1 and x_1 exponentially convergent and asymptotically stable (with respect to not $|x_1(t_0)|$ but $\|x(t_0)\|$).

It follows from (3.87) that

$$\lim_{t \to \infty} [u_1(t) e^{\zeta(t-t_0)}] = -\zeta c_2,$$

which is non-vanishing unless $\|x(t_0)\| = 0$. Hence, the same state transformation of (3.83) can be applied to render Transformed System 3.82 with $c_1 = \zeta c_2 k_1$, to result in control $u_2(t)$ in (3.84), and to yield stability and convergence.

To apply Controls 3.86 and 3.84 to the fourth-order chained system in the form of (3.2), let us choose $k_1 = 5.77$, $\zeta = 0.15$ and $Q_2 = 10I_{2 \times 2}$. For the initial condition of $x(t_0) = [0, -2, 3, 6]^T$, we can determine that the corresponding values of $u_1(t_0)$ and $c_2(x_0)$ are 7 and 1.2448, respectively. Solution P_2 to (3.85) is

$$P_2 = \begin{bmatrix} 53.7723 & -28.2099 & 5.7674 \\ -28.2099 & 16.6800 & -3.9419 \\ 5.7674 & -3.9419 & 1.2133 \end{bmatrix}.$$

Performance of the system under Controls 3.86 and 3.84 is shown in Fig. 3.11.

It is worth noting that control component u_1 in (3.86) is a pure-feedback control law and that, based on the discussion in Section 3.1.4, there is no pure-feedback control law for u_2. Given the fact that the overall system is non-linear, Algebraic Riccati Equation 3.85 may depend upon initial condition $x(t_0)$ through constant c_1. Although Transformation 3.83 is globally well defined for any initial condition of $x(t_0) \neq 0$, Control 3.84 has the computational shortcoming that the ratio of two infinitesimals of $z_i/e^{-(n-1-i)\zeta(t-t_0)}$ must be calculated. Such a computation is numerically unstable but unavoidable in order to achieve exponential convergence. In what follows, the time scaling approach is presented as a remedy to overcome this computational problem, and it achieves asymptotic stabilization rather than exponential convergence.

A Time-scaling Design

Consider the following dynamic feedback control of u_1:

$$\dot{u}_1 = -2\lambda(t)u_1 - \omega^2 x_1, \quad \lambda(t) \triangleq \frac{1}{t - t_0 + 1}, \qquad (3.88)$$

where $u_1(t_0) = c_0\|x(t_0)\|$, $\omega > 0$ and $c_0 > 0$ are design parameters. The solutions to the first sub-system of (3.25) and to Dynamic Control 3.88 can be found by direct computation as

$$x_1(t) = \lambda(t)\eta_1(t, x_0), \quad u_1(t) = \lambda(t)\eta_2(t, x_0), \qquad (3.89)$$

where time functions $\eta_i(t, x_0)$ are defined by

$$\eta_1(t, x_0) \triangleq x_1(t_0) \cos(\omega t - \omega t_0) + \frac{c_0\|x(t_0)\| + x_1(t_0)}{\omega} \sin(\omega t - \omega t_0),$$

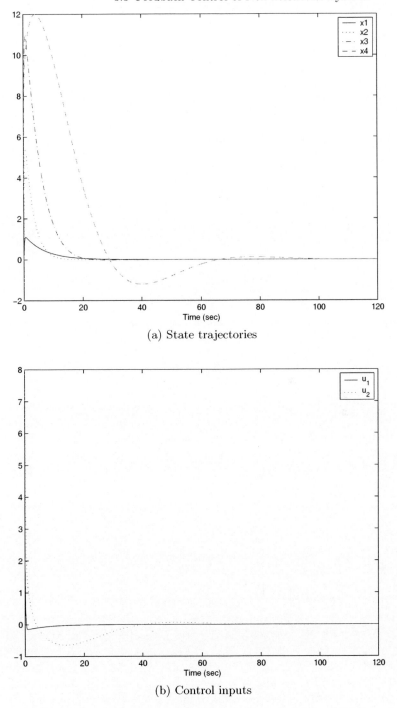

(a) State trajectories

(b) Control inputs

Fig. 3.11. Performance of Stabilizing Controls 3.86 and 3.84

and

$$\eta_2(t, x_0) \triangleq \{c_0\|x(t_0)\| + [1 - \lambda(t)]x_1(t_0)\} \cos(\omega t - \omega t_0)$$
$$- \left\{\lambda(t)\frac{c_0\|x(t_0)\|}{\omega} + \left[\frac{\lambda(t)}{\omega} + \omega\right] x_1(t_0)\right\} \sin(\omega t - \omega t_0), \quad (3.90)$$

and both are uniformly non-vanishing and uniformly bounded in proportion to $\|x(t_0)\|$. Accordingly, $x_1(t)$ is asymptotically stable, and the time-scaling method in Example 3.10 can be applied.

Under time mapping $\tau = \ln(t - t_0 + 1)$, we can use (3.89) to rewrite the second sub-system of (3.25) as

$$\frac{dz'}{d\tau} = \eta_2'(\tau, x_0)A_c z' + B_c u_2', \quad (3.91)$$

where $z'(\tau) = z(e^\tau + t_0 - 1)$, $\eta_2'(\tau, x_0) = \eta_2(e^\tau + t_0 - 1, x_0)$, and $u_2'(\tau) = e^\tau u_2(e^\tau + t_0 - 1)$. Since System 3.91 is linear time-varying and uniformly completely controllable, it is asymptotically stabilized under the control

$$u_2'(\tau) = -\frac{1}{r_2}B_c P_2'(\tau)z', \quad (3.92)$$

where matrix Q_2 is positive definite, and $P_2'(\tau)$ over $\tau \in [0, \infty)$ is the positive definite solution to the differential Riccati equation:

$$0 = \frac{dP_2'(\tau)}{d\tau} + \eta_2'(\tau, x_0)P_2'(\tau)A_c + A_c^T P_2'(t)\eta_2'(\tau, x_0) - \frac{1}{r_2}P_2'(\tau)B_c B_c^T P_2'(\tau) + Q_2.$$
$$(3.93)$$

Exponential stability of z' can be shown using Lyapunov function $V_2 = (z')^T P_2'(\tau)z'$. Hence, state vector $z(t) = z'(\ln(t - t_0 + 1))$ is asymptotically stable under the following control equivalent to (3.92):

$$u_2(t) = \lambda(t)u_2'(\ln(t - t_0 + 1)) = -\frac{\lambda(t)}{r_2}B_c^T P_2'(\ln(t - t_0 + 1))z \triangleq -\frac{\lambda(t)}{r_2}B_c^T P_2(t)z,$$
$$(3.94)$$

where $\lambda(t)$ is that in (3.88), and $P_2(t)$ is positive definite and uniformly bounded.

For the fourth-order chained system in the form of (3.2) and with initial condition $x(t_0) = [0, -2, 3, 6]^T$, time-scaling control inputs of (3.88) and (3.94) are simulated with the choices of $r_2 = 1$, $Q_2 = 10I_{2\times2}$, $\omega = 1$ and $u_1(t_0) = 10$, and their performance is shown in Fig. 3.12.

The asymptotically stabilizing control components of (3.88) and (3.94) do not have the computational problem suffered by Controls 3.81, 3.86 and 3.84. As a tradeoff, Control 3.94 requires the solution to a differential Riccati equation while Control 3.84 is based on an algebraic Riccati equation. Nonetheless, since $\eta_2(t, x_0)$ becomes periodic as the time approaches infinity, Lyapunov matrix $P_2(t)$ also becomes periodic in the limit and hence can be found using one of the well-established methods [3]. The two pairs of controls, Control 3.86 and 3.84 as well as Control 3.88 and 3.94, can also be shown to be inversely optimal [204].

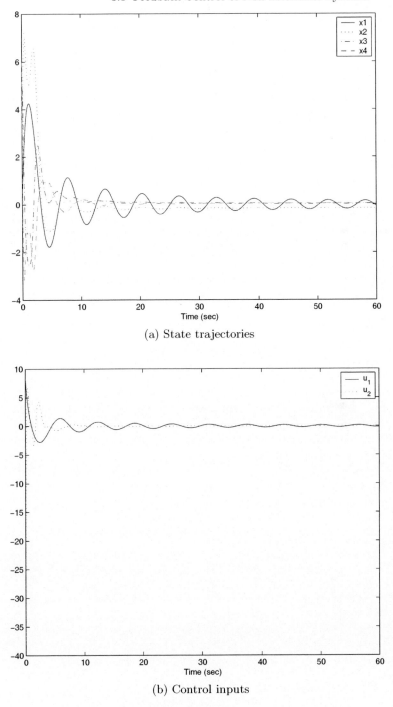

(a) State trajectories

(b) Control inputs

Fig. 3.12. Performance under Controls 3.88 and 3.94

3.3.3 Other Feedback Designs

Besides the quadratic Lyapunov designs based on uniform complete control-lability, there are other approaches to design either discontinuous or time-varying feedback controls. In what follows, several representative designs are outlined.

Discontinuous Controls

For feedback stabilization of Chained System 3.2, a discontinuous control can be designed using the partial feedback linearization in Section 3.1.3, also called σ-process [9]. Specifically, in the case that $x_1(t_0) \neq 0$, control component u_1 is set to be that in (3.22) which ensures $x_1(t) \neq 0$ for all finite time; then Transformation 3.20 is applied to yield the linear time-invariant system of (3.23) for which control component u_2 can easily be designed (using either pole placement [9] or algebraic Riccati equation [143]). In the case that $x_1(t_0) = 0$ (but $\|x(t_0)\| \neq 0$), a separate control can be designed over time interval $[t_0, t_0 + \delta t)$ (for instance, $u_1(t) = \pm c$ with $c > 0$ and $u_2(t) = 0$) such that $x_1(t_0 + \delta t) \neq 0$, and the control can then be switched to the previous one since $x_1(t_0 + \delta t) \neq 0$.

Because the solution of $x_1(t)$ to Control 3.22 is $x_1(t) = x_1(t_0)e^{-k(t-t_0)}$, any choice of control $u_2(t)$ based on Transformation 3.20 suffers from the same computational issue as Control 3.84. Meantime, Transformation 3.20 has the singularity hyperplane of $x_1(t)=0$, which can cause excessive chattering in the presence of noises and disturbances. To limit the chattering, the control magnitude can be saturated and the resulting stability becomes semi-global [140]. Also, robustification against certain types of noises and disturbances can be achieved [100, 191].

Instead of employing the σ-process, discontinuous stabilizing controls can also be designed using a piecewise-analytic state feedback design [22], a sliding mode design [21], and a series of nested invariant manifolds [139].

Time-varying Controls

A 2π-periodic stabilizing control law can be synthesized through constructing a 2π-periodic Lyapunov function. To expose the basic idea, let us consider the third-order chained system

$$\dot{x}_1 = u_1, \quad \dot{x}_2 = u_2, \quad \dot{x}_3 = x_2 u_1. \tag{3.95}$$

For System 3.95, the smooth time-varying control is given by

$$u_1 = -(x_2 + x_3 \cos t)x_2 \cos t - (x_2 x_3 + x_1),$$
$$u_2 = x_3 \sin t - (x_2 + x_3 \cos t).$$

It follows that Lyapunov function

$$V(t, x) = \frac{1}{2}(x_2 + x_3 \cos t)^2 + \frac{1}{2}x_1^2 + \frac{1}{2}x_3^2$$

is 2π-periodic and that its time derivative along the system trajectory is

$$\dot{V} = -(x_2 + x_3 \cos t)^2 - [(x_2 + x_3 \cos t)x_2 \cos t + (x_2 x_3 + x_1)]^2,$$

which is negative semi-definite and also 2π-periodic. Then, asymptotic stability can be shown by extending LaSalle's invariant set theorem to a 2π-periodic Lyapunov stability argument, which together with the constructive design for Chained System 3.2 can be found in [189].

The idea of sinusoidal steering control can be extended to yield time-varying stabilizing control design by using center manifold theory and averaging. To illustrate the concept, consider again System 3.95, and choose the sinusoidal feedback control as

$$u_1 = -x_1 - x_3^2 \sin t, \quad u_2 = -x_2 - x_3 \cos t.$$

The sinusoidal time functions can be generated from the following exogenous system:

$$\dot{w}_1 = w_2, \quad \dot{w}_2 = -w_1,$$

where $w_1(0) = 0$ and $w_2(0) = 1$. Hence, the time-varying control can be rewritten as

$$u_1 = -x_1 - x_3^2 w_1, \quad u_2 = -x_2 - x_3 w_2.$$

Stability of the augmented system (including the exogenous system) can be analyzed using center manifold theory [108]. It is straightforward to show that the local center manifold is given by

$$x_1 \approx -\frac{1}{2}x_3^2(w_1 - w_2), \quad x_2 \approx -\frac{1}{2}x_3(w_1 + w_2),$$

and that the dynamics on the center manifold is approximated by

$$\dot{x}_3 \approx -\frac{1}{4}x_3^3(w_1 - w_2)^2,$$

which is asymptotically stable using the averaging analysis. For nth-order non-holonomic systems, a sinusoidal stabilizing control can be designed for Power Form 3.34, and local stability can be concluded using center manifold and averaging analysis [258].

For systems in the Skew-symmetric Chained Form 3.36, a time-varying feedback control is given by

$$u_1 = -k_u z_1 + h(Z_2, t), \quad w_2 = -k_w z_n,$$

where k_u and k_w are positive constants, $Z_2 = [z_2, z_3, \cdots, z_n]^T$, and $h(Z_2, t) = \|Z_2\|^2 \sin t$. Global asymptotical stability of the closed-loop system can be shown [222] by using Barbalat lemma and a Lyapunov-like argument with the Lyapunov function

$$V(Z_2) = \frac{1}{2}\left(z_2^2 + \frac{1}{k_1}z_3^2 + \frac{1}{k_1 k_2}z_4^2 + \cdots + \frac{1}{\prod_{j=1}^{n-2} k_j}z_n^2 \right),$$

where $k_i > 0$ are constants.

Additional time-varying feedback controls are also available, for instance, those designed using homogeneous Lyapunov function [77, 153, 162]. Asymptotic stabilization can also be cast into the so-called ρ-exponential stabilization, a weaker concept of exponential stability [154, 162].

Hybrid Control

A hybrid control is a set of analytic feedback control laws that are piecewise switched according to a discrete event supervising law. For instance, a hybrid feedback control for the third-order Chained System 3.95 is, within the interval $t \in [2\pi k,\ 2\pi(k+1))$,

$$u_1(t) = -x_1 + |\alpha_k(x)| \cos t, \quad u_2(t) = -x_2 + \alpha_k(x)\sin t,$$

where $\alpha_k(x)$ is the piecewise-switching function defined by

$$\alpha_k(x) = \begin{cases} \alpha_{k-1} & \text{if } x_3(2\pi k)\alpha_{k-1} \geq 0 \\ \gamma|\alpha_{k-1}|\text{sign}(x_3(2\pi k)) & \text{if } x_3(2\pi k)\alpha_{k-1} < 0 \end{cases},$$

$\gamma > 0$ is a constant, $\alpha_0 = 0$ if $x(t_0) = 0$, and $\alpha_0 \neq 0$ if $x(t_0) \neq 0$. More details on hybrid control designs can be found in [22, 111, 112, 239].

3.4 Control of Vehicle Systems

Consider the control problem for a vehicle system which consists of n_r vehicles. The vehicles are indexed by set $\Omega = \{1, \cdots, n_r\}$, and they have configuration outputs q_i respectively. Individually, each of the vehicles can be controlled using the path planning and feedback control algorithms discussed in the preceding sections. In this section, we discuss the issues arising from controlling n_r vehicles as a group. In particular, we investigate the problems of how to achieve a vehicle group formation and how to ensure that all the vehicles avoid collision both among the group and with any of the obstacles in their vicinities. In the study, we assume that output information of each vehicle are available to all others and for all time. In practice, the information is collected or transmitted *via* a sensing and communication network, and the corresponding analysis and design will be carried out in the subsequent chapters.

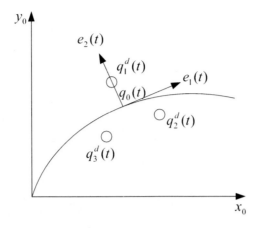

Fig. 3.13. Illustration of a formation in 2-D

3.4.1 Formation Control

Let $q_0(t) \in \Re^l$ be the desired trajectory of the virtual leader specifying the group motion of these vehicles. A formation specifies desired geometric requirements on relative positions of the vehicles while they move, and it can be expressed in terms of the coordinates in a motion frame $\mathcal{F}(t)$ that moves according to $q_0(t)$. Without loss of any generality, we assume $\mathcal{F}(t) \in \Re^l$, set the origin of $\mathcal{F}(t)$ to be at $q_0(t)$, and choose $[e_1, e_2, \cdots, e_l]$ to be its l orthonormal basis vectors. Then, for any given formation consisting of n_r vehicles, their desired positions in frame $\mathcal{F}(t)$ can be expressed as $\{P_1, \cdots, P_{n_r}\}$, where

$$P_i = \sum_{j=1}^{l} \alpha_{ij} e_j. \tag{3.96}$$

Alternatively, the formation can also be characterized by its edge vectors: for any $i \neq k$,

$$P_i - P_k = \sum_{j=1}^{l} (\alpha_{ij} - \alpha_{kj}) e_j. \tag{3.97}$$

Note that α_{ij} are constant if the formation is rigid over time and that α_{ij} are time functions if otherwise. Figure 3.13 illustrates a formation and its frame $F(t)$ for the case of $l = 2$ and $n_r = 3$. In this case, it follows from $q_0(t) = \begin{bmatrix} x_0(t) & y_0(t) \end{bmatrix}^T$ that

$$e_1(t) = \begin{bmatrix} \dfrac{\dot{x}_0(t)}{\sqrt{[\dot{x}_0(t)]^2 + [\dot{y}_0(t)]^2}} \\ \dfrac{\dot{y}_0(t)}{\sqrt{[\dot{x}_0(t)]^2 + [\dot{y}_0(t)]^2}} \end{bmatrix}, \quad e_2(t) = \begin{bmatrix} -\dfrac{\dot{y}_0(t)}{\sqrt{[\dot{x}_0(t)]^2 + [\dot{y}_0(t)]^2}} \\ \dfrac{\dot{x}_0(t)}{\sqrt{[\dot{x}_0(t)]^2 + [\dot{y}_0(t)]^2}} \end{bmatrix}.$$

The formation control problem is to control the vehicles such that they move together according to $q_0(t)$ and asymptotically form and maintain the desired formation. In what follows, several approaches of formation control are described.

Formation Decomposition

Based on the desired formation in (3.96) and on the virtual leader $q_0(t)$, the desired trajectories of individual vehicles can be calculated as

$$q_i^d(t) = q_0(t) + \sum_{j=1}^{l} \alpha_{ij} e_j, \quad i = 1, \cdots, n_r. \tag{3.98}$$

Through the decomposition, the formation control problem is decoupled into the standard tracking problem, and the design in Section 3.3.1 can readily be applied to synthesize control u_i to make q_i track q_i^d.

Example 3.16. To illustrate the formation decomposition and its corresponding control, consider a group of three differential-drive vehicles that follow a given trajectory while maintaining the formation of a right triangle. Suppose that the 2-D trajectory of the virtual leader is circular as

$$q_0(t) = [2\cos t, 2\sin t]^T,$$

and hence the moving frame is given by

$$e_1(t) = [-\sin t, \cos t]^T, \quad e_2(t) = [-\cos t, -\sin t]^T.$$

For a right triangle, formation parameters are

$$\alpha_{11} = 0, \quad \alpha_{12} = 0, \quad \alpha_{21} = -1, \quad \alpha_{22} = 1, \quad \alpha_{31} = -1, \quad \alpha_{32} = -1.$$

It then follows from (3.98) that the desired individual trajectories are

$$q_1^d = \begin{bmatrix} 2\cos t \\ 2\sin t \end{bmatrix}, \quad q_2^d = \begin{bmatrix} \cos t + \sin t \\ \sin t - \cos t \end{bmatrix}, \quad q_3^d = \begin{bmatrix} 3\cos t + \sin t \\ 3\sin t - \cos t \end{bmatrix},$$

respectively. In addition to vehicle positions in the formation, assume that the desired orientation of the vehicles are aligned as specified by $\theta_i^d = t + \pi/2$.

As shown in Section 3.1, models of the vehicles can be expressed in the chained form as, given $q_i = [x_{i1} \ x_{i2}]^T$,

$$\dot{x}_{i1} = u_{i1}, \quad \dot{x}_{i2} = x_{i3} u_{i1}, \quad \dot{x}_{i3} = u_{i2}.$$

Through Transformations 3.8 and 3.9, the desired trajectories can be mapped into those for the state of the above chained form:

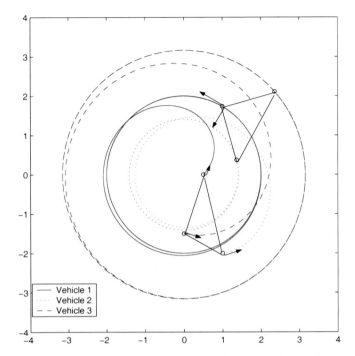

Fig. 3.14. Performance under formation decomposition and control

$$\begin{cases} x_{11}^d = t + \frac{\pi}{2}, \\ x_{12}^d = 2, \\ x_{13}^d = 0, \\ u_{11}^d = 1, \ u_{12}^d = 0, \end{cases} \qquad \begin{cases} x_{21}^d = t + \frac{\pi}{2}, \\ x_{22}^d = 1, \\ x_{23}^d = -1, \\ u_{21}^d = 1, \ u_{22}^d = 0, \end{cases} \qquad \begin{cases} x_{31}^d = t + \frac{\pi}{2}, \\ x_{32}^d = 3, \\ x_{33}^d = -1, \\ u_{31}^d = 1, \ u_{32}^d = 0. \end{cases}$$

Then, the tracking control presented in Section 3.3.1 can be readily applied. Simulation is done for initial configurations $[x_i(t_0), y_i(t_0), \theta_i(t_0)]^T$ given by $[0.5, 0, \frac{\pi}{4}]$, $[1, -2, \frac{\pi}{6}]$ and $[0, -1.5, 0]$, respectively; and phase portraits of the vehicle trajectories are shown in Fig. 3.14. $\diamondsuit$

Desired trajectory in (3.98) requires calculation of $\dot{q}_0(t)$, and the high-order derivatives of $q_0(t)$ may also be needed to write down appropriate tracking error systems for the tracking control design. If time derivatives of $q_0(t)$ are not available, standard numerical approximations can be used.

While this approach is the simplest, there is little coordination between the vehicles. As such, the formation is not robust either under disturbance or with noisy measurements or in the absence of synchronized timing because any formation error cannot be corrected or attenuated. In certain applications, $\dot{q}_0(t)$ may be known only to some of the vehicles but not all. To overcome these shortcomings, certain feedback mechanism among the vehicles should be present.

Leader-follower Design

Let $\Omega_i \subset \Omega$ be a non-empty sub-set that contains the indices of the leader vehicles for the ith vehicle to track. Then, it follows from (3.97) that the desired trajectory of the ith vehicle can be expressed as

$$q_i^d(t) = \sum_{k\in\Omega_i} d_{ik}\left[q_k^d(t) + \sum_{j=1}^{l}(\alpha_{ij} - \alpha_{kj})e_j\right], \tag{3.99}$$

where $d_{ik} > 0$ are positive weighting coefficients such that $\sum_{k\in\Omega_i} d_{ik} = 1$. As long as $\cup_{i=1}^{n_r}\Omega_i = \Omega$, a tracking control design can be applied to make q_i track the desired trajectory in (3.99) and in turn to achieve the desired formation asymptotically.

Formation 3.99 requires that $e_j(t)$ and hence $q_0(t)$ be explicitly available to all the vehicles. To overcome this restriction, we can adopt a leader-follower hierarchy and decompose the formation in (3.98) or (3.99) further in terms of n_r moving frames attached to individual vehicles (rather than just one moving frame attached to the virtual leader), that is,

$$q_i^d(t) = \sum_{k\in\Omega_i} d'_{ik}\left[q_k^d(t) + \sum_{j=1}^{l}\alpha'_{ij}e'_{kj}\right], \quad i = 1,\cdots,n_r, \tag{3.100}$$

where $\{e'_{kj} : j = 1,\cdots,l\}$ are the orthonormal basis vectors of moving frame $\mathcal{F}_k^d(t)$ at the kth vehicle, α'_{ij} for $j = 1,\cdots,l$ are the desired coordinates of the ith vehicle with respect to frame $\mathcal{F}_k^d(t)$ and according to the desired formation, and $d'_{ik} > 0$ are positive weighting coefficients such that $\sum_{k\in\Omega_i} d'_{ik} = 1$.

Example 3.17. To illustrate the leader-follower design, consider the formation control problem that a leader-following structure is imposed among three differential-drive vehicles and that each vehicle is to follow its leader(s) with certain desired offset(s) while the virtual leader for the group moves along a circular path of $q_0(t) = [2\cos t, 2\sin t]^T$.

Suppose the first vehicle is to follow the virtual leader with zero offset, that the second vehicle is to track the first with the offset values $\alpha'_{21} = -\sqrt{2}$ and $\alpha'_{22} = 2 - \sqrt{2}$, and that the third vehicle is to follow the second with the offset values $\alpha'_{31} = -\sqrt{2}$ and $\alpha'_{32} = 2 - \sqrt{2}$. It follows (3.100) that

$$q_1^d(t) = q_0(t), \quad q_2^d = q_1^d + \alpha'_{21}e'_{11} + \alpha'_{22}e'_{12}, \quad q_3^d = q_2^d + \alpha'_{31}e'_{21} + \alpha'_{32}e'_{22},$$

where e'_{kj} (with $j = 1,\cdots,l$) are determined along the trajectory of $q_k^d(t)$. That is, we have that

$$e'_{11}(t) = \begin{bmatrix} -\sin t \\ \cos t \end{bmatrix}, \quad e'_{12}(t) = \begin{bmatrix} -\cos t \\ -\sin t \end{bmatrix} \implies q_2^d = \sqrt{2}\begin{bmatrix} \cos t + \sin t \\ \sin t - \cos t \end{bmatrix},$$

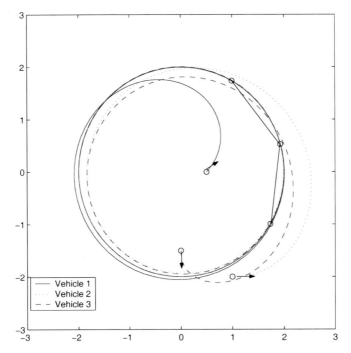

Fig. 3.15. Performance under a leader-follower formation control

and that

$$e'_{21}(t) = \left[\begin{array}{c} \frac{-\sin t + \cos t}{\sqrt{2}} \\ \frac{\cos t + \sin t}{\sqrt{2}} \end{array} \right], \quad e'_{22}(t) = \left[\begin{array}{c} -\frac{\sin t + \cos t}{\sqrt{2}} \\ \frac{\cos t - \sin t}{\sqrt{2}} \end{array} \right] \quad \Longrightarrow \quad q^d_3 = \left[\begin{array}{c} 2\sin t \\ -2\cos t \end{array} \right].$$

Once again, the desired orientation of all vehicles are assumed to be $\theta^d_i = t + \pi/2$.

Then, as did in Example 3.16, the desired trajectories can be mapped into the chained form as

$$\left\{ \begin{array}{l} x^d_{11} = t + \frac{\pi}{2}, \\ x^d_{12} = 2, \\ x^d_{13} = 0, \\ u^d_{11} = 1, \ u^d_{12} = 0, \end{array} \right. \quad \left\{ \begin{array}{l} x^d_{21} = t + \frac{\pi}{2}, \\ x^d_{22} = \sqrt{2}, \\ x^d_{23} = -\sqrt{2}, \\ u^d_{21} = 1, \ u^d_{22} = 0, \end{array} \right. \quad \left\{ \begin{array}{l} x^d_{31} = t + \frac{\pi}{2}, \\ x^d_{32} = 0, \\ x^d_{33} = -2, \\ u^d_{31} = 1, \ u^d_{32} = 0, \end{array} \right.$$

and the tracking control in Section 3.3.1 can be applied. Performance of the formation control is shown in Fig. 3.15 for the vehicles whose initial configurations of $[x_i(t_0), y_i(t_0), \theta_i(t_0)]^T$ are given by $[0.5, 0, \frac{\pi}{4}]$, $[1, -2, \frac{\pi}{6}]$ and $[0, -1.5, 0]$, respectively. $\qquad \diamond$

To make the leader-follower hierarchy robust and to avoid calculation of $q^d_i(t)$ for most of the vehicles, Desired Trajectory 3.100 can further be changed to

$$\hat{q}_i^d(t) = \sum_{k \in \Omega_i} d'_{ik} \left[q_k(t) + \sum_{j=1}^{l} \alpha'_{ij} \hat{e}'_{kj} \right], \quad i = 1, \cdots, n_r, \tag{3.101}$$

where $\hat{e}'_{kj}$ are the basis vectors of frame $\mathcal{F}_k(t)$ located at $q_k(t)$ (rather than its desired frame $\mathcal{F}_k^d(t)$ located at $q_k^d(t)$). Accordingly, time derivatives of $q_k(t)$ would appear in $\hat{e}'_{kj}$, in the tracking error system of the ith vehicle, and in the resulting control u_i. However, time derivatives of $q_k(t)$ are usually not available to the ith vehicle. If we choose to eliminate those derivatives by substituting the model and its control of the kth vehicle, we will have more complicated error dynamics, and the resulting control u_i will require state feedback of x_k and possibly u_k as well. Consequently, Ω_i should be chosen such that $\cap_{i=1}^{n_r} \Omega_i$ is empty in order to avoid the potential of algebraic loop in control design. Alternately, to simplify the design of u_i, standard observers can be used to estimate time derivatives of $q_k^{(j)}(t)$ from $q_k(t)$.

By designating the layers of leaders and followers, the vehicle system has a fixed structure of command and control. While uncertainties and disturbances can now be compensated for, its rigid structure means that, if one of the vehicles becomes disabled, all the vehicles supposed to follow that vehicle (among other leader vehicles) are too impacted to keep the rest of the formation intact. In order to make the formation robust, a vehicle should follow the prevailing behavior of its peers. That is, to maintain the formation as much as possible, every vehicle is commanded to track majority of the vehicles in its vicinity, which leads to the flexible structure of neighboring feedback design.

Neighboring Feedback

Let $N_i(t) \subset \Omega$ be the index set of those vehicles neighboring the ith vehicle at time t. In the spirit of (3.101), the desired trajectory can be set to be

$$\hat{q}_i^d(t) = \sum_{k \in N_i(t)} d''_{ik}(t) \left[q_k(t) + \sum_{j=1}^{l} \alpha'_{ij} \hat{e}'_{kj} \right], \tag{3.102}$$

where $d''_{ik}(t) \geq 0$ and $\sum_{k \in N_i} d''_{ik}(t) = 1$. In essence, the neighboring approach makes the ith vehicle adjusts its motion according to its neighbors. Based on (3.102), control u_i can be designed to make q_i track $\hat{q}_i^d$ provided that $N_i(t)$ changes seldom. In the general case, $N_i(t)$ could change quickly and even unpredictably, Desired Trajectory 3.102 would become discontinuous and not differentiable and hence the corresponding error system could not be predetermined. Formation control under these settings will be accomplished using cooperative control in Sections 5.3.4 and 6.5.

3.4.2 Multi-objective Reactive Control

One way to make the vehicle system become pliable to environmental changes and to ensure no collision among the vehicles in the system is to update both $q_0(t)$ and $q_i^d(t)$ using a real-time path planning and collision avoidance algorithm such as that in Section 3.2. However, this approach has several shortcomings in applications. First, it usually requires too much information, is too time consuming, and may even be unsolvable in many scenarios to replan $q_0(t)$ so that the whole formation avoids all the obstacles. Second, if the ith vehicle replans its desired trajectory, Geometrical Requirement 3.98 needs to be relaxed but also be maintained as close as possible. Third, if all the vehicles replan their desired trajectories, these trajectories must be collision-free among each other, which becomes more involved and may require centralized coordination. Finally, in the case that there are many vehicles and (moving) obstacles in a congested area, it would be more efficient for each vehicle to navigate its local traffic while attempting to relate itself to the virtual leader $q_0(t)$.

As the alternative that addresses these issues, a multi-objective reactive feedback control is preferred for each vehicle to track either the virtual leader $q_0(t)$ or its leader among the vehicle group while avoiding obstacles. The control design is to synthesize reactive forces based on potential field functions: attractive force from an attractive potential field associated with the virtual leader, and repulsive forces from repulsive potential field functions built around obstacles and the vehicles. The intuition is that, by having the two kinds of potential field functions interact with each other, the multiple control objectives are integrated into one by which obstacle avoidance is the priority during the transient and goal tracking is to be achieved asymptotically.

Composite Potential Field Function

Consider the following composite potential field function for the ith vehicle:

$$P(q_i - q_i^d, q_i - q_{o_j}) = P_a(q_i - q_i^d) + \sum_{j \in N_i} P_{r_j}(q_i - q_{o_j}), \qquad (3.103)$$

where $q \in \Re^l$, $N_i(t)$ is the set containing all the indices of the obstacles (including other vehicles) in a neighborhood around the ith vehicle (according to its sensing range), $P_a(\cdot)$ is the attractive potential field function, $P_{r_j}(\cdot)$ is the repulsive potential field function around the jth obstacle, $q_i^d(t)$ represents its desired position (which may be calculated according to (3.102)), and $q_{o_j}(t)$ is the location of the jth obstacle. In what follows, both $P_a(s)$ and $P_r(s)$ are assumed to be differentiable up to the second-order.

To generate an attractive force which moves the vehicle toward its goal, the attractive potential field function should have the following properties:

$$\begin{cases} \text{(i)} \ \ P_a(s) \text{ is positive definite and locally uniformly bounded,} \\ \text{(ii)} \ \ s^T \dfrac{\partial P_a(s)}{\partial s} \text{ is positive definite and locally uniformly bounded.} \end{cases} \qquad (3.104)$$

In (3.104), Property (i) is the basic characteristics of an attractive potential field function, and Property (ii) implies that the negative gradient $-\partial P_a(s)/\partial s$

can be used a feedback control force to make the value of P_a decrease over time. On the other hand, for collision avoidance, the repulsive potential field functions should be chosen such that

$$\begin{cases} \text{(i)} \ P_{r_j}(s) = +\infty \text{ if } s \in \Omega_{o_j}, \text{ and } \lim_{s \to \Omega_{o_j}, \ s \notin \Omega_{o_j}} \left\| \frac{\partial P_{r_j}(s)}{\partial s} \right\| = +\infty, \\ \text{(ii)} \ P_{r_j}(s) = 0 \text{ for } s \notin \overline{\Omega}_{o_j}, \text{ and } P_{r_j}(s) \in (0, \infty) \text{ for all } s \in \overline{\Omega}_{o_j} \cap \Omega_{o_j}^c, \end{cases}$$
$$(3.105)$$

where $\Omega_{o_j} \subset \Re^l$ is the compact set representing the envelope of the jth obstacle, $\Omega_{o_j}^c$ is the complement set of Ω_{o_j} in $\Re^l$, and compact set $\overline{\Omega}_{o_j}$ being the enlarged version of Ω_{o_j} represents the region in which the repulsive force becomes active. In (3.105), Property (ii) states that $P_{r_j}(\cdot)$ is positive semidefinite, and Property (i) implies that configuration q cannot enter region Ω_{o_j} without an infinite amount of work being done. In digital implementation, a maximum threshold value on $P_{r_j}(\cdot)$ may be imposed, in which case collision avoidance can be ensured by either an adaptive sampling scheme (that is, as the value of $P_{r_j}(\cdot)$ increases, the sampling period is reduced) or a velocity-scaled repulsive potential field function (*i.e.*, the higher the speed of the vehicle, the larger compact set $\overline{\Omega}_{o_j}$ while the maximum value on $P_{r_j}(\cdot)$ is fixed).

Composite Potential Field Function 3.103 will be used to generate a control Lyapunov function based on which a reactive control is designed. It should be noted that, to ensure asymptotical convergence, additional conditions on q_i^d, q_{o_j} and $P(\cdot)$ will be required.

Assumption 3.18 *Desired trajectory q_i^d can be tracked* asymptotically, *that is, after some finite time instant t^*, $[q_i^d(t) - q_{o_j}(t)] \notin \Omega_{o_j}$ for all $t \geq t^*$ and for all $j \in N_i(t)$. Accordingly, repulsive potential field functions P_{r_j} should be chosen such that $[q_i^d(t) - q_{o_j}(t)] \notin \overline{\Omega}_{o_j}$ for all $t \geq t^*$ and for all $j \in N_i(t)$.*

Assumption 3.19 *In the limit of $t \to \infty$, Composite Potential Field Function 3.103 has $q_i^d(t)$ as the only local minimum around $q_i(t)$.*

Obviously, Assumption 3.18 is also necessary for q_i to approach q_i^d asymptotically. In general, there are more than one "stationary" solution q_i^* to the algebraic equation $\partial P(q_i^* - q_i^d, q_i^* - q_{o_j})/\partial q_i^* = 0$. It follows from Assumption 3.18 and the properties in (3.104) that $q_i^* = q_i^d$ is the global minimum of function $P(q_i^* - q_i^d, q_i^* - q_{o_j})$. If the obstacles form a trap encircling the ith vehicle but not its desired trajectory, Assumption 3.19 does not hold. Neither does Assumption 3.19 have to hold during the transient. Without loss of any generality, we assume that $\overline{\Omega}_{o_j} \cap \overline{\Omega}_{o_k}$ be empty for $j \neq k$. Hence, we can solve the algebraic equation

$$\frac{\partial P(q_i^* - q_i^d, q_i^* - q_{o_j})}{\partial q_i^*} = \frac{\partial P_a(q_i^* - q_i^d)}{\partial q_i^*} + \frac{\partial P_{r_j}(q_i^* - q_{o_j})}{\partial q_i^*} = 0 \qquad (3.106)$$

for any pair of $\{P_a,\ P_{r_j}\}$ and analyze the property of solution q_i^* during the transient as well as in the limit.

For instance, consider the following typical choices of potential field functions:

$$P_a(q - q^d) = \lambda_a[(x - x^d)^2 + (y - y^d)^2], \qquad (3.107)$$

and

$$P_r(q - q_o) = \begin{cases} +\infty & \text{if } \|q - q_o\|^2 \leq r_o^2, \\ \lambda_r \left[\ln \dfrac{(r_o + \epsilon_o)^2 - r_o^2}{\|q - q_o\|^2 - r_o^2} - \dfrac{(r_o + \epsilon_o)^2 - \|q - q_o\|^2}{(r_o + \epsilon_o)^2 - r_o^2}\right] \\ & \text{if } r_o^2 \leq \|q - q_o\|^2 \leq (r_o + \epsilon_o)^2, \\ 0 & \text{if } \|q - q_o\|^2 \geq (r_o + \epsilon_o)^2, \end{cases} \qquad (3.108)$$

where $q = [x\ y]^T \in \Re^2$, $q^d = [x^d\ y^d]^T$, $q_o = [x_o\ y_o]^T$, $\lambda_a, \lambda_r > 0$ are constant gains, $\Omega_o = \{q:\ \|q - q_o\| \leq r_o\}$, and $\overline{\Omega}_o = \{q:\ \|q - q_o\| \leq r_o + \epsilon_o\}$ with $r_o, \epsilon_o > 0$. In the case that $y^d = y_o$ and $x^d - x_o > r_o + \epsilon_o$, the corresponding total potential force of $\partial P(q - q^d, q - q_o)/\partial q$ has its vector field shown in Fig. 3.16. Clearly, $q = q^d$ is the only global minimum, and there is a saddle point on the left of the obstacle.

In general, it follows from (3.106) that stationary point q^* exists only at location(s) where repulsive and attractive force balance each other as

$$\frac{\partial P_a(q^* - q^d)}{\partial q^*} = -\frac{\partial P_r(q^* - q_o)}{\partial q^*}. \qquad (3.109)$$

Hence, if $q^d \notin \overline{\Omega}_o$, there is no stationary point outside $\overline{\Omega}_o$ except for the global minimum at q^d. Inside $\overline{\Omega}_o$, additional stationary point(s) may exist. The following lemma provides the condition on the stationary point being a saddle point, which ensures Assumption 3.19.

Lemma 3.20. *Suppose that the level curves of $P_a(q - q^d) = c_a$ and $P_r(q - q_o) = c_r$ are closed and convex, that these two families of level curves are tangent to each other along a straight-line segment, and that $q^d \notin \overline{\Omega}_o \subset \Re^2$. Then, $q^* \in \overline{\Omega}_o$ is a saddle point if the curvature of $P_a(q - q^d) = c_a$ at $q = q^*$ is smaller than that of $P_r(q - q_o) = c_r$ at $q = q^*$.*

Proof: Consider any stationary point $q^* \in \overline{\Omega}_o$ and, at q^*, establish a local coordinate system of $q' = [x'\ y']^T$ such that its origin is at $q = q^*$ and its y' axis is along the direction of $\partial P_a(q^* - q^d)/\partial q^*$. That is, according to (3.109), axis y' points toward the obstacle and also the goal q^d further along, it is parallel to the normal vector of both level curves of $P_a(q - q^d) = c_a$ and $P_r(q - q_o) = c_r$ at $q' = 0$, and the x' axis is along the tangent direction of both level curves of $P_a(q - q^d) = c_a$ and $P_r(q - q_o) = c_r$ at $q' = 0$. Therefore, we know from the geometry that $q = Rq' + q^*$ for some rotational matrix R, that

$$\left.\frac{\partial P_a}{\partial x'}\right|_{q'=0} = \left.\frac{\partial P_r}{\partial x'}\right|_{q'=0} = 0 \qquad (3.110)$$

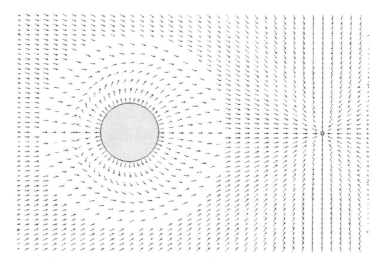

Fig. 3.16. Vector field of $\partial P(q - q^d, q - q_o)/\partial q$ under (3.107) and (3.108)

and

$$\left.\frac{\partial P_a}{\partial y'}\right|_{q'=0} = -\left.\frac{\partial P_r}{\partial y'}\right|_{q'=0} > 0, \tag{3.111}$$

and that, since the level curves are closed and convex,

$$\left.\frac{\partial^2 P_a}{\partial(y')^2}\right|_{q'=0} < 0, \qquad \left.\frac{\partial P_r}{\partial(y')^2}\right|_{q'=0} < 0. \tag{3.112}$$

By assumption and according to (3.110), the two sets of level curves are tangent along the y' axis and consequently equations

$$\left.\frac{\partial P_a}{\partial x'}\right|_{q'=[0\ \delta y']^T} = \left.\frac{\partial P_r}{\partial x'}\right|_{q'=[0\ \delta y']^T} = 0$$

hold for all small $\delta y'$, which in turn implies

$$\left.\frac{\partial^2 P}{\partial y' \partial x'}\right|_{q'=0} = \left.\frac{\partial^2 P}{\partial x' \partial y'}\right|_{q'=0} = \lim_{\delta y' \to 0} \frac{\left.\frac{\partial P}{\partial x'}\right|_{q'=[0\ \delta y']^T} - \left.\frac{\partial P}{\partial x'}\right|_{q'=0}}{\delta y'} = 0. \tag{3.113}$$

Stability of stationary point q^* is determined by Hassian matrix $\partial^2 P/\partial(q')^2$ at that point. Therefore, it follows from (3.112) and (3.113) that q^* is a saddle point if

$$\left.\frac{\partial^2 P_a}{\partial(x')^2}\right|_{q'=0} + \left.\frac{\partial P_r}{\partial(x')^2}\right|_{q'=0} > 0, \tag{3.114}$$

and it is a stable node if the above inequality is reversed.

On the other hand, the curvature of function $y = f(x)$ is $(d^2y/dx^2)/[1 + (dy/dx)^2]^{3/2}$. Taking the first-order derivative along the level curve of $P_a(q - q^d) = c_a$ yields

$$\left.\frac{dy_a}{dx'}\right|_{q'=0} = -\left.\frac{\frac{\partial P_a}{\partial x'}}{\frac{\partial P_a}{\partial y'}}\right|_{q'=0} = 0,$$

in which (3.110) is used, and $y_a = y_a(x')$ is the solution to the implicit equation of $P_a(q - q^d) = c_a$. Using the above result and taking the second-order derivative along the level curve of $P_a(q - q^d) = c_a$ yields

$$\left.\frac{dy_a^2}{d(x')^2}\right|_{q'=0} = -\left.\frac{\frac{\partial^2 P_a}{\partial(x')^2}}{\frac{\partial P_a}{\partial y'}}\right|_{q'=0}, \tag{3.115}$$

whose absolution value is the curvature of level curve $P_a(q - q^d) = c_a$. Similarly, it follows that the curvature of level curve $P_r(q - q_o) = c_r$ is the absolute value of

$$\left.\frac{dy_r^2}{d(x')^2}\right|_{q'=0} = -\left.\frac{\frac{\partial^2 P_r}{\partial(x')^2}}{\frac{\partial P_r}{\partial y'}}\right|_{q'=0}. \tag{3.116}$$

Recalling that the level curves are convex and hence have opening upwards in the coordinate of q' and around $q' = 0$, we know that $dy_a^2/d(x')^2$ and $dy_r^2/d(x')^2$ defined by (3.115) and (3.116) are both positive. It follows from (3.111) that

$$\left.\frac{\partial^2 P_a}{\partial(x')^2}\right|_{q'=0} < 0, \quad \left.\frac{\partial P_r}{\partial(x')^2}\right|_{q'=0} > 0.$$

Hence, Inequality 3.114 can be concluded by comparing curvatures.   □

Lemma 3.20 can be visualized graphically, and its proof also implies that q^* is a stable node (and hence a local minimum) if the curvature of $P_a(q-q^d) = c_a$ at $q = q^*$ is larger than that of $P_r(q - q_o) = c_r$ at $q = q^*$. Presence of stable node(s) in $\overline{\Omega}_o$ should be avoided by properly selecting the repulsive potential field function for any given obstacle. The global minimum solution of q^d characterizes the tracking control objective, and the rest of stationary solutions q^* introduced inevitably by the presence of repulsive potential field functions should be saddle points in order to achieve both the objectives of collision avoidance and goal tracking. Nonetheless, the presence of saddle points q_i^* needs to be accounted for in the subsequent reactive control design.

Reactive Control

To simplify the technical development, let us consider the case that dynamic feedback linearization has been applied (as in Section 3.1.3) to transfer the vehicle model into

$$\ddot{q}_i = u_i, \tag{3.117}$$

where $q_i \in \Re^l$ is the coordinate of configuration output variables, and u_i is the control to be designed. Then, the multi-objective reactive control can be chosen to be

$$
\begin{aligned}
u_i = & -\frac{\partial P(q_i - q_i^d, q_i - q_{o_j})}{\partial q_i} - \xi_i(q_i - q_i^d, q_i - q_{o_j})(\dot{q}_i - \dot{q}_i^d) + \ddot{q}_i^d \\
& -k \sum_{j \in N_i} \frac{\partial^2 P_{r_j}(q_i - q_{o_j})}{\partial (q_i - q_{o_j})^2}(\dot{q}_i - \dot{q}_{o_j})\|\dot{q}_i^d - \dot{q}_{o_j}\|^2 \\
& -2k \sum_{j \in N_i} \frac{\partial P_{r_j}(q_i - q_{o_j})}{\partial (q_i - q_{o_j})}(\dot{q}_i^d - \dot{q}_{o_j})^T(\ddot{q}_i^d - \ddot{q}_{o_j}) \\
& +\varphi_i(q_i^*, \dot{q}_i - \dot{q}_i^d, \dot{q}_i - \dot{q}_{o_j})
\end{aligned}
\tag{3.118}
$$

where $\xi_i(\cdot) \geq 0$ is the non-linear damping gain which vanishes as $(q_i - q_{o_j})$ approaches set Ω_{o_j} and is uniformly positive anywhere else (i.e., $\xi_i = \xi_i(q_i - q_i^d) > 0$ if $(q_i - q_{o_j}) \notin \overline{\Omega}_{o_j}$), $k > 0$ is a gain associated with the terms that resolve the potential conflict between asymptotic tracking and collision avoidance, and $\varphi_i(\cdot)$ is a uniformly bounded perturbation term which is non-zero only if the vehicle stays idle (in the sense that $\dot{q}_i - \dot{q}_i^d = \dot{q}_i - \dot{q}_{o_j} = 0$) at saddle point q_i^* (which can be solved from (3.106)). At the saddle point, a simple choice of $\varphi_i(\cdot)$ would be a bounded force along the tangent direction of the level curve of $P_{r_j}(q_i - q_{o_j}) = P_{r_j}(q_i^* - q_{o_j})$.

Collision avoidance is established in the following theorem by showing that, under Control 3.118, the vehicle never enters the moving set Ω_{o_j} around q_{o_j}. Asymptotic tracking convergence is also shown under Assumptions 3.18 and 3.19.

Theorem 3.21. *Suppose that Potential Field Function 3.103 satisfies Properties 3.104 and 3.105, where $[\dot{q}_i^d(t) - \dot{q}_{o_j}(t)]$ and $q_i^d(t)$ are uniformly bounded. Then, as long as $q_i(t_0) \notin \Omega_{o_j}$, System 3.117 under Control 3.118 is collision-free. Furthermore, if Assumptions 3.18 and 3.19 hold, q_i converges asymptotically to $q_i^d(t)$.*

Proof: Should moving sets $\overline{\Omega}_{o_j}$ and $\overline{\Omega}_{o_k}$ overlap for some $j \neq k$ and during certain time period, the two obstacles can be combined into one obstacle. Thus, we need only consider in the proof of collision avoidance that compact moving sets $\overline{\Omega}_{o_j}$ are all disjoint in the configuration space.

Consider first the simpler case that $\ddot{q}_i^d - \ddot{q}_{o_j} = 0$. In this case, Control 3.118 with $\varphi_i(\cdot) = 0$ reduces to

$$
u_i = -\frac{\partial P(q_i - q_i^d, q_i - q_{o_j})}{\partial q_i} - \xi_i(q_i - q_i^d, q_i - q_{o_j})(\dot{q}_i - \dot{q}_i^d) + \ddot{q}_i^d. \tag{3.119}
$$

Choose the Lyapunov function candidate

$$V_1(t) = \frac{1}{2}\|\dot{q}_i - \dot{q}_i^d\|^2 + V_0(t),$$

where

$$V_0(t) = P_a(q_i - q_i^d) + \sum_j P_{r_j}(q_i - q_{o_j}).$$

It is obvious that $V_1(t_0)$ is finite for any finite and collision-free initial conditions. It follows that, along the trajectory of System 3.117 and under Control 3.119,

$$\dot{V}_1 = (\dot{q}_i - \dot{q}_i^d)^T(\ddot{q}_i - \ddot{q}_i^d) + (\dot{q}_i - \dot{q}_i^d)^T\frac{\partial P_a(q_i - q_i^d)}{\partial(q_i - q_i^d)}$$

$$+ \sum_{j\in N_i(t)} (\dot{q}_i - \dot{q}_{o_j})^T\frac{\partial P_{r_j}(q_i - q_{o_j})}{\partial(q_i - q_{o_j})}$$

$$= -\xi_i(q_i - q_i^d, q_i - q_{o_j})\|\dot{q}_i - \dot{q}_i^d\|^2 + \sum_{j\in N_i(t)} (\dot{q}_i^d - \dot{q}_{o_j})^T\frac{\partial P_{r_j}(q_i - q_{o_j})}{\partial(q_i - q_{o_j})}$$

$$= -\xi_i(q_i - q_i^d, q_i - q_{o_j})\|\dot{q}_i - \dot{q}_i^d\|^2,$$

which is negative semi-definite. Thus, $V_1(t)$ is non-increasing as $V_1(t) \leq V_1(t_0)$ for all t, $P_{r_j}(q_i - q_{o_j}) < \infty$ for all t and for all j, and hence the vehicle is collision-free.

For the general case of $\dot{q}_i^d - \dot{q}_{o_j} \neq 0$, let us consider the Lyapunov function

$$V_2(t) = \frac{1}{2}\left\| \dot{q}_i - \dot{q}_i^d + k\sum_j \frac{\partial P_{r_j}(q_i - q_{o_j})}{\partial(q_i - q_{o_j})}\|\dot{q}_i^d - \dot{q}_{o_j}\|^2 \right\|^2 + V_0(t).$$

Similarly, $V_2(t_0)$ is finite for any finite and collision-free initial conditions. It follows from (3.117) and (3.118) that

$$\dot{V}_2 = \left[\dot{q}_i - \dot{q}_i^d + k\sum_{j\in N_i} \frac{\partial P_{r_j}(q_i - q_{o_j})}{\partial(q_i - q_{o_j})}\|\dot{q}_i^d - \dot{q}_{o_j}\|^2\right]^T [\ddot{q}_i - \ddot{q}_i^d$$

$$+k\sum_{j\in N_i} \frac{\partial^2 P_{r_j}(q_i - q_{o_j})}{\partial(q_i - q_{o_j})^2}(\dot{q}_i - \dot{q}_{o_j})\|\dot{q}_i^d - \dot{q}_{o_j}\|^2$$

$$+2k\sum_{j\in N_i} \frac{\partial P_{r_j}(q_i - q_{o_j})}{\partial(q_i - q_{o_j})}(\dot{q}_i^d - \dot{q}_{o_j})^T(\ddot{q}_i^d - \ddot{q}_{o_j})\right] + (\dot{q}_i - \dot{q}_i^d)^T\frac{\partial P_a(q_i - q_i^d)}{\partial(q_i - q_i^d)}$$

$$+ \sum_{j\in N_i} (\dot{q}_i - \dot{q}_{o_j})^T\frac{\partial P_{r_j}(q_i - q_{o_j})}{\partial(q_i - q_{o_j})}$$

$$= -k\sum_{j\in N_i} \left\|\frac{\partial P_{r_j}(q_i - q_{o_j})}{\partial(q_i - q_{o_j})}\right\|^2 \|\dot{q}_i^d - \dot{q}_{o_j}\|^2 - \xi_i(q_i - q_i^d, q_i - q_{o_j})\|\dot{q}_i - \dot{q}_i^d\|^2$$

$$+w(q_i - q_i^d, q_i - q_{o_j}, \dot{q}_i - \dot{q}_i^d, \dot{q}_i - \dot{q}_{o_j}), \tag{3.120}$$

where

$$
\begin{aligned}
& w(q_i - q_i^d, q_i - q_{o_j}, \dot{q}_i - \dot{q}_i^d, \dot{q}_i - \dot{q}_{o_j}) \\
& = \sum_{j \in N_i} (\dot{q}_i^d - \dot{q}_{o_j})^T \frac{\partial P_{r_j}(q_i - q_{o_j})}{\partial(q_i - q_{o_j})} \\
& \quad -k \sum_{j \subset N_i} \left(\frac{\partial P_{r_j}(q_i - q_{o_j})}{\partial(q_i - q_{o_j})} \right)^T \frac{\partial P_a(q_i - q_i^d)}{\partial(q_i - q_i^d)} \| \dot{q}_i^d - \dot{q}_{o_j} \|^2 \\
& \quad -k \sum_{j \in N_i} \left(\frac{\partial P_{r_j}(q_i - q_{o_j})}{\partial(q_i - q_{o_j})} \right)^T \xi_i(q_i - q_i^d, q_i - q_{o_j})(\dot{q}_i - \dot{q}_i^d) \| \dot{q}_i^d - \dot{q}_{o_j} \|^2.
\end{aligned}
$$

It follows from (3.104) and from $q_i^d(t)$ being uniformly bounded that, for any q_i satisfying $(q_i - q_{o_j}) \in \overline{\Omega}_{o_j}$, $\partial P_a(q_i - q_i^d)/\partial(q_i - q_i^d)$ is uniformly bounded. Hence, we know from $(\dot{q}_i - \dot{q}_{o_j})$ being uniformly bounded that, for some constants $c_1, c_2 \geq 0$,

$$
\begin{aligned}
& |w(q_i - q_i^d, q_i - q_{o_j}, \dot{q}_i - \dot{q}_i^d, \dot{q}_i - \dot{q}_{o_j})| \\
& \leq c_1 \sum_{j \in N_i} \left\| \frac{\partial P_{r_j}(q_i - q_{o_j})}{\partial(q_i - q_{o_j})} \right\| \| \dot{q}_i^d - \dot{q}_{o_j} \| \\
& \quad + c_2 \sum_{j \in N_i} \xi_i(q_i - q_i^d, q_i - q_{o_j}) \left\| \frac{\partial P_{r_j}(q_i - q_{o_j})}{\partial(q_i - q_{o_j})} \right\| \| \dot{q}_i^d - \dot{q}_{o_j} \| \| \dot{q}_i - \dot{q}_i^d \|.
\end{aligned}
$$

Substituting the above inequality into (3.120), we can conclude using the property of $\xi_i(q_i - q_i^d, q_i - q_{o_j})$ that, as $(q_i - q_{o_j}) \to \Omega_{o_j}$, $\dot{V}_2 < 0$. Thus, $V_2(t)$ will be finite in any finite region and hence there is no collision anywhere. Since perturbation force $\varphi_i(\cdot)$ is uniformly bounded and is only active at saddle points, its presence does not have any impact on collision avoidance.

To show asymptotic convergence under Assumptions 3.18 and 3.19, we note from Lemma 3.20 (and as illustrated by Fig. 3.16) that after $t = t^*$, the vehicle cannot stay in $\overline{\Omega}_{o_j}$ indefinitely and upon leaving $\overline{\Omega}_{o_j}$ the tracking error dynamics of System 3.117 and under Control 3.118 reduce to

$$
\dot{e}_{i1} = e_{i2}, \quad \dot{e}_{i2} = -\frac{\partial P_a(e_{i1})}{\partial e_{i1}} - \xi_i(e_{i1})e_{i2},
$$

where $e_{i1} = q_i - q_i^d$ and $e_{i2} = \dot{q}_i - \dot{q}_i^d$. Adopting the simple Lyapunov function

$$
V_3(t) = P_a(e_{i1}) + \frac{1}{2} \| e_{i2} \|^2
$$

we have

$$
\begin{aligned}
\dot{V}_3 & = e_{i2}^T \frac{\partial P_a(e_{i1})}{\partial e_{i1}} + e_{i2}^T \left[-\frac{\partial P_a(e_{i1})}{\partial e_{i1}} - \xi_i(e_{i1})e_{i2} \right] \\
& = -\xi_i(e_{i1}) \| e_{i2} \|^2,
\end{aligned}
$$

from which asymptotic convergence can be concluded using Theorem 2.21. □

It is worth noting that trajectory planning and reactive control can be combined to yield better performance toward both collision avoidance and goal tracking. Specifically, if moving sets $\overline{\Omega}_{o_j}$ are all relatively far apart from each other, Control 3.118 can be simplified to be

$$u_i = -\frac{\partial P(q_i - \hat{q}_i^d, \hat{q}_i^d - \dot{q}_{o_j})}{\partial q_i} - \xi_i(q_i - \hat{q}_i^d, q_i - q_{o_j})(\dot{q}_i - \dot{\hat{q}}_i^d) + \ddot{\hat{q}}_i^d. \quad (3.121)$$

where $\hat{q}_i^d$ is the re-planned version of q_i^d such that $\hat{q}_i^d$ never enters set $\overline{\Omega}_{o_j}$ around q_{o_j} (although q_i^d may enter temporarily), $\dot{\hat{q}}_i^d - \dot{q}_{o_j} = 0$ holds whenever $(q_i^d - q_{o_j})$ enters set $\overline{\Omega}_{o_j}$, and $\lim_{t\to\infty} \hat{q}_i^d - q_i^d = 0$. In other words, when $(q_i^d - \dot{q}_{o_j})$ enters moving set $\overline{\Omega}_{o_j}$, $\hat{q}_i^d(t)$ is kept outside by choosing $\hat{q}_i^d$ such that $(\hat{q}_i^d - \dot{q}_{o_j})$ is constant; and when $(q_i^d - \dot{q}_{o_j})$ exits set $\overline{\Omega}_{o_j}$, $\hat{q}_i^d(t)$ is replanned to track q_i^d asymptotically or in finite time. Comparing (3.121) and (3.119), we know that Control 3.121 is collision-free. It follows from the proof of Theorem 3.21 that Control 3.121 also makes $(q_i^d - q_i^d)$ asymptotically convergent. The uniformly bounded detouring force $\varphi_i(\cdot)$ added in Control 3.118 is to move the vehicle away from any saddle equilibrium point so that asymptotic tracking becomes possible, and hence it should also be added into (3.121). Finally, in the proof of Theorem 3.21, a control Lyapunov function can be found to replace V_3 once the expression of attractive potential field function $P_a(\cdot)$ is available.

To illustrate performance of the multi-objective reactive control in (3.118), consider the 2-D motion that one vehicle is in the form of (3.117), has its initial position at $q(t_0) = [2.5\ 1]^T$, and is to follow the desired trajectory given by

$$q^d(t) = \left[\, 2\cos\tfrac{\pi}{20}t\ \ 2\sin\tfrac{\pi}{20}t\,\right]^T.$$

In this case, Reactive Control 3.118 is designed with potential field functions

$$P_a(q) = \frac{1}{2}\lambda_a\|q - q^d\|^2$$

$$P_{r_j}(q) = \begin{cases} \dfrac{1}{2}\lambda_{r_j}\left(\dfrac{1}{\|q - q_{o_j}\| - \rho_j} - \dfrac{1}{\rho_j + \epsilon_j}\right)^2 \\ \qquad\qquad \text{if } \rho_j < \|q - q_{o_j}\| \le \rho_j + \epsilon_j, \\ 0 \qquad\qquad \text{if } \|q - q_{o_j}\| > \rho_j + \epsilon_j, \end{cases} \quad (3.122)$$

where λ_a, λ_{r_j}, ρ_j and ϵ_j are positive constants. Two scenarios are simulated. In the first scenario, there is only one static obstacle located at $q_{o_1} = [-2\ 0]^T$. Control 3.118 is implemented with design parameters $\xi = 2$, $k = 20$, $\varphi_i = 0$, $\lambda_a = 5$, $\lambda_{r_1} = 10$, and $\rho_1 = \epsilon_1 = 0.2$, and the simulation result is shown in Fig. 3.17. In the second scenario, there is additionally one moving obstacle which oscillates between the two points of $(1, 0)$ and $(3, 0)$, parallel to the x axis,

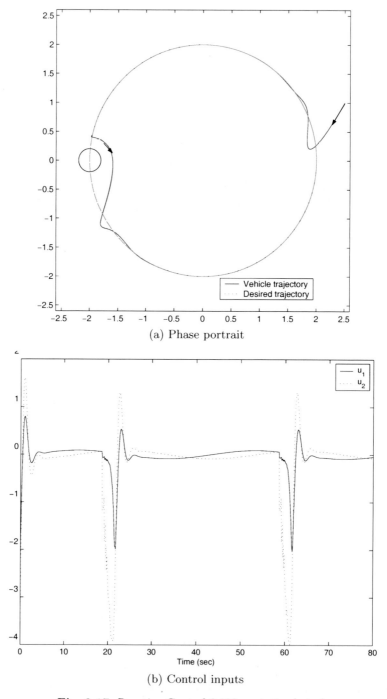

(a) Phase portrait

(b) Control inputs

Fig. 3.17. Reactive Control 3.118: a static obstacle

and with a constant velocity of $\pm 1/40$. Control 3.118 is then implemented with additional design parameters of $\lambda_{r_2} = 30$ and $\rho_2 = \epsilon_2 = 0.2$, and the simulation results are given in Figs. 3.18 and 3.19. In the figures, the obstacles are represented by the larger circles, and the vehicle is denoted by the smaller circle. Figure 3.18(a) shows all the trajectories over the time period of $t \in [0, 160]$; that is, the vehicle passes through the points numbered 1 up to 14 and in the direction along the marked arrows, while the moving obstacle oscillates once from the starting point $(3, 0)$ to point $(1, 0)$ and back. As illustrated by the snapshots in Fig. 3.19(a) and (b), the vehicle successfully avoids both obstacles while tracking the desired trajectory.

3.5 Notes and Summary

The chained form is the canonical model for analyzing non-holonomic systems [123, 168]. By their nature, non-holonomic systems are small-time controllable but not uniformly completely controllable in general, and they can be partially dynamically feedback linearizable but cannot be asymptotically stabilized under any smooth time-invariant feedback control. Accordingly, the open-loop and closed-loop control problems of non-holonomic systems are interesting, and several approaches to solve these problems are presented in this chapter. Specifically, open-loop steering controls can be parameterized in terms of basis functions (such as sinusoidal functions, piecewise-constant functions, and polynomial functions). To follow a desired trajectory generated under a non-vanishing input, a feedback tracking control can be designed using the Lyapunov direct method and Riccati equation. For asymptotic stabilization with respect to a fixed-point, a feedback control should be either discontinuous or time-varying or both. Among the feedback control design approaches, the technique of recovering uniform complete controllability and then solving for quadratic Lyapunov function from the Riccati equation has the advantages of simplicity and inverse optimality.

Without considering obstacles, motion planning under non-holonomic constraints is to generate appropriate steering controls, and among the available approaches are differential geometry [250, 251], differential flatness [67], input parameterization [159, 170, 259], the spline method [59, 126, 188], and optimal control [64]. It is shown through recasting the non-holonomic motion planning problem as an optimal control problem that the feasible and shortest path between two boundary conditions is a concatenation of arcs and straight-line segments belonging to a family of 46 different types of three-parameter steering controls [211, 251]. Trajectories planned for non-holonomic systems need to be both collision-free and feasible. For obstacles moving with known constant velocities, collision avoidance can be achieved using the concept of velocity cone [66]. In [206, 279], a pair of time and geometrical collision avoidance criteria are derived, and an optimized trajectory planning algorithm is proposed to handle both non-holonomic constraints of a vehicle and its dy-

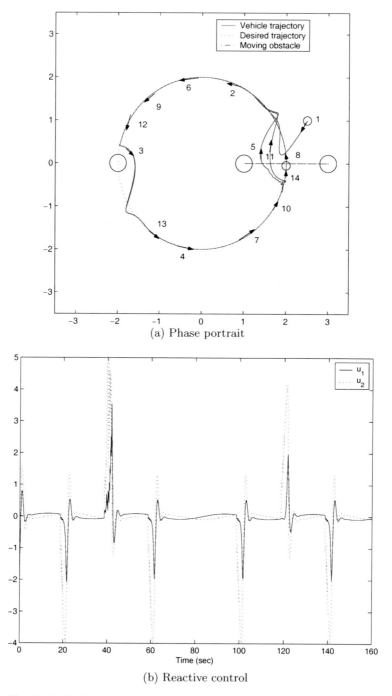

(a) Phase portrait

(b) Reactive control

Fig. 3.18. Performance of reactive control: static and moving obstacles

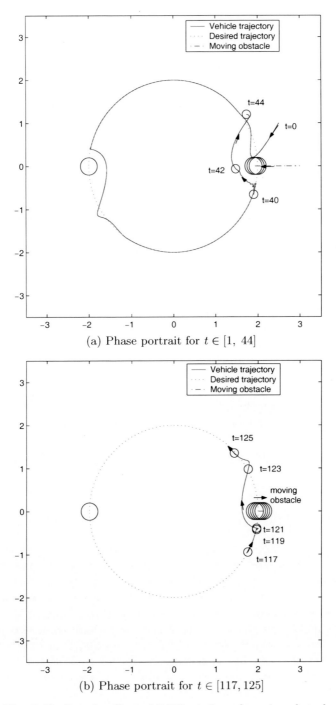

(a) Phase portrait for $t \in [1, \ 44]$

(b) Phase portrait for $t \in [117, 125]$

Fig. 3.19. Reactive Control 3.118: static and moving obstacles

namically changing environment. On the other hand, standard approaches of motion planning deal with holonomic systems in the presence of static obstacles [122], and they include potential field [36, 95, 109, 218], vector field histogram [26], exhaustive search methods [15, 58, 93, 125], the iterative method [53, 106], dynamic programming [55] and optimal control [124, 249]. These methods could also be used to generate feasible trajectories by embedding either an integration or approximation component, both of which may be computationally involved.

For a vehicle system, formation control of multiple vehicles can be designed using several approaches [6]. For instance, navigation and exploration can be pursued using cellular robot systems [17], or relative geometry approach [269], or artificial intelligence methods [167], or a probabilistic approach [69], or computer architectural method [183]. There is the behavior-based approach [7, 13, 31, 150] by which a set of primitive behaviors such as moving-to-goal, avoiding-obstacle and maintaining-formation are prescribed, their importance are represented by a set of weights, and the combined behavior is determined by aggregating the weights. And, basic artificial behaviors can be improving by setting up a reward function and its learning mechanism, for example, reinforcement learning [151]. The notion of string stability [253] can be used for line formations. To achieve precisely a general formation, one option is to impose a virtual structure on geometrical relationship among the vehicles, and a centralized formation control can be designed to minimize vehicles' errors with respect to the virtual structure [131]. The virtual structure can also be broken down using a set of moving frames so the formation problem becomes a set of individual tracking problems [105, 231]. The other option of maintaining a formation is to adopt a leader-following strategy [50, 257] by which the formation control problem is reduced to a tracking control problem. Artificial potential field functions have been used for distributed control of a multi-vehicle formation. In [128, 177], virtual leaders and artificial potentials are used to maintain a group geometry, although a unique desired formation may not be achieved and several virtual vehicles may be required to yield a composite potential field function with a unique minimum. In [179], a formation graph and its structural potential field functions are used to design distributed formation control for achieving a desired formation, and avoidance of static obstacles may be achieved if two kinds of special-purpose agents can be physically added through flocking protocols [180]. Cooperative control of unicycle robots have been studied by employing the combination of discontinuous control, non-smooth Lyapunov theory and graph theory [52], or by using the combination of Frenet-Serret model, time-varying control and average theory [135, 164], or by assuming constant driving velocity [227, 278]. By imposing several conditions including freezing the reference trajectory within avoidance zone, collision avoidance and formation control of differential-drive vehicles are studied in [148, 246], and tracking errors are ensured to be bounded. In Section 3.4.1 and later in Section 5.3.4, the formation control problem is formulated toward designing a neighboring-feedback cooperative and pliable formation control. Explicit conditions on how to choose an appropriate composite potential field function are found, and a reactive control is provided for each vehicle to asymptotically track its goal while avoiding static objects, moving obstacles and other vehicles. The multi-objective reactive control design, originally proposed in [42], makes the vehicles pliable to their dynamical surroundings. In Section 6.6, this pliable control will be combined into cooperative control to achieve any cooperative and pliable formation of vehicles in a dynamically changing environment, while the consensus problem of non-holonomic systems will be addressed in Section 6.5.

4

Matrix Theory for Cooperative Systems

In this chapter, non-negative matrices and their properties are introduced. Among the important concepts to characterize non-negative matrices are irreducibility and reducibility, cyclicity and primitiveness, and lower-triangular completeness. The so-called Perron-Frobenius theorem provides the fundamental results on eigenvalues and eigenvectors of a non-negative matrix.

Non-negative matrices arise naturally from systems theory, in particular, non-negative matrices for discrete-time positive systems and/or cooperative systems, row-stochastic matrices for Markov chains and cooperative systems, Metzler matrices for continuous-time positive systems and/or cooperative systems, and M-matrices for asymptotically stable systems. Geometrical, graphical and physical meanings of non-negative matrices are explicitly shown.

Analysis tools and useful properties of the aforementioned matrices are detailed for their implications in and applications to dynamical systems, especially cooperative systems. Specifically, eigenvalue properties are explored, existence of and explicit solution to Lyapunov function are determined, and convergence conditions on matrix sequences are obtained.

4.1 Non-negative Matrices and Their Properties

Consider two matrices/vectors $E, F \in \Re^{r_1 \times r_2}$. The notations of $E = F$, $E \geq F$, and $E > F$ are defined with respect to all their elements as, for all i and j, $e_{ij} = f_{ij}$, $e_{ij} \geq f_{ij}$, and $e_{ij} > f_{ij}$, respectively. Operation $E = |F|$ of any matrix F is defined element-by-element as $e_{ij} = |f_{ij}|$. Matrix/vector E is said to be *non-negative* if $E \geq 0$ and *positive* if $E > 0$. The set of all non-negative matrices is denoted by $\Re_+^{r_1 \times r_2}$.

Matrix $\mathbf{J}_{r \times r} \in \Re_+^{r \times r}$ and vector $\mathbf{1}_r \in \Re_+^r$ are the special positive matrix and vector, respectively, whose elements are all 1. Matrix E is said to be *binary* if its elements are either 0 or 1. Matrix $E \in \Re_+^{r \times r}$ is said to be *diagonally positive* if $e_{ii} > 0$ for all $i = 1, \cdots, r$. Matrix $E \in \Re_+^{r_1 \times r_2}$ is said to be *row-*

stochastic if $E1_{r_2} = 1_{r_1}$, and it is said to be *column-stochastic* if E^T is row-stochastic.

4.1.1 Reducible and Irreducible Matrices

Non-negative matrices have special properties, and the most distinctive is irreducibility or reducibility that captures physical connectivity or topological structure of any dynamical system associated with the given matrix.

Definition 4.1. *A non-negative matrix $E \in \Re^{r \times r}$ with $r \geq 2$ is said to be reducible if the set of its indices, $\mathcal{I} \triangleq \{1, 2, \cdots, r\}$, can be divided into two disjoint non-empty sets $\mathcal{S} \triangleq \{i_1, i_2, \cdots, i_\mu\}$ and $\mathcal{S}^c \triangleq \mathcal{I}/\mathcal{S} = \{j_1, j_2, \cdots, j_\nu\}$ (with $\mu + \nu = r$) such that $e_{i_\alpha j_\beta} = 0$, where $\alpha = 1, \cdots, \mu$ and $\beta = 1, \cdots, \nu$. Matrix E is said to be irreducible if it is not reducible.*

Permutation matrix corresponds to reordering of the indices, and its application as a coordinate (or state) transformation rearranges matrix rows and columns. It follows from Definition 4.1 and its associated permutation matrix that the following lemma provides the most basic properties of both a reducible matrix and an irreducible matrix. Hence, the *lower triangular structure* of matrix $F_\searrow$ in (4.1) is the canonical form for reducible matrices.

Lemma 4.2. *Consider matrix $E \in \Re_+^{r \times r}$ with $r \geq 2$. If E is reducible, there exist an integer $p > 1$ and a permutation matrix T such that*

$$T^T E T = \begin{bmatrix} F_{11} & 0 & \cdots & 0 \\ F_{21} & F_{22} & \cdots & 0 \\ \vdots & \vdots & \ddots & \vdots \\ F_{p1} & F_{p2} & \cdots & F_{pp} \end{bmatrix} \triangleq F_\searrow, \tag{4.1}$$

where $F_{ii} \in \Re^{r_i \times r_i}$ is either square and irreducible sub-matrices of dimension higher than 1 or a scalar, and $\sum_{i=1}^p r_i = r$. If E is irreducible, vector $z' = (I_{r \times r} + E)z$ has more than η positive entries for any vector $z \geq 0$ containing exactly η positive entries, where $1 \leq \eta < r$ and $I_{r \times r} \in \Re^{r \times r}$ is the identity matrix.

The following corollaries can directly be concluded from Lemma 4.2, and they can be used as the two simple tests on irreducibility or reducibility.

Corollary 4.3. *Consider matrix $E \in \Re_+^{r \times r}$. Then, if and only if E is irreducible, inequality $\gamma z \geq Ez$ with constant $\gamma > 0$ and vector $z \geq 0$ implies either $z = 0$ or $z > 0$.*

Corollary 4.4. *Consider matrix $E \in \Re_+^{r \times r}$. Then, E is irreducible if and only if $(cI_{r \times r} + E)^{r-1} > 0$ for any scalar $c > 0$. If all the matrices in sequence*

$\{E(k)\}$ *are irreducible and diagonally positive,* $E(k+\eta)\cdots E(k+1)E(k) > 0$ *for some* $1 \le \eta \le r-1$ *and for all* k.

An irreducible matrix needs to be analyzed as a whole; and by utilizing the lower triangular structure in (4.1), analysis of a reducible matrix could be done in terms of both irreducible blocks on the diagonal and blocks in the lower triangular portion.

Definition 4.5. *A reducible matrix* E *is said to be* lower triangularly complete *if, in its canonical form of (4.1) and for every* $1 < i \le p$, *there exists at least one* $j < i$ *such that* $F_{ij} \ne 0$. *It is said to be* lower triangularly positive *if, in (4.1),* $F_{ij} > 0$ *for all* $j < i$.

Physical and geometrical meanings of a lower triangularly complete matrix will be provided in Section 4.2, and its role in matrix sequence convergence will be studied in Section 4.4.

Example 4.6. Binary matrix

$$A_1 = \begin{bmatrix} 0 & 1 & 0 \\ 0 & 0 & 1 \\ 1 & 0 & 0 \end{bmatrix}$$

is irreducible, which can be easily verified by Definition 4.1. Alternatively, this conclusion can be shown using either Corollary 4.3 or Corollary 4.4; that is, $z = A_1 z$ yields $z = z_1 \mathbf{1}$, or

$$(I + A_1)^2 > 0.$$

Identity matrix I of dimension higher than 1 and its permutated versions of dimension higher than 2 are all reducible. Again, this can be shown by Definition 4.1 or by one of the two tests.

Matrix

$$A_2 = \begin{bmatrix} 1 & 0 & 0 \\ a_{21} & 1 & 0 \\ a_{31} & a_{32} & 1 \end{bmatrix}$$

is reducible for any $a_{21}, a_{31}, a_{32} \ge 0$. Non-negative matrix A_2 is lower triangularly complete if $a_{21} > 0$ and $a_{31} + a_{32} > 0$. $\diamond$

4.1.2 Perron-Frobenius Theorem

The *Perron-Frobenius theorem*, given below as Theorem 4.8, is the fundamental result on eigenvalue and eigenvector analysis of non-negative matrices, positive matrices, and irreducible matrices. If A is reducible, A can be permuted into its lower triangular canonical form A', and then Statement (c) can be applied to the resulting diagonal blocks A'_{ii}, and the spectrum of A is the union of the spectra of the A'_{ii}'s. Proof of Perron-Frobenius theorem is facilitated by the following Brouwer's fixed-point theorem.

Theorem 4.7. *Let $K \subset \Re^n$ be compact and convex, and let $f : K \to K$ be continuous. Then, f has a fixed-point, i.e., there exists $x \in K$ such that $f(x) = x$.*

Theorem 4.8. *Consider $A \in \Re_+^{n \times n}$. Then,*

(a) *Spectrum radius $\rho(A) \geq 0$ is an eigenvalue and its corresponding eigenvector is non-negative.*

(b) *If A is positive, $\rho(A) > 0$ is a simple eigenvalue of A (of multiplicity 1, both algebraically and geometrically), and its eigenvector is positive.*

(c) *If A is irreducible, there is a positive vector v such that $Av = \rho(A)v$, and $\rho(A)$ is a simple eigenvalue of A.*

Proof: It follows from Section 2.1 that, if $x \geq 0$, $\|x\|_1 = \mathbf{1}^T x$ and that, if $A \geq 0$, $\|A\|_\infty = \|A^T \mathbf{1}\|_\infty$.

(a) Let $\lambda \in \sigma(A)$ have the maximum modulus and v be its corresponding eigenvector with $\|v\|_1 = \sum_i |v_i| = 1$. Then, we have

$$\rho(A)|v| = |\lambda v| = |Av| \leq A|v|. \tag{4.2}$$

We need to find λ^* (among those λ's) such that $\lambda^* = \rho(A)$ and $v^* \geq 0$. To this end, define

$$K \overset{\triangle}{=} \{x \in \Re^n : x \geq 0, \sum_{i=1}^n x_i = 1, Ax \geq \rho(A)x\}.$$

It is apparent that K is compact and convex. Also, (4.2) shows $|v| \in K$ and hence K is not empty. Two distinct cases are analyzed.

The first case is that there exists $x \in K$ such that $Ax = 0$. It follows from $0 = Ax \geq \rho(A)x$ that $\rho(A) = 0$, and Statement (a) is established in this case.

In the second case, $Ax \neq 0$ for all $x \in K$. Define the mapping

$$f(x) = \frac{1}{\|Ax\|_1} Ax, \quad x \in K.$$

It is apparent that $f(x)$ is continuous, $f(x) \geq 0$, $\|f(x)\|_1 = 1$, and

$$Af(x) = \frac{1}{\|Ax\|_1} AAx \geq \frac{1}{\|Ax\|_1} A\rho(A)x = \rho(A)f(x).$$

These properties show that $f(x)$ is a continuous mapping from K to K. Hence, by Theorem 4.7, there exists $v^* \in K$ such that $f(v^*) = v^*$, that is, with $\lambda^* = \|Av^*\|_1$,

$$\lambda^* v^* = Av^* \geq \rho(A)v^*,$$

which implies $v^* \geq \rho(A)$ and in turn $v^* = \rho(A)$.

(b) Since $A > 0$, it follows from (a) and from $Av = \rho(A)v$ that $v > 0$.
Next, let us prove that $\rho(A)$ has algebraic multiplicity 1, that is, eigen-vector v is the only eigenvector associated with $\rho(A)$. Suppose that there were another, v' which is linearly independent of v. Since $v > 0$, it is pos-sible to find a linear combination $w = cv + v'$ such that $w \geq 0$ but $w \not> 0$. However, since $Aw = \rho(A)w$ is strictly positive, there is a contradiction. Therefore, algebraical multiplicity of $\rho(A)$ is 1.
Geometrical multiplicity of $\rho(A)$ being 1 is also shown by contradiction. Suppose there were a Jordan chain of length at least two associated with $\rho(A)$. That is, there is a vector z such that $[A - \rho(A)I]z = v$ while $Av = \rho(A)v$. Now let w be the positive eigenvector of A^T corresponding to $\rho(A)$. It follows that

$$0 = w^T[A - \rho(A)I]z = \rho(A)w^T v > 0,$$

which is contradictory.

(c) According to Corollary 4.4, $(\alpha I + A)^k > 0$ for some $k > 0$ and any constant $\alpha > 0$. Since eigenvalues and eigenvectors of A can be computed from those of $(\alpha I + A)^k > 0$, we know that $\rho(A)$ is simple and v is positive.

$\square$

Example 4.9. Consider the two matrices

$$A_1 = \begin{bmatrix} 0 & 1 & 0 \\ 0 & 0 & 1 \\ 1 & 0 & 0 \end{bmatrix}, \quad A_2 = \begin{bmatrix} 1 & 0 & 0 \\ 1 & 1 & 0 \\ 0 & 1 & 1 \end{bmatrix}.$$

Matrix A_1 has eigenvalues 1 and $-0.5000 \pm 0.8660j$, and $\rho(A_1) = 1$ (with eigenvector $\mathbf{1}$) is unique since A_1 is irreducible. Matrix A_2 has identical eigen-values of 1, and $\rho(A_2) = 1$ is not unique since A_2 is reducible.

In general, a non-negative matrix A can have complex eigenvalues, and some of its eigenvalues (other than $\rho(A)$) may be in the left half plan. Sym-metric (non-negative) matrices have only real eigenvalues. While lower tri-angular matrices are only triangular in blocks and may not be strictly tri-angular, eigenvalues of strictly triangular non-negative matrices are all real, non-negative, and equal to those elements on the diagonal.

Matrix $\mathbf{J}_{n \times n}$ has special properties. First, it is symmetric and hence di-agonalizable. Second, $\mathbf{J} = \mathbf{1}\mathbf{1}^T$ is of rank 1, and $\mathbf{J}\mathbf{1} = n\mathbf{1}$. Third, there are column vectors s_i for $i = 1, \cdots, n-1$ such that s_i are orthonormal ($s_i^T s_i = 1$ and $s_i^T s_j = 0$ for $i \neq j$) and that $\mathbf{1}^T s_i = 0$, i.e., $\mathbf{J}s_i = 0$. Fourth, choosing T to be the unitary matrix with columns s_i and with $1/\sqrt{n}$ as its last column, we have

$$T^{-1}\mathbf{J}T = T^T\mathbf{J}T = T^T\mathbf{1}(T^T\mathbf{1})^T = \text{diag}\{0, \cdots, 0, n\}.$$

Hence, all the eigenvalues of $\mathbf{J}$ are zero except for the unique eigenvalue of $\rho(\mathbf{J}) = n$.

$\diamond$

4.1.3 Cyclic and Primitive Matrices

Statement (c) of Perron-Frobenius theorem also has a non-trivial second part: if A is irreducible and if $p \geq 1$ eigenvalues of A are of modulus $\rho(A)$, then these eigenvalues are all distinct roots of polynomial equation $\lambda^p - [\rho(A)]^p = 0$, and there exists a permutation matrix P such that

$$
PAP^T = \begin{bmatrix}
0 & E_{12} & 0 & \cdots & 0 \\
0 & 0 & E_{23} & \cdots & 0 \\
\vdots & \vdots & \vdots & \ddots & \vdots \\
0 & 0 & 0 & \cdots & E_{(p-1)p} \\
E_{p1} & 0 & 0 & \cdots & 0
\end{bmatrix},
\tag{4.3}
$$

where the zero blocks on the diagonal are square. In (4.3), index p is called the *cyclicity index* of A, or equivalently, matrix A is called *p-cyclic*. The above statement has the following partial inverse: Given matrix E in (4.3), matrix E (and hence A) is irreducible if the product of $E_{12}E_{23} \cdots E_{(p-1)p}E_{p1}$ is irreducible.

A non-negative matrix E is said to be *primitive* if there exists a positive integer k such that $E^k > 0$. The following lemma shows that a primitive matrix has (up to scalar multiples) only one non-negative eigenvector.

Lemma 4.10. *Let E be a primitive matrix with a non-negative eigenvector v associated with eigenvalue λ. Then, $v > 0$ is unique and $\lambda = \rho(E)$.*

Proof: Since E is primitive, we have $k > 0$ such that $E^k > 0$ and thus $E^k v > 0$. It follows from $E^k v = \lambda^k v$ that $v > 0$ and $\lambda > 0$. Consider eigenvalue $\rho(E)$ of E and its eigenvector w. It follows from $E^k w = \rho^k(E)w$ and from Perron theorem, (b) in Theorem 4.8, that $w > 0$ and hence can be scaled by a positive constant such that $0 < w < v$. Therefore, we have that, for all k,

$$0 < \rho^k(E)w = E^k w \leq E^k v = \lambda^k v,$$

which is impossible if $\lambda < \rho(E)$. Thus, $\lambda = \rho(E)$.   □

Corollary 4.4 suggests there is a relationship between E being irreducible and primitive. Indeed, matrix A is irreducible and 1-cyclic if and only if A is primitive, which is applied in the following example.

Example 4.11. Consider the following matrices and their eigenvalues:

$$A_1 = \begin{bmatrix} 0 & 2 \\ 1 & 1 \end{bmatrix}, \quad \lambda(A_1) = 2, -1;$$

$$A_2 = \begin{bmatrix} 0 & 4 \\ 1 & 0 \end{bmatrix}, \quad \lambda(A_2) = 2, -2;$$

$$A_3 = \begin{bmatrix} 1 & 1 \\ 0 & 1 \end{bmatrix}, \quad \lambda(A_3) = 1, 1.$$

Among the three matrices, only A_1 is primitive.   ◇

In addition to the above necessary and sufficient condition for a matrix to be primitive, a sufficient condition is that, if A is irreducible and diagonally positive, it is also primitive. This is due to the fact that A being irreducible implies $(\epsilon I + A)^{n-1} > 0$ for any $\epsilon > 0$ and, if A is diagonally positive, ϵ can be set to be zero. The following example shows that matrix A being irreducible does not necessarily imply $A^k > 0$.

Example 4.12. Consider irreducible matrix

$$A = \begin{bmatrix} 0 & 0 & 1 \\ 1 & 0 & 0 \\ 0 & 1 & 0 \end{bmatrix}.$$

It follows that

$$A^2 = \begin{bmatrix} 0 & 1 & 0 \\ 0 & 0 & 1 \\ 1 & 0 & 0 \end{bmatrix}$$

is also irreducible, but $A^3 = I$ is reducible. Matrix A is 3-cyclic, and its power is never positive. $\diamond$

In general, for an irreducible and p-cyclic matrix A, A^p is block diagonal. If $p > 1$, A is not primitive, and the whole spectrum $\sigma(A)$ contains points of magnitude $\rho(A)$ and angles $\theta = 2\pi/p$ apart. For instance, matrix

$$A = \begin{bmatrix} 0 & 0 & 2 & 0 \\ 0 & 0 & 0 & 1 \\ 0 & 1 & 0 & 0 \\ 2 & 0 & 0 & 0 \end{bmatrix}$$

is irreducible, is 4-cyclic, and has the spectrum

$$\sigma(A) = \{\sqrt{2},\ -\sqrt{2},\ \sqrt{2}j,\ -\sqrt{2}j\}.$$

If matrix A is either diagonal or in Jordan form, it is reducible and not primitive. In this case, $\rho(A)$ may not be a simple eigenvalue, and the corresponding Perron eigenvector may not be strictly positive.

As an interesting application of Theorem 4.8, we can study the limit of powers of a primitive matrix. Consider the matrix

$$E = \begin{bmatrix} 1 & 3 \\ 2 & 2 \end{bmatrix},$$

which has left eigenvector $w = [2\ 3]$ and right eigenvector $v = [1\ 1]^T$ associated with spectral radius $\rho(E) = 4$. The other eigenvalue is $\lambda = -1$. It follows that

$$\frac{1}{wv}vw = \frac{1}{5}\begin{bmatrix} 2 & 3 \\ 2 & 3 \end{bmatrix},$$

which is of rank 1. It is not difficult to verify that

$$E^n = 4^n \begin{bmatrix} \frac{2}{5} & \frac{3}{5} \\ \frac{3}{5} & \frac{3}{5} \\ \frac{2}{5} & \frac{2}{5} \end{bmatrix} + (-1)^n \begin{bmatrix} \frac{3}{5} & -\frac{3}{5} \\ -\frac{2}{5} & \frac{2}{5} \end{bmatrix},$$

and that

$$\lim_{n \to \infty} \frac{E^n}{\rho^n(E)} = \frac{1}{wv} vw. \tag{4.4}$$

The general result illustrated by the example is given below.

Corollary 4.13. *Let E be a primitive matrix with left eigenvector w and right eigenvector v associated with eigenvalue $\rho(E)$. Then, the limit in (4.4) holds, and the convergence is exponential.*

Proof: Since E is primitive, $\rho(E)$ is a simple eigenvalue, and any other eigenvalue λ has the property that $|\lambda| < \rho(E)$. Letting $A = vw/(wv)$, we can establish (4.4) by showing the following four equations:

$$Av = v = \frac{Ev}{\rho(E)},$$

$$wA = w = \frac{wE}{\rho(E)},$$

$$A\beta_r = 0 = \lim_{n \to \infty} \frac{E^n \beta_r}{\rho^n(E)},$$

$$\beta_l A = 0 = \lim_{n \to \infty} \frac{\beta_l E^n}{\rho^n(E)},$$

where β_l is any of left eigenvectors of E but not associated with $\rho(E)$, and β_r is any of right eigenvectors of E and associated with $\lambda \neq \rho(E)$. The first two are obvious. To show the third equation, we know that

$$\rho^n(E)A\beta_r = \frac{1}{wv} v[\rho^n(E)w]\beta_r = \frac{1}{wv} vwE^n \beta_r = \frac{1}{wv} vw\lambda^n \beta_r = \lambda^n A\beta_r,$$

and that

$$\frac{1}{\rho^n(E)} E^n \beta_r = \frac{\lambda^n}{\rho^n(E)} \beta_r,$$

which implies that

$$A\beta_r = \lim_{n \to \infty} \frac{\lambda^n}{\rho^n(E)} A\beta_r = 0 = \lim_{n \to \infty} \frac{1}{\rho^n(E)} E^n \beta_r.$$

The fourth equation can be established similarly. $\qquad\square$

4.2 Importance of Non-negative Matrices

We begin with the following definitions:

Definition 4.14. *Matrices whose off-diagonal elements are non-negative are said to be* Metzler *matrices.*

Definition 4.15. $\mathcal{M}_1(c_0)$ *denotes the set of matrices with real entries such that the sum of the entries in each row is equal to the real number c_0. Let $\mathcal{M}_2(c_0)$ be the set of Metzler matrices in $\mathcal{M}_1(c_0)$. A matrix D is said to be row-stochastic if $D \geq 0$ and $D \in \mathcal{M}_1(1)$.*

A Metzler matrix A can be expressed as

$$A = -sI + E, \quad s \geq 0, \quad E \geq 0. \tag{4.5}$$

Therefore, properties of Metzler matrix A are analogous to those of non-negative matrix E. It also follows that $\mathcal{M}_2(c_0)$ are those matrices of A such that $\alpha I + \beta A$ is row-stochastic for some positive numbers α and β. Thus, properties of those matrices in $\mathcal{M}_2(c_0)$ can be derived from properties of row-stochastic matrices. The following lemma restates parts of Perron-Frobenius theorem, Theorem 4.8.

Lemma 4.16. *Consider two matrices $A_1 \in \mathcal{M}_1(c_0)$ and $A_2 \in \mathcal{M}_2(c_0)$. Then,*

(a) c_0 is an eigenvalue of A_1 with eigenvector $\mathbf{1}$.
(b) Real parts of all the eigenvalues of A_2 are less than or equal to c_0, and those eigenvalues with real part equal to c_0 are real.
(c) If A_2 is also irreducible, then c_0 is a simple eigenvalue of A_2.

As an application of non-negative row-stochastic matrix, consider a homogeneous discrete-time Markov chain $\{X_k\}$ described by its transition probability matrix $P = [p_{ij}]$, where

$$p_{ij} = Pr\{X_k = j | X_{k-1} = i\}$$

is the probability of a transition from state i to state j at any time step. If the number of possible states is finite, stochastic matrix P is square and finite dimensional, and $P\mathbf{1} = \mathbf{1}$. On the other hand, a row vector

$$p = \begin{bmatrix} p_1 & p_2 & \cdots & p_n \end{bmatrix}$$

is called a probability distribution vector if $p \geq 0$ and $p\mathbf{1} = 1$. Let p^0 be the initial probability distribution. Then, the probability distribution at step k is

$$p^k = p^{k-1}P = p^0 P^k, \quad k = 1, 2, \cdots$$

If the Markov chain is *ergodic*, there exists a steady state probability distribution independent of p^0, that is,

$$p^\infty = p^\infty P,$$

which says that p^∞ is the left eigenvector of P corresponding to the eigenvalue 1. In matrix terms, ergodicity of the homogeneous discrete-time Markov chain means that P is irreducible and primitive (and hence $\lambda = 1$ is a simple eigenvalue, it is the only eigenvalue on the unit circle, and $p^\infty > 0$). Application to non-homogeneous discrete-time Markov chain will be mentioned in Section 4.4.

In systems and control, non-negative matrices arise naturally from the class of linear positive systems defined below.

Definition 4.17. *A linear system with triplet $\{A,\ B,\ C\}$ is said to be positive if and only if, for every non-negative initial condition and for every non-negative input, its state and output are non-negative.*

Lemma 4.18. *A continuous-time linear system*

$$\dot{x} = A(t)x + B(t)u, \quad y = C(t)x$$

is positive if and only if A is a Metzler matrix, $B \geq 0$, and $C \geq 0$. A discrete-time linear system of triplet $\{A,\ B,\ C\}$ is positive if and only if $A \geq 0$, $B \geq 0$ and $C \geq 0$.

Proof: Consider first the continuous-time system, and necessary is proven by contradiction. Assume that matrix A contain $a_{ij} < 0$ for some $i \neq j$. If so, under zero input and under the initial conditions of $x_k(t_0) = 0$ for all k except that $x_j(t_0) > 0$, differential equation of $x_i(t)$ becomes

$$\dot{x}_i = a_{ij}(t)x_j, \quad x_i(t_0) = 0.$$

according to which $x_i(t_0 + \delta t)$ will leave set R_+^n. This result contradicts with Definition 4.17 and hence $a_{ij} \geq 0$ for all $i \neq j$, i.e., matrix A is Metzler. Similarly, we can show that $B \geq 0$ and $C \geq 0$ are also necessary. Sufficiency can be established by noting that, given Metzler matrix A and non-negative matrices B and C, $\dot{x} \geq 0$ along the boundary of R_+^n. The result on discrete-time system can be shown in a similar fashion. □

Among the class of positive systems, there are *cooperative systems* described by

$$\dot{x} = k_2[-k_1 I + D(t)]x, \tag{4.6}$$

where $k_1, k_2 > 0$ are scalar gains, and $D(t)$ is a non-negative and row-stochastic matrix. Cooperative systems will be studied in depth in Chapter 5. It is worth noting that, although matrix E in (4.5) is not necessarily row-stochastic, a linear time-invariant (or piecewise-constant) positive system can always be transformed into the cooperative system in (4.6) for the purpose of asymptotic stability analysis. To validate this point, consider any positive system $\dot{z} = [-sI + E]z$ with $E \geq 0$. If E is irreducible, it follows from Theorem 4.8 that, for some vector $v > 0$, $Ev = \rho(E)v$ where $\rho(E) > 0$. Letting T be the diagonal matrix such that $T\mathbf{1} = v$, we can introduce the transformed state $x = T^{-1}z$ and have

$$\dot{x} = [-sI + T^{-1}ET]x = \rho(E)\left[-\frac{s}{\rho(E)}I + \frac{1}{\rho(E)}T^{-1}ET\right]x \triangleq \rho(E)[-k_1 I + D]x,$$

in which matrix D is non-negative and row-stochastic since $T^{-1}ET\mathbf{1} = T^{-1}Ev = \rho(E)T^{-1}v = \rho(E)\mathbf{1}$. If E is reducible, the same process can be applied. For instance, let us say

$$E = \begin{bmatrix} E_{11} & 0 \\ E_{21} & E_{22} \end{bmatrix}$$

where E_{11} and E_{22} are irreducible. As before, we can find the diagonal state transformation $z = T^{-1}x$ with $T = \text{diag}\{T_1, T_2\}$, $T_i\mathbf{1} = v_i$, and $E_{ii}v_i = \rho(E_{ii})v_i$ such that the resulting system becomes

$$\dot{x} = \begin{bmatrix} -sI + \rho(E_{11})D_{11} & 0 \\ D_{21} & -sI + \rho(E_{22})D_{22} \end{bmatrix} x,$$

where both D_{11} and D_{22} are row-stochastic. Asymptotic stability of the above system is the same as that of system

$$\dot{x} = \begin{bmatrix} -sI + \rho(E_{11})D_{11} & 0 \\ 0 & -sI + \rho(E_{22})D_{22} \end{bmatrix} x,$$

Hence, we can use (4.6) as the canonical form to study asymptotic stability of positive systems. The following is such a result, and it comes directly from Lemma 4.16.

Corollary 4.19. *Consider System 4.6 with $k_1, k_2 > 0$ and with D being non-negative, row-stochastic, and constant. Then,*

(a) The system is asymptotically stable if $k_1 > 1$.
(b) The system is unstable if $k_1 < 1$.
(c) The system is Lyapunov stable if $k_1 = 1$ and if D is irreducible.

The above stability result, in particular, Statement (c) will be extended further in Section 4.3 to the case that matrix D is reducible. Similar results can be stated for discrete-time positive system $x_{k+1} = Ax_k$ with non-negative matrix A; that is, the system is unstable if $\rho(A) > 1$, it is Lyapunov stable if $\rho(A) = 1$ and if A is irreducible, and it is asymptotically stable if $\rho(A) < 1$.

A non-negative matrix associated with a positive system has explicit *physical meanings*. For illustration, let us use the cooperative system in (4.6) as an example and view each of its state variables as a separate entity. Then, dynamics of each entity consists of negative feedback about itself and positive feedback from others. The individual negative feedback is necessary to maintain stability for continuous-time systems, while positive feedback from other entities represent information exchanges and their impacts.

Irreducibility or reducibility is the structural property of a non-negative matrix, whose physical meaning is dynamical grouping or coupling among the entities. Should matrix D be irreducible, the value of (or the information about) any of the entities can propagate to any other entity through the system dynamics. We can see this fact from the following state solution to System 4.6 with constant matrix D:

$$x(t) = e^{-(k_1+1)k_2 t} e^{k_2(I+D)t} x(0) = e^{-(k_1+1)k_2 t} \sum_{j=0}^{\infty} \frac{k_2^j t^j}{j!} (I + D)^j x(0),$$

in which $(I + D)^j \geq 0$ for all j. Since $(I + D)^{n-1} > 0$ for any irreducible matrix D, $x(t) > 0$ for all finite time $t > 0$ under any initial condition on the boundary of $\Re_+^n$ but not at the origin. Hence, we know that all the entities corresponding to an irreducible matrix move together as one group.

On the other hand, if System 4.6 has a reducible matrix D, we can find its triangular canonical form shown in (4.1). Its permutation matrix tells us how the state variables can be separated into the groups corresponding to diagonal irreducible blocks of $D_\searrow$. If matrix D is lower triangularly complete, then the group of the entities corresponding to the first diagonal block of $D_\searrow$ acts as the unique leader group by sending its information to the rest of entities but receiving nothing back, and all the other groups of entities act as followers by receiving information from the leader. Thus, we know that the groups of entities corresponding to a reducible but lower triangularly complete matrix also move together as one larger composite group. If matrix D is not lower triangularly complete, matrix $D_\searrow$ contains two or more isolated diagonal blocks and hence there are at least two groups of entities that move independently from each other and from the rest of entities.

For discrete-time system $x_{k+1} = D x_k$, the same observation holds except for the differences that negative feedback about an entity itself is no longer needed for asymptotic stability and that $D^k > 0$ only if D is primitive. If D is cyclic (irreducible but not primitive), there are two or more groups of entities among which information is transmitted in such a way that they act each other in cycles. In the same fashion as that of continuous-time systems, grouping of discrete entities depends upon whether $D_\searrow$ is lower triangularly complete if D is reducible.

It is because of the above physical implications that non-negative matrices and their irreducibility or lower triangular completeness play a critical role in analysis and synthesis of cooperative systems. Discussions in the next two subsections reinforce these implications on dynamical systems.

4.2.1 Geometrical Representation of Non-negative Matrices

Let $K \subset \Re^n$ be a set. Then, set K is said to be *convex* if K contains the linear segment between any two points therein, and set K is said to be a *cone* if K contains all finite non-negative linear combinations of its elements. The simplest convex cones are $K = \Re^n$ and $K = \Re^n_+$, and any square matrix A can be viewed as a mapping of $A : \Re^n \to \Re^n$. It follows that $A \in \Re^{n \times n}_+$ if and only if $A(\Re^n_+) \subset \Re^n_+$.

Convex cone K is said to be *solid* if the interior of K, denoted by $int\ K$, is not empty; K is *pointed* if $K \cap (-K) = \{0\}$; and K is called a *proper* cone if it is solid, pointed, and closed. Among proper cones is $\Re^n_+$. Given a proper cone K, F is said to be a *face* of K if $F \subset K$ is closed and pointed and if

$$x \in F,\ x - y \in K \implies y \in F.$$

The face F is non-trivial if $F \neq \{0\}$ and $F \neq K$. As an example, the faces of R^n_+ are of the form

$$F_I = \{x \in R^n_+ :\ x_i = 0 \text{ if } i \notin I\},$$

where $I \subset \{1, \cdots, n\}$.

Non-negativeness, positiveness, and irreducibility can be defined equivalently but geometrically as follows: For any given proper cone K,

(a) Matrix $A \in \Re^{n \times m}$ is said to be K-non-negative if it leaves K invariant, i.e., $A(K) \subset K$.

(b) Matrix A is said to be K-positive if it maps non-zero elements of K into the interior, i.e., $A(K - \{0\}) \subset int\ K$.

(c) Matrix A is said to be K-irreducible if and only if faces of K that it leaves invariant are $\{0\}$ and K itself. Matrix A is said to be K-reducible if it leaves invariant a non-trivial face of K.

Geometrically, the Perron-Frobenius theorem tells us the following:

(a) If A is K-non-negative, $\rho(A)$ is an eigenvalue, and K contains an eigenvector of A corresponding to $\rho(A)$.

(b) Matrix A is K-irreducible if and only if no eigenvector of A lies on the boundary of K. Matrix A is K-irreducible if and only if A has exactly one (up to scale multiples) eigenvector in K, and this vector is in its interior $int\ K$.

(c) Matrix A is K-irreducible if and only if $(I + A)^{n-1}$ is K-positive.

Letting $K = \Re^{n \times n}_+$ renders geometrical meanings about the results in Section 4.1.

(a)   (b)

Fig. 4.1. Examples of graphs

4.2.2 Graphical Representation of Non-negative Matrices

Given any matrix $A \in \Re^{n \times n}$, we can associate it with a *directed graph* or *digraph* $\mathcal{G}(A) = (\mathcal{I}, \mathcal{E})$ with vertex set (or nodes) $\mathcal{I} \triangleq \{1, 2, \cdots, n\}$ and their arc set (or edges) $\mathcal{E} \subset \mathcal{I} \times \mathcal{I}$, where $(i, j) \in \mathcal{E}$ if and only if $a_{ij} \neq 0$. Conversely, given a directed graph $\mathcal{G} = (\mathcal{I}, \mathcal{E})$, there is an adjacency matrix A whose elements are binary and defined by

$$a_{ij} = \begin{cases} 1, & \text{if } (i, j) \in \mathcal{E} \\ 0, & \text{else} \end{cases}.$$

The out-degree (or in-degree) of a node is the number of arcs originating (or terminating) at the node. A node is balanced if its in-degree and out-degree are equal, and a digraph is *balanced* if all of its nodes are balanced, or equivalently in terms of adjacency matrix A, $A^T \mathbf{1} = A\mathbf{1} = c\mathbf{1}$ for some $c \in \Re_+$. If matrix A is symmetrical and row-stochastic, it corresponds to an *undirected* graph (whose edges are all bi-directional), and hence the graph is balanced.

Example 4.20. Consider matrices:

$$A_1 = \begin{bmatrix} 0 & 0 & 1 \\ 1 & 0 & 0 \\ 0 & 1 & 1 \end{bmatrix}, \quad \text{and} \quad A_2 = \begin{bmatrix} 0 & 1 & 0 \\ 0 & 0 & 1 \\ 1 & 0 & 0 \end{bmatrix}.$$

Their corresponding graphs are shown in Fig. 4.1. Graph in Fig. 4.1(b) is balanced, but A_2 is not symmetrical and hence graph in Fig. 4.1(a) is not balanced.   ◇

A *path* from node i to j is a sequence of successive edges $\{(i, k_1), \cdots, (k_l, j)\}$ in $\mathcal{E}$ that connects node j to node i. Directed graph $\mathcal{G}$ is *strongly connected* if, for any pair of vertices $i, j \in \mathcal{I}$, there exists a path in $\mathcal{G}$ that starts at i and terminates at j. Directed graph $\mathcal{G}$ has a *global reachable node* or a *spanning tree* if it has a node to which there exists a path from every other node. A *loop* of length k in $\mathcal{G}$ is a path of length k which begins and ends at the same vertex.

The following are useful facts that are apparent from the above discussions: For any $A \geq 0$,

(a) The $(i,j)th$ element of matrix A^k is positive for some $k > 0$ if and only if there is a path from node i to node j.
(b) Matrix A is irreducible if and only if its graph $\mathcal{G}(A)$ is strongly connected.
(c) Matrix A is lower triangularly complete in its canonical form if and only if its graph $\mathcal{G}(A)$ has a global reachable node or a spanning tree.
(d) The cyclicity (period) of matrix A is the greatest common divisor of the lengths of all the loops in $\mathcal{G}(A)$.

As such, the matrix-theoretical results in the subsequent sections and chapters can be stated analogously and equivalently using the terminology of the graph-theoretical approach [19, 51, 78].

4.3 M-matrices and Their Properties

We begin with the following definition.

Definition 4.21. *Let $\mathcal{Z}$ denote the set of square matrices whose off-diagonal elements are non-positive, that is,*

$$\mathcal{Z} \triangleq \{A = [a_{ij}] \in \mathfrak{R}^{n \times n} : a_{ij} \leq 0,\ i \neq j\}.$$

Then, matrix A is called a non-singular (or singular) M-matrix if $A \in \mathcal{Z}$ and if all its principal minors are positive (or non-negative).

By definition, matrix $A \in \mathcal{Z}$ can also be expressed in the form of

$$A = sI - B, \quad s > 0, \quad B \geq 0; \tag{4.7}$$

that is, $-A$ is a Metzler matrix. If A is a non-singular M-matrix, $s > \rho(B)$ and $(-A)$ is both a Metzler matrix and a Hurwitz matrix. If A is a singular M-matrix, $s \geq \rho(B)$. A symmetrical non-singular (or singular) M-matrix must be positive definite (or positive semi-definite).

4.3.1 Diagonal Dominance

The following Gerschgorin's circle criterion, also referred to as Gerschgorin disc theorem [90], is typically used to estimate the locations of eigenvalues of a matrix.

Theorem 4.22. *All the eigenvalues of matrix* $A = [a_{ij}] \in \Re^{n \times n}$ *are located within the following union of n discs:*

$$\bigcup_{i=1}^{n} \left\{ z \in \mathcal{C} : \ |z - a_{ii}| \leq \sum_{j \neq i} |a_{ij}| \right\}.$$

Diagonal dominance of a matrix enables us to apply Gerschgorin's circle criterion and to conclude the eigenvalue properties in Corollary 4.24.

Definition 4.23. *Matrix* $E = [e_{ij}] \in \Re^{n \times n}$ *is said to be* diagonally dominant *if, for all i,*

$$e_{ii} \geq \sum_{j \neq i} |e_{ij}|.$$

Matrix E is said to be strictly diagonally dominant *if the above inequalities are all strict.*

Corollary 4.24. *If matrix $A \in \Re^{n \times n}$ is diagonally dominant (strictly diagonally dominant), then eigenvalues of A have non-negative (positive) real parts. Additionally, if $A \in \mathcal{Z}$, matrix A is a singular (non-singular) M-matrix.*

4.3.2 Non-singular M-matrices

The following theorem summarizes several useful properties of non-singular M-matrices, and a comprehensive list of 50 properties can be found in [18] (Theorem 2.3, p.134).

Theorem 4.25. *Matrix $A \in \mathcal{Z}$ is a non-singular M-matrix under one of the following conditions, and the conditions are all equivalent:*

(a) The leading principal minor determinants of A are all positive.
(b) The eigenvalues of A have positive real parts.
(c) A^{-1} exists and is non-negative.
(d) There exist vectors $x, y > 0$ such that both Ax and $A^T y$ are positive, denoted by $Ax, Ay > 0$.
(e) There exists a positive diagonal matrix S such that $AS + SA^T$ is strictly diagonally dominant and hence also positive definite, that is,

$$a_{ii}s_i > \sum_{j \neq i} |a_{ij}|s_j, \quad \forall \, i. \tag{4.8}$$

Proof: Equivalence between (a) and (b) is classical and well known.
Since $A \in \mathcal{Z}$, we can rewrite matrix A as

$$A = \lambda I - B$$

where B is non-negative. It follows from the property of non-negative matrices that spectrum $\rho(B) \geq 0$ is an eigenvalue of B and the corresponding eigenvector is $v \geq 0$. It follows that Property (b) is equivalent to $\rho(B) < \lambda$.

Letting $T = B/\lambda$, it follows that $A^{-1} = (I - T)^{-1}/\lambda$ and that $\rho(T) < 1$ if and only if (b) holds. On the other hand, let us consider the infinite series

$$(I - T)^{-1} \triangleq \sum_{k=1}^{\infty} T^k, \tag{4.9}$$

which exists and is non-negative provided that the series is convergent. Since the above series converges if and only if $\rho(T) < 1$, the equivalent between (b) and (c) is established.

Now, choosing $x = A^{-1}\mathbf{1}$, we know that $Ax = \mathbf{1} > 0$ and also $x > 0$ since A^{-1} is non-negative, invertible and hence cannot have a zero row. That is, we know that (c) implies (d). Letting $S = \mathrm{diag}\{x_1, \cdots, x_n\}$, we have

$$AS\mathbf{1} = Ax = \mathbf{1} > 0,$$

which means strict diagonal dominance of AS. Hence, (d) implies (e). On the other hand, letting $P = S^{-1}$, we know that

$$PA + A^T P = P[AS + SA^T]P$$

is positive definite, which means that (e) implies (a).   □

Statement (a) implies that, if $A \in \mathcal{Z}$ and if A is symmetrical and positive definite, A must be a non-singular M-matrix. If $A \in \mathcal{Z}$ and if $A^T + A$ is positive definite, A is also a non-singular M-matrix since $A + A^T \in \mathcal{Z}$. However, if A is a non-singular M-matrix, $A^T + A$ may not be positive definite. Instead, statement (e) tells us that, for any non-singular M-matrix, there exists a positive diagonal matrix P such that $AP + A^T P$ is positive definite.

In some literature, Inequality 4.8 is also referred to as *strict pseudo-diagonal dominance*, and it includes strict diagonal dominance as a special case. Since $-(AS + SA^T)$ is a symmetric and Metzler matrix, $P = S^{-1}$ is a diagonal Lyapunov matrix and, by Lyapunov direct method, Corollary 4.24 yields statement (a) from (e) of Theorem 4.25. In comparison, Theorem 4.25 establishes in matrix set $\mathcal{Z}$ the equivalence between strict pseudo-diagonal dominance and non-singular M-matrix.

Example 4.26. Matrix

$$A = \begin{bmatrix} 1 & 0 \\ -3 & 1 \end{bmatrix}$$

is a non-singular M-matrix. It follows that

$$A + A^T = \begin{bmatrix} 2 & -3 \\ -3 & 2 \end{bmatrix}$$

is not positive definite and that, under the choice of

$$S = \begin{bmatrix} 1 & 0 \\ 0 & 3 \end{bmatrix}, \quad AS + SA^T = \begin{bmatrix} 2 & -3 \\ -3 & 6 \end{bmatrix}$$

is positive definite.   ◇

4.3.3 Singular M-matrices

The following theorem summarizes several useful properties of singular M-matrices, i.e., matrices of form $A = \rho(B)I - B$ with $B \geq 0$.

Theorem 4.27. *Matrix $A \in \mathcal{Z}$ is called a singular M-matrix under one of the following equivalent conditions:*

(a) The leading principal minor determinants of A are all non-negative.
(b) The eigenvalues of A have non-negative real parts.
(c) $(A + D)^{-1}$ exists and is non-negative for each positive diagonal matrix D.
(d) There exist non-negative non-zero vectors $x, y \geq 0$ such that both Ax and $A^T y$ are non-negative, denoted by $Ax, Ay \geq 0$.
(e) There exists a non-negative diagonal matrix S such that $S \neq 0$ and

$$a_{ii}s_i \geq \sum_{j \neq i} |a_{ij}|s_j. \tag{4.10}$$

Proof: It follows from

$$A + \epsilon I = (\rho(B) + \epsilon)I - B$$

that $A + \epsilon I$ is a non-singular M-matrix for all $\epsilon > 0$ if and only if A is a singular M-matrix. Hence, statements (a), (b), (c), (d) and (e) are parallel to those in Theorem 4.25, and they can be shown using Theorem 4.25 by taking the limit of $\epsilon \to 0$.   □

If S is diagonal and positive definite, Inequality 4.10 is referred to as *pseudo-diagonal dominance*. Since S is only positive semi-definite, Statement (e) of Theorem 4.27 does not imply that $SA + SA^T$ is diagonally dominant or matrix $-A$ is Lyapunov stable. To ensure these properties, an additional condition needs to be imposed on A in addition to A being a singular M-matrix. To this end, the following definition is introduced.

Definition 4.28. *An M-matrix A is said to have "Property c" if A can be split as $A = \rho(B)I - B$ with $B \geq 0$ and $\rho(B) > \max_i a_{ii}$ and if $T = B/\rho(B)$ is semi-continuous in the sense that $\lim_{j \to \infty} T^j$ exists.*

Note that, as shown in (4.9), matrix sum $\sum_j T^j$ is convergent and $\lim_{j \to \infty} T^j = 0$ if and only if $\rho(T) < 1$. Hence, all non-singular M-matrices have "Property c," but not all singular M-matrices. Also, note that the split of $A = \rho(B)I - B$ is not unique but should be done such that $\rho(B) > \max_i a_{ii}$. For example, consider

$$A = \begin{bmatrix} 1 & -1 \\ -1 & 1 \end{bmatrix}.$$

It follows that, upon splitting A as $A = \lambda(I - T)$, $\lim_{j\to\infty} T^j$ exists for any $\lambda > 1$ but does not exist for $\lambda = 1$.

The following theorem provides the characterization of singular M-matrices having "Property c." Note that, in (c), Lyapunov matrix W is not necessarily diagonal.

Theorem 4.29. *For any matrix $A \in \mathcal{Z}$, the following statements are equivalent to each other:*

(a) A is a singular M-matrix with "Property c."
(b) Rank of A^k is invariant as rank $rank(A) = rank(A^2)$.
(c) There exists a symmetric positive definite matrix W such that $WA + A^T W$ is positive semi-definite; that is, matrix $-A$ is Lyapunov stable.

Proof: Recall from Perron-Frobenius theorem that, for any singular M-matrix A, $A = \rho(B)I - B$ with $B \geq 0$. The choice of $\rho(B) > \max_i a_{ii}$ ensures that $T = B/\rho(B)$ is diagonally positive, hence the cyclicity index of T is 1 and $\rho(T) = 1$ is the unique eigenvalue of modulus 1. Suppose that S is the non-singular transformation matrix under which

$$S^{-1}TS = \begin{bmatrix} J & 0 \\ 0 & H \end{bmatrix}$$

is the Jordan form for T, where J has 1 on its diagonal and $\rho(H) < 1$. It follows that

$$\lim_{j\to\infty} T^j = S^{-1} \lim_{j\to\infty} \begin{bmatrix} J^j & 0 \\ 0 & H^j \end{bmatrix} S = S^{-1} \begin{bmatrix} \lim_{j\to\infty} J^j & 0 \\ 0 & 0 \end{bmatrix} S,$$

which exists if and only if $J = I$. On the other hand, $J = I$ is true if and only if $rank(I - T) = rank((I - T)^2)$, and hence if and only if $rank(A) = rank(A^2)$. This shows that (a) and (b) are equivalent.

To show that (a) and (c) are equivalent, note that (a) is equivalent to

$$A = \rho(B)(I - T) = \rho(B)S^{-1} \begin{bmatrix} 0 & 0 \\ 0 & I - H \end{bmatrix} S,$$

where $\rho(H) < 1$. Hence, the above is true if and only if there exists symmetric positive definite matrix P_h such that $P_h(I - H) + (I - H)^T P_h$ is positive definite, thus if and only if there exists $W = S^T \text{diag}\{I, P_h\}S$ such that

$$WA + A^T W = \rho(B)S^T \begin{bmatrix} 0 & 0 \\ 0 & P_h(I - H) + (I - H)^T P_h \end{bmatrix} S$$

is positive semi-definite. $\qquad\square$

The following lemma provides an alternative condition.

Lemma 4.30. *If $A \in \mathcal{Z}$ and if there exists vector $x > 0$ such that $Ax \geq 0$, then A is a singular M-matrix with "Property c."*

Proof: Let $y = x/\rho(B)$. It follow that $y > 0$ and $(I - T)y \geq 0$ or simply $y \geq Ty$. Recalling that T is non-negative, we have

$$y \geq Ty \geq T(Ty) = T^2 y \geq \cdots \geq T^j y.$$

Since $y > 0$, the above inequality implies T^j is finite for all j. Proof of Theorem 4.25 also shows that A is a singular M-matrix with "Property c" if and only if the corresponding matrix T has the property that T^j is finite for all j. Thus, the proof is done.   □

4.3.4 Irreducible M-matrices

Irreducibility or reducibility of M-matrix $A = sI - B$ is defined in terms of that of non-negative matrix B. The following theorem shows that Lyapunov stability of $-A$ is always guaranteed for a singular but irreducible M-matrix A.

Theorem 4.31. *Let $A \in \mathcal{Z}$ be an nth-order irreducible singular M-matrix. Then,*

(a) A has rank $n - 1$.
(b) There exists a vector $x > 0$ such that $Ax = 0$.
(c) $Ax \geq 0$ implies $Ax = 0$.
(d) Each principal sub-matrix of A of order less than n is a non-singular M-matrix.
(e) Matrix A has "Property c."
(f) There exists a positive diagonal matrix P such that $PA + A^T P$ is positive semi-definite; that is, matrix A is pseudo-diagonally dominant.

Proof: Let $A = \rho(B)I - B$ with $B \geq 0$ and $\rho(B) > \max_i a_{ii}$. It follows that B is irreducible and, by Perron-Frobenius theorem, $\rho(B)$ is a simple eigenvalue of B. Then, matrix A has 0 as its simple eigenvalue and is of rank $n - 1$.

It follows from Perron-Frobenius theorem that there exists a vector $x > 0$ such that $Bx = \rho(B)x$, that is, $Ax = 0$.

Suppose that $Ax \geq 0$ for some x. It follows from (b) that there exists $y > 0$ such that $A^T y = 0$. Thus, $y^T(Ax) = (A^T y)^T x = 0$ for all x. If $Ax \neq 0$, $y^T(Ax) \neq 0$, which is a contradiction.

It follows from (b) that there exist $x, y > 0$ such that $Ax = 0$ and $A^T y = 0$. Defining $A_{adj} \overset{\triangle}{=} cxy^T$ for some $c \in \Re$, we know from $AA_{adj} = A_{adj}A = det(A)I = 0$ that A_{adj} is the adjoint matrix of A. It follows from (a) that A is rank $n - 1$ and hence $c \neq 0$. If $c < 0$, there would exist an $\epsilon > 0$ such that the adjoint matrix of $A + \epsilon I$ would be negative, which contradicts the proof of Theorem 4.27. Thus, $c > 0$ and hence $A_{adj} > 0$, from which all the principal minors of order $n - 1$ of A are positive. This in turn implies that all the principal minors of order less than n are positive.

Since $\rho(B)$ is a simple eigenvalue of B, 1 is a simple eigenvalue of $T = B/\rho(B)$, and it follows from the proof of Theorem 4.29 that A has "Property c."

Let $x, y > 0$ be the right and left eigenvector of B and associated with $\rho(B)$, that is, $Ax = 0$ and $A^T y = 0$. Defining $P = \text{diag}\{y_1/x_1, \cdots, y_n/x_n\}$, we have that $PA + A^T P \in Z$ and that

$$(PA + A^T P)x = P(Ax) + A^T(Px) = 0 + A^T y = 0,$$

from which $(PA + A^T P)$ being a singular M-matrix is concluded from (d) of Theorem 4.27. Since $(PA + A^T P)$ is symmetrical, it must be positive semi-definite. □

Under irreducibility, there is a simpler transition from singular M-matrix to non-singular M-matrix as stated below.

Lemma 4.32. *If $A \in Z$ is a singular but irreducible M-matrix, then matrix $[A + \text{diag}\{0, \cdots, 0, \epsilon\}]$ is a non-singular M-matrix, and matrix $[-A + \text{diag}\{0, \cdots, 0, \epsilon\}]$ is an unstable matrix, where $\epsilon > 0$.*

Proof: Partition matrix $A = [a_{ij}]$ as follows:

$$A = \begin{bmatrix} A_{11} & A_{12} \\ A_{21} & a_{nn} \end{bmatrix},$$

where $A_{11} \in \Re^{(n-1)\times(n-1)}$. Clearly, principal sub-matrices of matrix

$$(A + \text{diag}\{0, \cdots, 0, s\})$$

up to $(n - 1)$th-order are identical to those of A. By definition, we have the following property on determinant calculation:

$$\det(A + \text{diag}\{0, \cdots, 0, s\}) = \det(A) + \det\begin{bmatrix} A_{11} & A_{12} \\ 0 & s \end{bmatrix} = \det(A) + s\det(A_{11}).$$

Since A is a singular irreducible M-matrix, $\det(A) = 0$ and, by (d) of Theorem 4.31, $\det(A_{11}) > 0$. Thus, $\det(A + \text{diag}\{0, \cdots, 0, s\})$ is non-zero and has the same sign as s, from which the conclusions are drawn. □

It follows from (e) of Theorem 4.25 that, if $A \in \Re^{n\times n}$ is a non-singular M-matrix, so is matrix $(A + \text{diag}\{0, \cdots, 0, \epsilon\})$ with $\epsilon > 0$. The following corollaries can be concluded from Lemma 4.32 by mixing permutation with addition of matrix $\text{diag}\{0, \cdots, 0, \epsilon_i\}$ and by invoking (f) of Theorem 4.31. In Corollary 4.34, by requiring irreducibility, strict pseudo-diagonal dominance is relaxed from all the rows in (4.8) to just one of the rows.

Corollary 4.33. *If $A \in \mathcal{Z}$ is a singular but irreducible M-matrix, then matrix $[A + diag\{\epsilon_i\}]$ is a non-singular M-matrix, and matrix $[-A + diag\{\epsilon_i\}]$ is an unstable matrix, where $\epsilon_i \geq 0$ for $i = 1, \cdots, n$ and at least one of them is positive.*

Corollary 4.34. *Consider matrix $A \in \mathcal{Z}$. If A is irreducible and if there exists a positive diagonal matrix S such that*

$$a_{ii}s_i \geq \sum_{j \neq i} |a_{ij}|s_j, \quad \forall\, i,$$

in which at least one of the inequalities is strict, then matrix A is a non-singular M-matrix.

4.3.5 Diagonal Lyapunov Matrix

For any non-singular M-matrix, Statement (e) of Theorem 4.25 provides a quadratic Lyapunov function whose special expression is a weighted sum of squares of all the state variables, and its proof is constructive for finding P. The following is an alternative way to find the diagonal Lyapunov matrix P. Let $x, y > 0$ be the vectors such that $u = Ax > 0$ and $v = A^T y > 0$. Defining $P = diag\{y_1/x_1, \cdots, y_n/x_n\}$, we have that $PA + A^T P \in \mathcal{Z}$ and that

$$(PA + A^T P)x = P(Ax) + A^T(Px) = Pu + A^T y = Pu + v > 0,$$

from which $(PA + A^T P)$ is a non-singular M-matrix according to (d) of Theorem 4.25. Since $(PA + A^T P)$ is symmetrical, it must be positive definite. Should matrix $A = \lambda I - B$ also be irreducible, the above vectors u and v used in finding diagonal Lyapunov matrix P can simply be chosen to the right and left eigenvectors of B and associated with $\rho(B) > 0$ respectively, in which case $u, v > 0$, $Au = (\lambda - \rho(B))u$ and $v^T A = (\lambda - \rho(B))v^T$. In addition, it follows from (4.9) that A is a non-singular irreducible M-matrix if and only if $A^{-1} > 0$.

For a singular but irreducible matrix, Statement (f) of Theorem 4.31 provides a quadratic Lyapunov function also in terms of a diagonal Lyapunov matrix, and its proof is also constructive. For a singular and reducible M-matrix A, Statement (e) of Theorem 4.27 does not provide a Lyapunov function to study Lyapunov stability of $-A$. If singular M-matrix A has "Property c," Lyapunov stability of $-A$ is shown in Theorem 4.29, but the corresponding Lyapunov matrix is not necessarily diagonal. In general, a Lyapunov matrix does not exist without "Property c" since matrix $-A$ is not Lyapunov stable. The following example illustrates these points.

Example 4.35. Consider the following reducible and singular M matrices:

$$A_1 = \begin{bmatrix} \epsilon & 0 \\ -\epsilon & 0 \end{bmatrix}, \quad A_2 = \begin{bmatrix} 0 & 0 \\ -\epsilon & \epsilon \end{bmatrix}, \quad A_3 = \begin{bmatrix} 0 & 0 \\ -\epsilon & 0 \end{bmatrix},$$

where $\epsilon > 0$. Apparently, for any two-dimensional vector $x > 0$, neither can $A_1 x \geq 0$ or $A_2^T x \geq 0$ or $A_3 x \geq 0$ hold. Both matrices $-A_1$ and $-A_2$ are Lyapunov stable and also have "Property c", hence there are Lyapunov matrices P_i such that $P_i A_i + A_i^T P_i$ is positive semi-definite for $i = 1, 2$, but P_i are not diagonal. For instance, for matrix A_2, we can choose positive definite matrix

$$P_2 = \frac{1}{\epsilon} \begin{bmatrix} 2 & -1 \\ -1 & 1 \end{bmatrix}$$

and obtain

$$P_2 A_2 + A_2^T P_2 = \begin{bmatrix} 2 & -2 \\ -2 & 2 \end{bmatrix},$$

which is positive semi-definite. Nonetheless, there is no diagonal Lyapunov function P for matrix A_2 since

$$P = \begin{bmatrix} p_1 & 0 \\ 0 & p_2 \end{bmatrix} \implies PA_2 + A_2^T P = \begin{bmatrix} 0 & -\epsilon p_2 \\ -\epsilon p_2 & 2\epsilon p_2 \end{bmatrix}.$$

Matrix A_3 is unstable and thus Lyapunov matrix P_3 does not exist such that P_3 is positive definite and $P_3 A_3 + A_3^T P_3$ is positive semi-definite. Indeed, matrix A_3 does not have "Property c." $\diamond$

The above results show that, unless M-matrix A is both reducible and singular, its quadratic Lyapunov function exists and the corresponding Lyapunov matrix P can simply be chosen to be diagonal. These results can be applied directly to continuous-time positive system $\dot{x} = -Ax$, where $-A$ is Metzler. If A is both reducible and singular, one should proceed with analysis either by checking "Property c" or by first finding its canonical form of (4.1) and then utilizing properties of irreducible diagonal blocks.

For discrete-time positive system $x_{k+1} = Ax_k$, the following theorem provides a similar result on the existence of diagonal Lyapunov matrix. Once again, if A is reducible, one can proceed with analysis by employing the canonical form of (4.1).

Theorem 4.36. *If A is a non-negative and irreducible matrix and if $\rho(A) < 1$ (or $\rho(A) = 1$), then there exists a positive diagonal matrix P such that $P - A^T PA$ is positive definite (or positive semi-definite).*

Proof: It follows from Theorem 4.8 that there exists positive vectors $u, v > 0$ and eigenvalue $\rho(A) > 0$ such that

$$Au = \rho(A)u, \quad A^T v = \rho(A)v.$$

Letting $P = \text{diag}\{v_i / u_i\}$, we know that off-diagonal elements of matrix $(P - A^T PA)$ are all non-positive and that

$$(P - A^T PA)u = Pu - \rho(A)A^T Pu = v - \rho(A)A^T v = (1 - \rho^2(A))v \geq 0,$$

from which $(P - A^T P A)$ being a non-singular or singular M-matrix can be concluded from Theorems 4.25 or 4.27. Since $(P - A^T P A)$ is also symmetrical, it must be positive definite or positive semi-definite.   □

4.3.6 A Class of Interconnected Systems

Consider the following collection of sub-systems: for $i = 1, \cdots, q$,

$$\dot{x}_i = Ax_i + B(u + D_i y), \quad y_i = Cx_i, \tag{4.11}$$

where $x_i \in \Re^n$, $u \in \Re$, $y_i \in \Re$,

$$x = \begin{bmatrix} x_1^T & \cdots & x_q^T \end{bmatrix}^T, \quad y = \begin{bmatrix} y_1 & \cdots & y_q \end{bmatrix}^T, \quad B = \begin{bmatrix} 0 & 1 \end{bmatrix}^T,$$

$$C = \begin{bmatrix} 1 & 0 \end{bmatrix}, \quad D = \begin{bmatrix} D_1^T & \cdots & D_q^T \end{bmatrix}^T,$$

and matrices A, B, C and D are of proper dimensions. Individual dynamics of each sub-system are identical, and dynamic interaction among the sub-systems are through matrix D. But, different from a typical interconnected system, the above system has three distinct properties:

(a) Matrix $D \in \mathcal{M}_2(0)$ is a Metzler matrix with $D1_q = 0$, and it is also irreducible and symmetric. It follows from Theorem 4.31 that eigenvalues of D are $\lambda_1 \leq \cdots \leq \lambda_{q-1} < \lambda_q = 0$.
(b) All the sub-systems have the same control input.
(c) In designing the only control, the available feedback is limited to the mean of the sub-system states. That is, control u has to be of form

$$u = -\frac{1}{q} K \sum_{j=1}^{q} x_j, \tag{4.12}$$

where $K \in \Re^{1 \times n}$ is a feedback gain matrix.

The following lemma provides the corresponding stability condition.

Lemma 4.37. *Consider System 4.11 with an irreducible and symmetric matrix D in set $\mathcal{M}_2(0)$ and under Control 4.12. Then, the system is asymptotically stable if and only if matrices $(A - BK)$ and $(A + \lambda_i BC)$ are all Hurwitz for $i = 1, \cdots, q - 1$, where λ_i are negative real eigenvalues of D.*

Proof: It follows that

$$qBu = -BK \begin{bmatrix} I_{n \times n} & \cdots & I_{n \times n} \end{bmatrix} x$$
$$= -(1_q^T \otimes (BK))x$$

and that

$$BD_i y = (D_i \otimes B)\mathrm{diag}\{C, \cdots, C\}x$$
$$= (D_i \otimes (BC))x.$$

Hence, dynamics of the overall system can be written as

$$\dot{x} = (I_{q \times q} \otimes A)x - \frac{1}{q}(\mathbf{J}_{q \times q} \otimes (BK))x + (D \otimes (BC))x \triangleq \overline{A}x.$$

Based on Property (a), we know there exists a unitary matrix T with $1/\sqrt{q}$ as its last column such that

$$T^{-1}DT = T^T DT = \mathrm{diag}\{\lambda_1, \cdots, \lambda_q\}.$$

It has been shown in Example 4.9 that

$$T^{-1}\mathbf{J}_{q \times q}T = \mathrm{diag}\{0, \cdots, 0, q\}.$$

Thus, we have

$$(T \otimes I_{n \times n})^{-1}\overline{A}(T \otimes I_{n \times n})$$
$$= (I_{q \times q} \otimes A) - (\mathrm{diag}\{0, \cdots, 0, 1\} \otimes (BK)) + (\mathrm{diag}\{\lambda_1, \cdots, \lambda_{q-1}, 0\} \otimes (BC))$$
$$= \mathrm{diag}\{(A + \lambda_1 BC), \cdots, (A + \lambda_{q-1}BC), (A - BK)\},$$

from which the condition on asymptotic stability can be concluded. □

Lemma 4.37 proves to be very useful in friction control of nano-particles [82, 83]. It is worth noting that, by defining the average state $x_a = (\sum_{j=1}^{q} x_j)/q$ and by recalling $\mathbf{1}_q^T D = \mathbf{1}_q^T D^T = (D\mathbf{1}_q)^T = 0$, dynamics of the average state are described by

$$\dot{x}_a = \frac{1}{q}(\mathbf{1}_q^T \otimes I_{n \times n})\dot{x} = Ax_a - BKx_a + \frac{1}{q}((\mathbf{1}_q^T D) \otimes (BC))x = (A - BK)x_a,$$

which can always be stabilized as long as pair $\{A, B\}$ is controllable. Since all the sub-systems have the same average-feedback control, the average state is expected to be stabilized. Whether the first $q - 1$ states are all stabilized depends upon property of $(A + \lambda_j BC)$ for $j = 1, \cdots, q - 1$. Note that, if the fictitious system of $\dot{z} = Az + Bv$ and $y = Cz$ is stabilizable under any static output feedback control, matrices $(A + \lambda_j BC)$ are Hurwitz.

4.4 Multiplicative Sequence of Row-stochastic Matrices

Consider an infinite countable series of non-negative row-stochastic matrices $\{P_k : P_k \in \Re_+^{n \times n}, P_k\mathbf{1} = \mathbf{1}, k \in \aleph^+\}$, where $\aleph^+$ is the set of positive integers. In the subsequent discussion, let us denote that, for any $k > l$,

$$P_{k:l} \triangleq P_k P_{k-1} \cdots P_{l+2}P_{l+1}.$$

The objective of this section is to determine the conditions (including topological or structural conditions) under which the infinitely pre-multiplicative sequence of $P_{\infty:0}$ has the following limit: for some vector $c \in \Re^n$,

$$P_{\infty:0} \triangleq \lim_{k \to \infty} \prod_{j=1}^{k} P_j \triangleq \lim_{k \to \infty} P_k P_{k-1} \cdots P_2 P_1 = \mathbf{1} c^T. \qquad (4.13)$$

It will be shown in Chapter 5 that the convergence corresponds to the so-called *asymptotic cooperative stability*.

It is beneficial to begin convergence analysis with four simple facts. First, Limit 4.13 implies $c^T \mathbf{1} = 1$ since $P_j \mathbf{1} = \mathbf{1}$. Second, Limit 4.13 does not mean nor is it implied by irreducibility. For instance, it is shown in Example 4.12 that the infinite sequence of P^k does not have Limit 4.13 for any irreducible cyclic matrix P. Third, Limit 4.13 is different from Limit 2.35. Nonetheless, the general implications derived from the discussions in Section 2.4.2 do apply here. As a sufficient condition, if $A_k = \alpha_i P_i$ where $0 \leq \alpha_i \leq 1 - \epsilon$ for constant $\epsilon > 0$ and for all i, Limit 2.35 can be concluded using Limit 4.13. Fourth, for a non-homogeneous discrete-time Markov chain $\{X_k\}$ with transition probability matrix P_k, its ergodicity is characterized by the following limit of post-multiplicative sequence:

$$\lim_{k \to \infty} P_1 P_2 \cdots P_{k-1} P_k \triangleq \lim_{k \to \infty} \sqcap_{j=1}^{k} P_j = \mathbf{1} c^T.$$

Although the above sequence is different from $P_{\infty:0}$, both sequences can be analyzed using the same tools and their convergence conditions are essentially the same.

4.4.1 Convergence of Power Sequence

A special case of Limit 4.13 is that, for some $c \in \Re^n$,

$$\lim_{k \to \infty} P^k = \mathbf{1} c^T. \qquad (4.14)$$

The following theorem provides a necessary and sufficient condition, its proof is based on the next lemma (which is a restatement of Corollary 4.33 and can also be proven by directly showing that row sums of $E_{ii}^{r_i}$ are less than 1).

Lemma 4.38. *Suppose that non-negative row-stochastic matrix $E_\searrow$ is in the lower triangular canonical form of (4.1), where $E_{ii} \in \Re^{r_i \times r_i}$ are irreducible blocks on the diagonal. If $E_{ij} \neq 0$ for some $i > 1$ and for some $j < i$, then $(I - E_{ii})$ is a non-singular M-matrix, or simply $\rho(E_{ii}) < 1$.*

Theorem 4.39. *Let $E_\searrow$ in (4.1) be the lower triangular canonical form of row-stochastic matrix P. Then, Limit 4.14 exists if and only if $E_\searrow$ is lower triangularly complete and its E_{11} is primitive.*

Proof: Assume without loss of any generality that, in (4.1), $p = 2$ (if otherwise, the proof can be extended by induction). It follows that

$$E_{\searrow}^k = \begin{bmatrix} E_{11}^k & 0 \\ W_k & E_{22}^k \end{bmatrix},$$

where $W_k = 0$ if $E_{21} = 0$. Indeed, it follows from Lemma 4.38 that, if $E_{21} \neq 0$, $E_{22}^k \to 0$ and hence $W_k \mathbf{1} \to \mathbf{1}$. It follows from Corollary 4.13 and Example 4.12 that $E_{11}^k \to \mathbf{1}c_1^T$ if and only if E_{11} is primitive. Hence, combining the two conditions yields

$$E_{\searrow}^{2k} = \begin{bmatrix} E_{11}^{2k} & 0 \\ W_k E_{11}^k & E_{22}^{2k} \end{bmatrix} \to \begin{bmatrix} \mathbf{1}c_1^T & 0 \\ \mathbf{1}c_1^T & 0 \end{bmatrix}, \quad E_{\searrow}^{2k+1} \to E_{\searrow} \begin{bmatrix} \mathbf{1}c_1^T & 0 \\ \mathbf{1}c_1^T & 0 \end{bmatrix} = \begin{bmatrix} \mathbf{1}c_1^T & 0 \\ \mathbf{1}c_1^T & 0 \end{bmatrix},$$

which completes the proof. □

In fact, Lemma 4.38 and Corollary 4.13 also imply that, if Limit 4.14 exists, convergence is exponential. In [63], lower triangularly complete matrices are also said to be *indecomposable*. If E_{11} is primitive (while other E_{ii} may not be primitive), it has cyclicity 1 and hence is aperiodic. Therefore, a row-stochastic matrix having Limit 4.14 or satisfying Theorem 4.39 is also referred to as *stochastic-indecomposable-aperiodic* or *SIA* matrix [274] or *regular* matrix [85].

Lower triangularly complete and diagonally positive matrices are always SIA, and it will be shown in Subsection 4.4.4 that they are instrumental to ensuring Limit 4.13 arising from control systems. In general, matrices P_j being SIA do not guarantee Limit 4.13 as shown by the following example.

Example 4.40. Matrices

$$D_1 = \begin{bmatrix} 0 & 1 & 0 \\ 0 & 0 & 1 \\ 0.4 & 0 & 0.6 \end{bmatrix}, \quad D_2 = \begin{bmatrix} 0.6 & 0 & 0.4 \\ 1 & 0 & 0 \\ 0 & 1 & 0 \end{bmatrix}$$

are both irreducible and primitive, hence they are SIA. In fact, D_2 is a permuted version of D_1. However, it follows that

$$D_1 D_2 = \begin{bmatrix} 1 & 0 & 0 \\ 0 & 1 & 0 \\ 0.24 & 0.6 & 0.16 \end{bmatrix},$$

which is reducible, not lower triangularly complete, and hence not SIA. In other words, if sequence $P_{\infty:0}$ has Limit 4.13, simply inserting permutation matrices into the convergent sequence may make the limit void.   ◇

4.4.2 Convergence Measures

Existence of Limit 4.13 can be studied using the idea of contraction mapping discussed in Section 2.2.1. To this end, let us define the following two measures: given a squarely or rectangularly row-stochastic matrix $E \in \Re_+^{r_1 \times r_2}$,

$$\delta(E) = \max_{1 \leq j \leq r_2} \max_{1 \leq i_1, i_2 \leq r_1} |e_{i_1 j} - e_{i_2 j}|,$$

$$\lambda(E) = 1 - \min_{1 \leq i_1, i_2 \leq r_1} \sum_{j=1}^{r_2} \min(e_{i_1 j}, e_{i_2 j}). \tag{4.15}$$

It is obvious that $0 \leq \delta(E), \lambda(E) \leq 1$ and that $\lambda(E) = 0$ if and only if $\delta(E) = 0$. Both quantities measure how different the rows of E are: $\delta(E) = 0$ if all the rows of E are identical, and $\lambda(E) < 1$ implies that, for every pair of rows i_1 and i_2, there exists a column j (which may depend on i_1 and i_2) such that both $e_{i_1 j}$ and $e_{i_2 j}$ are positive.

For matrix $E \in \Re_+^{n \times n}$ with row vectors E_i, let us define vector minimum function $\min\{E_i^T, E_j^T\} \in \Re_+^n$ and their *convex hull* $\mathrm{co}(E) \subset \Re_+^n$ as

$$\min\{E_i^T, E_j^T\} \triangleq \left[\min\{e_{i1}, e_{j1}\}, \cdots, \min\{e_{in}, e_{jn}\} \right]^T,$$

and

$$\mathrm{co}(E) \triangleq \left\{ \sum_{j=1}^n a_j E_j^T : a_j \geq 0, \sum_{j=1}^n a_j = 1 \right\}.$$

Convex hull $\mathrm{co}(E)$ is a *simplex* (an n-dimensional analogue of a triangle) with vertices at E_j^T, and its size can be measured by its radius (or diameter) defined in p-norm by

$$\mathrm{rad}_p(E) \triangleq \frac{1}{2} \max_{i,j} \|E_i^T - E_j^T\|_p, \quad \mathrm{diam}_p(E) = 2\mathrm{rad}_p(E),$$

where $\|\cdot\|_p$ (including $\|\cdot\|_\infty$) is the standard vector norm. The following lemma provides both geometrical meanings and simple yet powerful inequalities in terms of norms and the aforementioned two measures.

Lemma 4.41. *Consider row-stochastic matrices $E, F, G \in \Re_+^{n \times n}$ with $G = EF$. Then,*

$$\delta(E) = \mathrm{diam}_\infty(E), \quad \lambda(E) = 1 - \min_{i,j} \| \min\{E_i^T, E_j^T\}\|_1, \quad \lambda(E) = \mathrm{rad}_1(E),$$
$$\tag{4.16}$$

and

$$\mathrm{diam}_p(G) \leq \lambda(E)\mathrm{diam}_p(F), \quad \text{or} \quad \mathrm{rad}_p(G) \leq \lambda(E)\mathrm{rad}_p(F), \tag{4.17}$$

or simply

$$\delta(G) \leq \lambda(E)\delta(F), \quad \lambda(G) \leq \lambda(E)\lambda(F), \tag{4.18}$$

where $\delta(\cdot)$ and $\lambda(\cdot)$ are the measures defined in (4.15).

Proof: The first two norm expressions in (4.16) are obvious from definition. To show the third expression, note that

$$\text{rad}_1(E) = \frac{1}{2} \max_{i,j} \sum_{k=1}^{n} |e_{ik} - e_{jk}|. \tag{4.19}$$

For any index pair (i, j) with $i \neq j$, the index set of $\{1, \cdots, n\}$ can be divided into two complementary and mutually-exclusive sets Ω_{ij} and Ω_{ij}^c such that $(e_{ik} - e_{jk})$ is positive for $k \in \Omega_{ij}$ and non-positive for $k \in \Omega_{ij}^c$. Hence,

$$\sum_{k=1}^{n} |e_{ik} - e_{jk}| = \sum_{k \in \Omega_{ij}} (e_{ik} - e_{jk}) - \sum_{k \in \Omega_{ij}^c} (e_{ik} - e_{jk}).$$

On the other hand, E being row-stochastic implies that

$$\sum_{k \in \Omega_{ij}} (e_{ik} - e_{jk}) + \sum_{k \in \Omega_{ij}^c} (e_{ik} - e_{jk}) = \sum_{k=1}^{n} (e_{ik} - e_{jk}) = 0,$$

and

$$1 = \sum_{k=1}^{n} e_{ik} = \sum_{k \in \Omega_{ij}} e_{ik} + \sum_{k \in \Omega_{ij}^c} e_{ik}.$$

Combining the above three equations yields

$$\frac{1}{2} \sum_{k=1}^{n} |e_{ik} - e_{jk}| = \sum_{k \in \Omega_{ij}} (e_{ik} - e_{jk})$$

$$= \sum_{k \in \Omega_{ij}} e_{ik} - \sum_{k \in \Omega_{ij}} e_{jk}$$

$$= 1 - \sum_{k \in \Omega_{ij}^c} e_{ik} - \sum_{k \in \Omega_{ij}} e_{jk}$$

$$= 1 - \sum_{k=1}^{n} \min\{e_{ik}, e_{jk}\},$$

which together with (4.19) and (4.15) yields a third expression in (4.16).

Next, consider again any index pair (i, j) with $i \neq j$. Let E_k, F_k, and G_k be the kth row of matrices E, F, and G, respectively. It follows that

$$G_i^T = \sum_{k=1}^{n} e_{ik} F_k^T, \quad G_j^T = \sum_{k=1}^{n} e_{jk} F_k^T.$$

Denoting $\beta_k = \min\{e_{ik}, e_{jk}\}$ or $[\beta_1, \cdots, \beta_n]^T \triangleq \min\{E_i^T, E_j^T\}$, we know from $E_i \mathbf{1} = E_j \mathbf{1} = 1$ that $\beta_{ij,k} \geq 0$ and, for some $\epsilon \geq 0$,

$$\sum_{k=1}^{n} \beta_k + \epsilon = 1,$$

in which $\epsilon = 0$ if and only of $e_{ik} = e_{jk} = \beta_k$ for all k. Let us now introduce matrix H whose lth row H_l is given by, for $l = 1, \cdots, n$,

$$H_l^T \overset{\triangle}{=} \epsilon F_l^T + \sum_{k=1}^{n} \beta_k F_k^T .$$

It is obvious that convex hull $co(H)$ is a simplex with vertices at H_j^T, that by the above definition $co(H) \subset co(F)$, and that

$$\text{diam}_p(H) = \left(1 - \sum_{k=1}^{n} \min\{e_{ik}, e_{jk}\}\right) \text{diam}_p(F) \qquad (4.20)$$

since, for any (l, m),

$$H_l^T - H_m^T = \epsilon(F_l^T - F_m^T) = \left(1 - \sum_{k=1}^{n} \min\{e_{ik}, e_{jk}\}\right)(F_l^T - F_m^T).$$

If $\epsilon = 0$, $G_i = G_j = H_l$, that is, $G_i, G_j \in co(H)$. If $0 < \epsilon \le 1$, we choose $a_k = (e_{ik} - \beta_k)/\epsilon$ and have that

$$a_k \ge 0, \quad \sum_{k=1}^{n} a_k = \frac{1}{\epsilon}\sum_{k=1}^{n}[e_{ik} - \beta_k] = \frac{1}{\epsilon} - \frac{1}{\epsilon}\sum_{k=1}^{n} \beta_k = 1,$$

and that

$$\sum_{l=1}^{n} a_l H_l^T = \epsilon \sum_{l=1}^{n} a_l F_l^T + \sum_{l=1}^{n} a_l \sum_{k=1}^{n} \beta_k F_k^T$$

$$= \epsilon \sum_{l=1}^{n} a_l F_l^T + \sum_{k=1}^{n} \beta_k F_k^T$$

$$= \sum_{m=1}^{n} [\epsilon a_m + \beta_m] F_m^T$$

$$= \sum_{m=1}^{n} e_{im} F_m^T = G_i^T ,$$

which implies $G_i^T \in co(H)$. Similarly, $G_j^T \in co(H)$. Combining the two cases on the value of ϵ, we know $G_i, G_j \in co(H)$ and hence

$$\|G_i^T - G_j^T\|_p \le \text{diam}_p(H). \qquad (4.21)$$

Since the pair (i, j) is arbitrary, Inequality 4.17 can be concluded by combining (4.20) and (4.21).

According to the alternative definitions in (4.16), the inequalities in (4.18) are special cases of Inequality 4.17 with $p = \infty$ and $p = 1$, respectively. $\quad\square$

Lemma 4.41 essentially shows that, for convergence of sequence $P_{\infty:0}$ to Limit 4.13, pre-multiplication of any matrix P_k with $\lambda(P_k) \le 1 - \epsilon$ for some $\epsilon > 0$ is a contraction mapping. As such, the following definition is introduced.

Definition 4.42. *Row-stochastic matrix E is said to be a* scrambling matrix *if $\lambda(E) < 1$. Sequence $E_{\infty:0}$ is said to be* sequentially scrambling *if there exists a scalar strictly-increasing sub-series $\{m_l \in \aleph^+ : l \in \aleph\}$ such that $\lambda(E_{m_l:m_{l-1}}) < 1$. Sequence $E_{\infty:0}$ is said to be* uniformly sequentially scrambling *if there exist a constant $\epsilon > 0$ and a scalar strictly-increasing sub-series $\{m_l \in \aleph^+ : l \in \aleph\}$ such that $\lambda(E_{m_l:m_{l-1}}) \leq 1 - \epsilon$ for all l.*

By the above definition, whether matrix E is scrambling or not depends solely on locations of its zero and non-zero entries. Hence, we can define the so-called *characteristic matrix* as the following binary matrix: for any non-negative matrix E,

$$\mathcal{B}(E) \triangleq [\text{sign}(e_{ij})], \qquad (4.22)$$

where $\text{sign}(\cdot)$ is the standard sign function (*i.e.*, $\text{sign}(0) = 0$ and $\text{sign}(\epsilon) = 1$ if $\epsilon > 0$). Obviously, there are at most 2^{n^2} choices for matrix $\mathcal{B}(\cdot)$; $\mathcal{B}(EF)$ is invariant for any pair of non-negative matrices (E, F) as long as $\mathcal{B}(E)$ and $\mathcal{B}(F)$ do not change; and, if $\mathcal{B}(E) \geq \mathcal{B}(F)$, $\lambda(F) < 1$ implies $\lambda(E) < 1$. The following lemma further provides useful results on both $\lambda(\cdot)$ and $\mathcal{B}(\cdot)$.

Lemma 4.43. *Consider row-stochastic matrices $E, F, G, W \in \Re_+^{n \times n}$ with $G = EF$.*

(a) *If E is diagonally positive, $\mathcal{B}(G) \geq \mathcal{B}(F)$. If F is diagonally positive, $\mathcal{B}(G) \geq \mathcal{B}(E)$. If both E and F are diagonally positive, then $\mathcal{B}(G) \geq \mathcal{B}(E + F)$.*

(b) *If W is diagonally positive and $\lambda(G) < 1$, then $\lambda(WEF) < 1$, $\lambda(EWF) < 1$, and $\lambda(EFW) < 1$. That is, scramblingness of a finite-length matrix product is invariant under an insertion of any diagonally positive matrix anywhere.*

(c) *If E is a SIA matrix and $\mathcal{B}(G) = \mathcal{B}(F)$, then $\lambda(F) < 1$ and $\lambda(G) < 1$. Similarly, if F is a SIA matrix and $\mathcal{B}(G) = \mathcal{B}(E)$, then both E and G are scrambling.*

Proof: It follows from

$$g_{ij} = \sum_{k=1}^n e_{ik} f_{kj}$$

that, if $e_{kk} > 0$, $f_{kj} > 0$ implies $g_{kj} > 0$ and that, if $g_{kk} > 0$, $e_{ik} > 0$ implies $g_{ik} > 0$. Hence, (a) is concluded.

It follows from (a) that, if W is diagonally positive, $\mathcal{B}(WG) \geq \mathcal{B}(G)$, $\mathcal{B}(GW) \geq \mathcal{B}(G)$, and $\mathcal{B}(EW) \geq \mathcal{B}(E)$ which implies $\mathcal{B}(EWF) \geq \mathcal{B}(EF)$. Thus, (b) is validated.

It follows from $\mathcal{B}(EF) = \mathcal{B}(F)$ that $\mathcal{B}(E^2F) = \mathcal{B}(EF)$ and hence $\mathcal{B}(E^2F) = \mathcal{B}(F)$. By induction, we have $\mathcal{B}(E^kF) = \mathcal{B}(F)$. It follows from the proof of Theorem 4.39 and from (4.15) that, since E is SIA, $\lambda(E^k) < 1$

for sufficiently large k. By Lemma 4.41, $\lambda(E^k F) \leq \lambda(E^k)\lambda(F) < 1$. Thus, Statement (c) can be claimed by recalling $\mathcal{B}(E^k F) = \mathcal{B}(G) = \mathcal{B}(F)$. □

By Definition 4.42, a lower triangularly positive and diagonally positive matrix is scrambling (and, by Theorem 4.39, it is also SIA). For the purpose of concluding that sequence $E_{\infty:0}$ is sequentially scrambling, the following lemma reduces the requirement of E_k being lower triangularly positive to that of E_k being lower triangularly complete.

Lemma 4.44. *Consider series $\{E_k \in \Re_+^{n \times n} : k \in \aleph\}$ of diagonally positive and row-stochastic matrices. If matrices E_k are in its lower triangular canonical form and are lower-triangularly complete, there exists a finite κ for every choice l such that $E_{(l+\kappa):l}$ is lower triangularly positive, and hence $E_{(l+\kappa):l}$ is both scrambling and SIA.*

Proof: Assume without loss of any generality that

$$E_k = \begin{bmatrix} E_{k,11} & 0 \\ E_{k,21} & E_{k,22} \end{bmatrix},$$

where $E_{k,11} \in \Re^{r_1 \times r_1}$ and $E_{k,22} \in \Re^{r_2 \times r_2}$ are irreducible and diagonally positive. If E_k contains more block rows, induction should be used to extend the subsequent analysis. Let us first consider the case that dimensions of $E_{k,11}$ and $E_{k,22}$ are independent of k. In this case, it follows that

$$E_{(l+1):(l-1)} = \begin{bmatrix} E_{(l+1),11}E_{l,11} & 0 \\ E_{l+1,21}E_{l,11} + E_{(l+1),22}E_{l,21} & E_{(l+1),22}E_{l,22} \end{bmatrix}$$
$$\triangleq \begin{bmatrix} E_{(l+1):(l-1),11} & 0 \\ E_{(l+1):(l-1),21} & E_{(l+1):(l-1),22} \end{bmatrix},$$

in which, according to Statement (a) of Lemma 4.43,

$$\mathcal{B}(E_{(l+1):(l-1),21}) \geq \mathcal{B}(E_{l+1,21} + E_{l,21}) \tag{4.23}$$

By Corollary 4.4, we know that, for all $m \geq \max\{r_1, r_2\} - 1$, diagonal products $E_{(l+jm-1):(l+jm-m-1),ii}$ are positive for $i = 1, 2$ and $j = 1, 2, 3$. It follows from (4.23) and from E_k being lower-triangularly complete that, by simply increasing the finite integer m, lower-triangular blocks $E_{(l+jm-1):(l+jm-m-1),21}$ become non-zero for $j = 1, 2, 3$. Furthermore, it follows from $E_{(l+2m-1):(l+m-1),ii}$ and $E_{(l+m-1):(l-1),ii}$ being positive and from

$$E_{(l+2m-1):(l-1),21} = E_{(l+2m-1):(l+m-1),21}E_{(l+m-1):(l-1),11}$$
$$+ E_{(l+2m-1):(l+m-1),22}E_{(l+m-1):(l-1),21}$$

that a positive entry in matrix $E_{(l+2m-1):(l+m-1),21}$ induces a positive row in matrix $E_{(l+2m-1):(l-1),21}$ and a positive entry in $E_{(l+2m-1):(l+m-1),21}$ induces a positive row in $E_{(l+2m-1):(l-1),21}$. Therefore, by taking another multiplication and repeating the argument of positive-entry propagation once more, we know that $E_{(l+3m-1):(l-1),21} > 0$. In the second case that dimensions of diagonal blocks of E_k vary as k increases, we know again from Statement (a) of Lemma 4.43 that, by computing products of finite length, sizes of irreducible blocks on the diagonal are all non-decreasing and also upper bounded. Thus, the second case can be handled by applying the result of the first case to the diagonal blocks of largest sizes appeared. In summary, the conclusion is drawn by choosing $\kappa \geq 3m$. □

If $\lambda(P_k)$ (or $\lambda(P_{m_l:m_{l-1}})$) is less than 1 but approaches 1 as l increases, sequence $P_{\infty:0}$ is sequentially scrambling but not uniformly sequentially scrambling, and it may not have the limit of (4.13). To illustrate this point, several matrix sequences are included in the following example.

Example 4.45. Matrix sequence $P_{\infty:0}$ with

$$P_k = \begin{bmatrix} 1 & 0 \\ \frac{1}{k+2} & \frac{k+1}{k+2} \end{bmatrix}, \quad k \in \aleph^+$$

is diagonally positive, lower triangularly positive and sequentially scrambling with $\lambda(P_k) = 1 - 1/(k+2)$. Direct computation yields $P_{\infty:0} = \mathbf{1}\begin{bmatrix} 1 & 0 \end{bmatrix}$. In contrast, matrix sequence $P'_{\infty:0}$ with

$$P'_k = \begin{bmatrix} 1 & 0 \\ \frac{1}{(k+2)^2} & \frac{(k+2)^2-1}{(k+2)^2} \end{bmatrix}, \quad k \in \aleph^+$$

is also diagonally positive, lower triangularly positive and sequentially scrambling with $\lambda(P_k) = 1 - 1/(k+2)^2$, but it does not have the limit of (4.13) as

$$P'_{\infty:0} = \begin{bmatrix} 1 & 0 \\ \frac{1}{3} & \frac{2}{3} \end{bmatrix}.$$

Sequences $P_{\infty:0}$ and $P'_{\infty:0}$ are not uniformly sequentially scrambling because lower triangular elements of P_k and P'_k are vanishing as k increases.

Next, consider the pair of matrix series $E_{\infty:0}$ and $E'_{\infty:0}$, where

$$E_k = \begin{bmatrix} 1 & 0 & 0 \\ \frac{1}{k+2} & \frac{k+1}{k+2} & 0 \\ 0 & \frac{(k+2)^2-1}{(k+2)^2} & \frac{1}{(k+2)^2} \end{bmatrix},$$

$$E'_k = \begin{bmatrix} 1 & 0 & 0 \\ \frac{k+1}{k+2} & \frac{1}{k+2} & 0 \\ 0 & \frac{1}{(k+2)^2} & \frac{(k+2)^2-1}{(k+2)^2} \end{bmatrix}.$$

Both sequences are sequentially scrambling but not uniformly sequentially scrambling as $E_{(k+2):k}$ and $E'_{(k+2):k}$ are lower triangularly positive but critical elements in their lower triangular forms as well as some of their diagonal elements are vanishing. It follows from direct computation that

$$E'_{\infty:0} = \begin{bmatrix} 1 & 0 & 0 \\ 1 & 0 & 0 \\ \frac{107}{465} & \frac{48}{465} & \frac{2}{3} \end{bmatrix},$$

while $E_{\infty:0} = \mathbf{1}\begin{bmatrix} 1 & 0 & 0 \end{bmatrix}$.   ◇

The following corollary further relaxes lower triangular completeness to sequential lower triangular completeness for diagonally positive and row-stochastic non-negative matrices. Matrix product $E_{(l+k):l}$ being lower triangularly complete is much less restrictive than any of E_k being lower triangularly complete. In light of Example 4.45, the uniform non-vanishing property is required in the definition of sequential lower triangular completeness. Under Definition 4.47, Lemma 4.44 reduces to Corollary 4.48.

Definition 4.46. *Given matrix series* $\{E_k \in \Re_+^{n \times n} : k \in \aleph\}$, *sequence* $E_{\infty:0}$ *is said to be* sequentially lower-triangular (or lower triangularly positive) *if there exists one permutation matrix that is independent of k and maps all E_k into the lower triangular canonical form of (4.1) (or their canonical forms are all lower-triangularly positive).*

Definition 4.47. *Sequence* $E_{\infty:0}$ *is said to be* sequentially lower-triangularly complete *if it is sequentially lower-triangular and if there exists a scalar strictly-increasing sub-series* $\{m_l \in \aleph^+ : l \in \aleph\}$ *such that products* $E_{m_l:m_{l-1}}$ *are all lower triangularly complete and diagonally positive and that both diagonal positiveness and lower triangular completeness are uniform[1] with respect to l. Sequence* $E_{\infty:0}$ *is said to be* uniformly sequentially lower-triangularly complete *if it is lower-triangularly complete and if the differences of $(m_l - m_{l-1})$ are all uniformly bounded with respect to l.*

Corollary 4.48. *Consider series* $\{E_k \in \Re_+^{n \times n} : k \in \aleph\}$ *of diagonally positive and row-stochastic matrices. If sequence* $E_{\infty:0}$ *is sequentially lower-triangularly complete, the sequence is uniformly sequentially scrambling. If sequence* $E_{\infty:0}$ *is uniformly sequentially lower-triangularly complete, there exist a finite integer $\kappa > 0$ and a scalar strictly-increasing sub-series* $\{m_l \in \aleph^+ : l \in \aleph\}$ *such that* $\lambda(E_{m_l:m_{l-1}}) \le 1 - \epsilon$ *(and hence sequence* $E_{\infty:0}$ *is uniformly sequentially scrambling) and* $m_l - m_{l-1} \le \kappa$ *for all l.*

In the next subsection, sufficient conditions of convergence to Limit 4.13 are presented, among which convergence of a uniformly sequentially scrambling sequence is shown. In Section 4.4.4, necessary and sufficient conditions of convergence are provided and, in particular, values of measure $\lambda(E_{m_l:m_{l-1}})$ need to be computed in order to determine convergence of a non-uniformly sequentially scrambling sequence $E_{\infty:0}$.

---

[1] The uniformity means that diagonal elements of $E_{m_l:m_{l-1}}$ are uniformly non-vanishing with respect to l and that, in every block row i of the lower triangular canonical form of $E_{m_l:m_{l-1}}$, there is at least one $j < i$ such that the corresponding block on the ith block row and the jth block column is uniformly non-vanishing with respect to l.

4.4.3 Sufficient Conditions on Convergence

The following theorem provides several sufficient conditions on sequence convergence as defined by (4.13).

Definition 4.49. *Sequence $F_{\infty:0}$ is said to be a* complete sub-sequence *of sequence $P_{\infty:0}$ if there exists a scalar sub-series $\{m_k : k \in \aleph, m_{k+1} > m_k, m_1 \geq 0\}$ of $\aleph$ such that $F_k = P_{m_{k+1}:m_k}$. Sequence $F_{\infty:0}$ is said to be a* sub-sequence *of sequence $P_{\infty:0}$ if there exists a scalar sub-series $\{m_k : k \in \aleph, m_{k+1} > m_k, m_1 \geq 0\}$ of $\aleph$ such that $F_k = P_{m_{2k+1}:m_{2k}}$ while $(m_{2k+2} - m_{2k+1})$ (and hence the maximum number of consecutive matrices being removed) is upper bounded.*

Theorem 4.50. *Consider series $\{P_k : k \in \aleph^+\}$ of squarely row-stochastic matrices. Then, sequence $P_{\infty:0}$ is guaranteed to have the limit in (4.13) if one of the following conditions holds:*

(a) Sequence $F_{\infty:0}$ is sequentially lower-triangularly complete.
(b) $P_{j_2:j_1} = \mathbf{J}/n$ (i.e., $\lambda(P_{j_2:j_1}) = 0$) for some pair of finite values of $j_1 < j_2$.
(c) Matrices P_j are chosen from a set of finite many distinct matrices and if all finite length products of these matrices are SIA.
(d) Matrices P_j are chosen from a set of finite many distinct matrices that, under the same permutation matrix, are transformed into the canonical form of (4.1) and are lower triangularly complete and diagonally positive.

Proof: It follows from Lemma 4.41 that inequality

$$\lambda\left(\prod_{j=1}^{k} P_{n_{j+1}:n_j}\right) \leq \lambda\left(\prod_{j=1}^{k} P_{n_{j+1}:n_j}\right) \leq \prod_{j=1}^{k} \lambda(P_{n_{j+1}:n_j}) \tag{4.24}$$

holds for any $k \in \aleph^+$, where $n_1 < n_2 < \cdots < n_k$. Statement (a) is obvious by applying Inequality 4.24 to a uniformly sequentially scrambling sequence in (4.13) and by invoking Corollary 4.48.

Statement (b) is also obvious since P_j is row-stochastic and hence, for any choice of P_j with $j > j_2$, $P_{j:j_1} = \mathbf{J}/n$.

To show (c), consider the finite-length product $P_{k:j} = P_{k:l}P_{l:j}$, where $k > l + 1$ and $l > j$. Since characteristic matrix $\mathcal{B}(\cdot)$ assumes a finite number of matrix values, $\mathcal{B}(P_{k:j}) = \mathcal{B}(P_{l:j})$ must hold for some l when $(k - j)$ becomes sufficiently large. By assumption, $P_{k:l}$ is SIA. Hence, it follows from Statement (c) of Lemma 4.43 that $\lambda(P_{k:j}) < 1$, and uniformity of $\lambda(P_{k:j}) < 1$ is ensured by the fact that only a finite number of seed matrices are used. Thus, Statement (c) is concluded by applying Statement (a) to the complete sub-sequence consisting of $P_{k:j}$.

Statement (d) is obviously from (c) since any finite-length product of lower triangular and diagonally positive matrices are always lower triangular and diagonally positive and since, by Theorem 4.39, a lower triangularly complete and diagonally positive matrix is SIA. □

As evidenced in Example 4.40, Condition (c) of Theorem 4.50 requires that all combinations of sufficiently long matrix products be tested for the SIA property. In comparison, the rest of conditions in Theorem 4.50 are in terms of individual matrices and hence more desirable for applications. Nonetheless, all the conditions in Theorem 4.50 are sufficient but not necessary in general.

4.4.4 Necessary and Sufficient Condition on Convergence

The following theorem provides necessary and sufficient conditions on convergence to Limit 4.13. The trivial case, Statement (b) of Theorem 4.50, is excluded in the development of the following necessary and sufficient conditions. Among the conditions, (d) and (e) depend only on structural (or topological) properties of matrices P_j and hence do not involve any numerical computation unless some of positive entries in matrices P_j are vanishing as j increases but they determine the aforementioned properties.

Definition 4.51. *Sequence $P_{\infty:0}$ is said to be* convergent *to the limit in (4.13) if, given any integer j and any constant $\epsilon' > 0$, there exist $\kappa(\epsilon', j) \in \aleph$ and $c_j \in \Re^n$ such that, for all $k \geq \kappa + j$,*

$$\|P_{k:j} - \mathbf{1}c_j^T\|_\infty \leq \epsilon'.$$

Sequence $P_{\infty:0}$ is said to be uniformly convergent *if $\kappa(\epsilon', j) = \kappa(\epsilon')$.*

Definition 4.52. *Sequence $P_{\infty:0}$ is said to be* (uniformly) sequentially complete *if one of its sub-sequences is (uniformly) sequentially lower-triangularly complete.*

Theorem 4.53. *Consider sequence $P_{\infty:0}$ of squarely row-stochastic matrices and, in light of (b) in Theorem 4.50, assume without loss of any generality that $\lambda(P_{j_2:j_1}) \neq 0$ for any $j_2 > j_1$. Then,*

(a) *Sequence $P_{\infty:0}$ is convergent to the limit in (4.13) if and only if one of its complete sub-sequences is.*

(b) *Sequence $P_{\infty:0}$ is convergent to the limit in (4.13) if and only if $P_{\infty:0}$ is uniformly sequentially scrambling.*

(c) *Sequence $P_{\infty:0}$ is convergent to the limit in (4.13) if and only if there exist a constant $\epsilon > 0$ and a scalar sub-series $\{m_k : k \in \aleph, m_{k+1} > m_k, m_1 \geq 0\}$ such that*

$$\prod_{k=1}^{\infty} \lambda(P_{m_{k+1}:m_k}) = 0.$$

(d) Sequence $P_{\infty:0}$ is convergent to the limit in (4.13) if and only if there exists a scalar sub-series $\{m_k : k \in \aleph, \ m_{k+1} > m_k, \ m_1 \geq 0\}$ such that

$$\sum_{k=1}^{\infty} [1 - \lambda(P_{m_{k+1}:m_k})] = +\infty.$$

(e) If sequence $P_{\infty:0}$ is sequentially lower triangular and diagonally positive, it is uniformly convergent to the limit in (4.13) if and only if it is uniformly sequentially lower triangularly complete.

(f) If matrices P_k are diagonally positive uniformly respect to k, sequence $P_{\infty:0}$ is uniformly convergent to the limit in (4.13) if and only if it is uniformly sequentially complete.

Proof: Necessity for (a) is trivial. To show sufficiency, assume that complete sub-sequence $F_{\infty:0}$ be convergent, that is, by Definition 4.51, inequality

$$\|F_{k:j} - \mathbf{1}c_j^T\|_{\infty} \leq \epsilon'$$

holds for all $k \geq \kappa + j$. Recall that $\|A\|_{\infty} = 1$ for any non-negative and row-stochastic matrix A. It follows that, for any (l', l) satisfying $l' < m_j < m_k < l$, $P_{l:l'} = P_{l:m_k} F_{k:j} P_{m_j:l'}$ and

$$\begin{aligned}
\|P_{l:l'} - \mathbf{1}(P_{m_j:l'}c)^T\|_{\infty} &= \|P_{l:m_k}[F_{k:j} - \mathbf{1}c^T]P_{m_j:l'}\|_{\infty} \\
&\leq \|P_{l:m_k}\|_{\infty} \cdot \|F_{k:j} - \mathbf{1}c^T\|_{\infty} \cdot \|P_{m_j:l'}\|_{\infty} \\
&\leq \epsilon',
\end{aligned}$$

from which convergence of sequence $P_{\infty:0}$ becomes obvious.

Sufficiency for (b) is obvious from (a) and from (a) of Theorem 4.50. To show necessity, we note from (4.16) that

$$\lambda(P_{k:l}) = \frac{1}{2} \max_{i,j} \|P_{k:l,i}^T - P_{k:l,j}^T\|_1 \leq \max_i \|P_{k:l,i}^T - c_l^T\|_1 = \|P_{k:l} - \mathbf{1}c_l^T\|_{\infty}, \tag{4.25}$$

where $P_{k:l,i}$ is the ith row of $P_{k:l}$. It follows from (4.24) and from Definition 4.51 that the value of $\lambda(P_{k:l})$ is monotone decreasing with respect to k and that, if sequence $P_{\infty:0}$ is convergent, $\lambda(P_{k:l}) \to 0$ for any l as $k \to \infty$. Hence, for any finite m_j and for any $0 < \epsilon < 1$, there must be a finite m_{j+1} such that $\lambda(P_{m_{j+1}:m_j}) \leq 1 - \epsilon$, and necessity is shown.

Sufficiency for (c) follows from (a) of Theorem 4.50, and necessity for (c) is apparent from (b).

Equivalence between (c) and (d) can be established by noting that, for any $0 \leq \beta_k < 1$,

$$\prod_{k=1}^{\infty}(1 - \beta_k) = 0 \quad \Longleftrightarrow \quad \sum_{k=1}^{\infty} \beta_k = \infty. \tag{4.26}$$

The above relationship is trivial if $\beta_k \not\to 0$. If $\beta_k \to 0$, the above relationship can be verified by using Taylor expansion of

$$\log \left(\prod (1 - \beta_k) \right) = \sum \log(1 - \beta_k) \approx - \sum \beta_k.$$

Sufficiency of (e) is obvious from Corollary 4.48 and from (a) of Theorem 4.50. It follows from (4.25) that, since the sequence is uniformly convergent, $\lambda(P_{k:l}) \leq 1 - \epsilon$ for some $\epsilon > 0$, for all $k > \kappa + l$, and all l. Thus, the sequence must be uniformly sequentially lower triangularly complete, and necessity of (e) is shown.

To show necessity of (f), note that, if $P_{\infty:0}$ is not uniformly sequentially complete, none of its sub-sequences can be uniformly sequentially lower triangularly complete. On the other hand, uniform convergence implies that $\lambda(P_{(l+\kappa):l}) \leq 1 - \epsilon$ holds for some $\kappa, \epsilon > 0$ and for all l and, since there are only a finite number of permutation matrices, at least one sub-sequence of $P_{\infty:0}$ is uniformly sequentially lower triangularly complete. This apparent contradiction establishes the necessity. To show sufficiency of (f), assume without loss of any generality that $E_{\infty:0}$ with $E_v = P_{k_{2v}+2:k_{2v}+1}$ and $v \in \aleph$ denote the uniformly sequentially lower-triangularly complete sub-sequence contained in $P_{\infty:0}$. According to Corollary 4.48, for any $\epsilon > 0$, there exists κ such that $\lambda(E_{v+\kappa} \cdots E_{v+1}) \leq 1 - \epsilon$. It follows that

$$\prod_{k=0}^{\infty} P_k = \prod_{v=1}^{\infty} [E_v F_v],$$

where $k_0 = 0$, $F_v = P_{k_{2v+1}:k_{2v}}$, $F_0 = I$ if $k_1 = 0$, and $F_0 = P_{k_1:0}$ if $k_1 > 0$. By Definition 4.49, F_v are diagonally positive uniformly with respect to v. Invoking (b) of Lemma 4.43, we know that $\lambda(E_{v+\kappa}F_{v+\kappa} \cdots E_{v+1}F_{v+1}) \leq 1 - \epsilon'$, where $\epsilon' > 0$ is independent of v. Thus, uniform convergence of sequence $P_{\infty:0}$ to the limit in (4.13) can be concluded using (4.24).   □

The following example uses two-dimensional row-stochastic matrix sequences to illustrate uniform convergence and non-uniform convergence to Limit 4.13 as well as non-convergence.

Example 4.54. Consider matrix series $\{E_k : k \in \aleph\}$, where $0 \le \beta_k \le 1$,

$$E_k = \begin{bmatrix} 1 & 0 \\ \beta_k & 1 - \beta_k \end{bmatrix},$$

and matrix E_k is lower triangularly complete. Matrix sequence $E_{\infty:0}$ provides the solution to dynamic system

$$\begin{bmatrix} x_1(k+1) \\ x_2(k+1) \end{bmatrix} = E_k \begin{bmatrix} x_1(k) \\ x_2(k) \end{bmatrix}.$$

It follows that $x(k+1) = E_{k:0}x(0)$, that the limit in (4.13) exists if and only if $[x_1(k+1) - x_2(k+1)] \to 0$ as $k \to \infty$, and that

$$[x_1(k+1) - x_2(k+1)] = (1 - \beta_k)[x_1(k) - x_2(k)].$$

Hence, it follows from (4.26) that sequence $E_{\infty:0}$ is convergent to the limit in (4.13) if and only if the sum of $\sum_k \beta_k$ is divergent.

Clearly, if β_k does not vanish as k increases, say $\beta_k = 1/3$, $\sum_k \beta_k = \infty$, sequence $E_{\infty:0}$ is uniformly sequentially lower triangularly complete and hence is uniformly convergent to the limit in (4.13). On the other hand, β_k can vanish as k increases, in which case $\{\beta_k : k \in \aleph\}$ is a Cauchy series and lower triangular completeness of matrix E_k is not uniform. Should the Cauchy series be divergent as $\sum_k \beta_k = \infty$, for instance,

$$\beta_k = \frac{1}{k+1},$$

sequence $E_{\infty:0}$ is not sequentially lower triangularly complete but it is convergent to the limit in (4.13). If the Cauchy series is convergent as $\lim_{k \to \infty} \sum_k \beta_k < \infty$, for instance,

$$\beta_k = \frac{1}{3^k},$$

sequence $E_{\infty:0}$ does not have the limit in (4.13).

Now, consider matrix series $\{P_k : k \in \aleph\}$ where

$$P_k = \begin{cases} I_{2\times 2} & \text{if } k \ne m_j \\ E_{m_j} & \text{if } k = m_j \end{cases},$$

where $\{m_j : j \in \aleph\}$ is a scalar sub-series, and $\{E_{m_j} : j \in \aleph\}$ is the sub-series of $\{E_k : k \in \aleph\}$. Should the sequence of $\{E_{m_j} : j \in \aleph\}$ be sequentially lower triangularly complete, so is sequence $P_{\infty:0}$. On the other hand, if sequence $E_{\infty:0}$ is not sequentially lower triangularly complete but is convergent and if E_{m_j} is lower triangularly complete (say, $\beta_k = 1/(k+1)$), the choice of scalar sub-series $\{m_j : j \in \aleph\}$ determines whether sequence $P_{\infty:0}$ is convergent. For instance, if $\beta_k = 1/(k+1)$ and $m_j = 3^j - 1$, sequence $P_{\infty:0}$ is not convergent since

$$\beta_{m_j} = \frac{1}{3^j};$$

and if $\beta_k = 1/(k+1)$ and $m_j = 2j$, sequence $P_{\infty:0}$ is convergent.   ◇

Theorem 4.39 on convergence of power sequence implies that lower triangular completeness of certain-length matrix products is needed for sequence $P_{\infty:0}$ to converge to the limit in (4.13), and diagonal positiveness required in (e) of Theorem 4.53 is also necessary in general because, due to the possibility of arbitrary permutation, there is no other alternative to ensure primitiveness for the first diagonal block in the canonical form of P_k. Matrix sequences arising from control systems can be made diagonally positive since any entity can always have (output) feedback from itself. The property of sequential lower triangular completeness or sequential completeness is structural (or topological), and it ensures convergence of sequence $P_{\infty:0}$. The proof of (d) and (e) in Theorem 4.53 indicates that convergence of Sequence 4.13 is exponential if it is uniformly sequentially complete. Without sequential completeness, it is evidenced by Examples 4.45 and 4.54 that the values of measure $\lambda(\cdot)$ have to be computed for certain-length matrix products, and convergence needs to be established in general by verifying that $P_{\infty:0}$ is uniformly sequentially scrambling.

4.5 Notes and Summary

This chapter introduces non-negative matrices (as well as Metzler matrices and M-matrices) and presents their properties that are most relevant to stability, Lyapunov function, and sequence convergence. Extensive discussions on non-negative matrices can be found in [14, 18, 74, 157, 262].

Perron-Frobenius theorem is the fundamental result on non-negative matrices. Theorem 4.7 is the famous fixed-point theorem by Brouwer [32], and the elegant proof of Statement (a) of Theorem 4.8 is given in [1]. Statement (b) of Theorem 4.8 is due to Perron [186]. The simple dominant eigenvalue $\rho(A)$ and its eigenvector are also known as Perron root and eigenvector of A, respectively. Later, Frobenius extended the result to irreducible matrices, Statement (c) of Theorem 4.8. The original proof [73] used expansions of determinants and was rather involved. The proof on (c) of Theorem 4.8 is due to Wielandt [273] and is based on the min-max property of $\rho(A)$ [74].

Matrix sequences arise naturally in the study of probability theory and systems theory. Convergence measures in Section 4.4.2 and the first inequality in (4.18) are classical, and they, together with Statements (b) and (c) of Theorem 4.53, are due to Hajnal [85, 86]. Seen as a generalization, the second inequality in (4.18) is reported in [54]. Statement (c) of Lemma 4.43 and Statement (d) of Theorem 4.50 were first reported in [274]. A historical review on early results on sequence convergence can be found in [225, 226]. General convergence of pre-multiplicative matrix sequence (that may not be row-stochastic) is studied in [48]. Diagonally-positive matrix sequences are studied in [200, 201], Statements (b) and (d) of Lemma 4.43 and (c) and (d) of Theorem 4.53 are developed in [202].

Applications of non-negative matrices have been found in such important areas as probability and Markov chains [63, 70, 226], input-output models in economics [129, 254], criticality studies of nuclear reactors [20], population models and mathematical bioscience [35], iterative methods in numerical analysis [262], data mining (DNA, recognition of facial expression, *etc.*) [127], web search and retrieval [120], positive systems [60], and cooperative control of dynamical systems [202].

5

Cooperative Control of Linear Systems

In this chapter, the cooperative control problem of linear systems is studied by first analyzing properties of linear cooperative systems. The results on multiplicative sequence of non-negative matrices from Chapter 4 naturally render the cooperative stability condition for linear systems. Based on the condition, a class of linear cooperative controls are designed, and cooperative controllability can be determined by the cumulative connective property of sensing/communication network topologies over time. As applications, cooperative control is used to solve the problems of consensus, rendezvous, tracking of a virtual leader, formation control, and synchronization. The issue of maintaining network connectivity is also addressed.

In the cooperative control problem, information are collected or transmitted through a sensing/communication network, and the cooperative control law is designed to account for topological changes of the network. Availability of the feedback needed for implementing decentralized controls is not continuous or continual or even predictable, which is the main distinction from a standard control problem. This distinction is obviously desirable for applications and positive for implementation, but it makes stability analysis much more involved. To establish the connection back to the standard methods and to describe the cumulative effect of topology changes, an average system is introduced and derived for cooperative systems. The average system enables us to determine Lyapunov function and control Lyapunov function for cooperative systems, although common Lyapunov function is not expected to exist in general. Robustness of cooperative systems against measurement noises and disturbance is also studied.

5.1 Linear Cooperative System

Dynamics of a continuous-time linear cooperative system can be expressed as

$$\dot{x} = [-I + D(t)]x, \quad x(t_0) \text{ given}, \quad t \geq t_0, \tag{5.1}$$

where $x \in \Re^n$, and matrix $D(t) \in \Re^{n \times n}$ is assumed to have the following properties:

(a) Matrix $D(t)$ is non-negative and row-stochastic.
(b) Changes of matrix $D(t)$ are discrete and countable. That is, matrix $D(t)$ is time-varying but piecewise-constant.
(c) Matrix $D(t)$ is diagonally positive[1].

The properties come from the fact that, as explained in Section 1.1, sub-systems of a cooperative system are to achieve the same objective while their environment keeps changing. Specifically, the sub-systems moving toward the same objective imply that the cooperative system must be a positive system and, as shown in Section 4.2, $D(t)$ should be non-negative and could be row-stochastic. In the sequel, further analysis shows that Property (a) captures the essence of cooperativeness. As to Property (b), the presence of a sensing/communication network yields discrete and intermittent changes in the feedback topology and hence the same type of changes in System Dynamics 5.1. In the subsequent analysis, $\{t_k : t_k > t_{k-1}, \ k \in \aleph\}$ denotes the sequence of time instants at which changes occur in matrix $D(t)$. Property (c) can always be assumed since every sub-system can have feedback continuously about itself.

5.1.1 Characteristics of Cooperative Systems

Basic characteristics of Cooperative System 5.1 can be seen from the following observations.

Set of equilibrium points: It follows from $D(t)$ being row-stochastic that $1/n$ is the right eigenvector associated with eigenvalue $\rho(D) = 1$ and that, for any state $x^e \in \Omega^e$ with $\Omega^e \triangleq \{x \in \Re^n : x = c\mathbf{1}, \ c \in \Re\}$,

$$\dot{x}^e = [-I + D(t)]x^e = -x^e + cD(t)\mathbf{1} = -x^e + c\mathbf{1} = 0.$$

Hence, $x^e = c\mathbf{1}$ includes the origin as the special case of $c = 0$, points x^e are among the *equilibria* of System 5.1, and set Ω^e of all the state variables being equal provides a mathematical description of the cooperative system reaching the common objective. By Theorem 4.31, Ω^e is the set of all the equilibrium points of System 5.1 if $D(t) = D$ is irreducible. If $D(t) = D$ is reducible, System 5.1 may have other equilibrium points outside Ω^e. It follows from (5.1) that, given $D(t) = D$, all the equilibrium points are the solution to the matrix equation

$$Dx_e = x_e, \quad \text{or equivalently,} \quad D^\infty x_e = x_e.$$

---

[1] This property is not necessarily needed for continuous-time systems but is required for discrete-time systems.

We know from Theorem 4.39 (if Property (c) holds) together with Theorem 4.31 (if Property (c) is not imposed) that there is no equilibrium point other than those in Ω^e if and only if matrix D is lower triangularly complete.

Stationary center of mass: Within the time interval $[t_k, t_{k+1})$, matrix $D(t) = D(k)$ is constant. Letting η_k be the left eigenvector of $D(t_k)$ associated with eigenvalue $\rho(D) = 1$ and defining the *center of mass* $\mathcal{O}_k(t) \triangleq \eta_k^T x(t)$, we know that

$$\frac{d\mathcal{O}_k(t)}{dt} = \eta_k^T[-I + D(t)]x = -\eta_k^T x + \eta_k^T D(t)x = -\eta_k^T x + \eta_k^T x = 0, \quad (5.2)$$

and hence $\mathcal{O}_k(t)$ is a also piecewise-constant time function over time intervals $t \in [t_k, t_{k+1})$. According to Theorem 4.8, $\eta_k > 0$ and $\mathcal{O}_k(t)$ is defined in terms of the whole state if $D(t)$ is irreducible, and $\mathcal{O}_k(t)$ may become degenerate if $D(t)$ is reducible. Over time, $\mathcal{O}_k(t)$ generally jumps from one point to another according to k. If $D(t)$ is also column-stochastic, $\eta_k \equiv 1/n$, and $\mathcal{O}_k(t)$ reduces to the average of the state variables and is invariant over time.

Uniform and non-decreasing minimum: Let $x_{min}(t) \triangleq \min_i x_i(t)$ denote the element-wise minimum value over time, and let $\Omega_{min}(t)$ be the timed index set such that $x_i(t) = x_{min}(t)$ if $i \in \Omega_{min}(t)$. It follows that, for any $i \in \Omega_{min}(t)$,

$$\dot{x}_i = -x_i + \sum_{j=1}^n d_{ij}x_j = \sum_{j=1}^n d_{ij}(x_j - x_i) \geq 0,$$

since $d_{ij} \geq 0$ and $(x_j - x_i) = [x_j(t) - x_{min}(t)] \geq 0$. Hence, System 5.1 is *minimum preserving* in the sense that $x_i(t) \geq x_{min}(t_0)$ for all i and for all $t \geq t_0$.

Uniform and non-increasing maximum: Let $x_{max}(t) \triangleq \max_i x_i(t)$ denote the element-wise maximum value over time, and let $\Omega_{max}(t)$ be the timed index set such that $x_i(t) = x_{max}(t)$ if $i \in \Omega_{max}(t)$. It follows that, for any $i \in \Omega_{max}(t)$,

$$\dot{x}_i = -x_i + \sum_{j=1}^n d_{ij}x_j = \sum_{j=1}^n d_{ij}(x_j - x_i) \leq 0,$$

since $d_{ij} \geq 0$ and $(x_j - x_i) = [x_j(t) - x_{max}(t)] \leq 0$. Hence, System 5.1 is *maximum preserving* in the sense that $x_i(t) \leq x_{max}(t_0)$ for all i and for all $t \geq t_0$.

Combining the last two properties, we know that $\|x(t)\|_\infty \leq \|x(t_0)\|_\infty$ and thus System 5.1 is ∞-*norm preserving* and Lyapunov stable. Due to equilibrium points of $x^e = c\mathbf{1}$, System 5.1 is not locally asymptotically stable (with respect to the origin). Stability of equilibrium points $x^e = c\mathbf{1}$ will be analyzed in the next subsection.

It should be noted that System 5.1 is the special case of System 4.6 with $k_1 = k_2 = 1$ and that the presence of $k_2 > 0$ and $k_2 \neq 1$ does not affect stability analysis. As demonstrated in Section 2.4.2, eigenvalue analysis cannot be applied in general to either System 5.1 or System 4.6 due to the presence of time-varying matrix $D(t)$. Nonetheless, the above analysis has established Statement (c) of Corollary 4.19 for Time-varying System 4.6, and it can also be used to show that Statements (a) and (b) of Corollary 4.19 hold in general for System 4.6. The subsequent analysis is to determine the condition(s) under which Time-varying System 5.1 converges to equilibrium set Ω^e.

5.1.2 Cooperative Stability

We begin with the following definition. It follows from the ∞-norm preserving property that System 5.1 is always cooperatively stable. Hence, asymptotic cooperative stability remains to be investigated for System 5.1. It is worth mentioning that, if $k_1 > 1$, System 4.6 is asymptotically stable and hence Definition 5.1 is satisfied with $c = 0$. This is the trivial case of asymptotic cooperative stability and not of interest in the subsequent discussions.

Definition 5.1. *System $\dot{x} = \mathcal{F}(x,t)$ is said to be cooperatively stable if, for every given $\epsilon > 0$, there exists a constant $\delta(t_0, \epsilon) > 0$ such that, for initial condition $x(t_0)$ satisfying $\|x_i(t_0) - x_j(t_0)\| \leq \delta$, $\|x_i(t) - x_j(t)\| \leq \epsilon$ for all i, j and for all $t \geq t_0$. The system is said to be* asymptotically cooperatively stable *if it is cooperatively stable and if $\lim_{t \to \infty} x(t) = c\mathbf{1}$, where the value of $c \in \Re$ depends upon the initial condition $x(t_0)$ and changes in the dynamics. The system is called* uniformly asymptotically cooperatively stable *if it is asymptotically cooperatively stable, if $\delta(t_0, \epsilon) = \delta(\epsilon)$, and if the convergence of $\lim_{t \to \infty} x(t) = c\mathbf{1}$ is uniform.*

In control theory, it is typical to convert a convergence problem into a standard stability problem. To determine the convergence condition of $\lim_{t \to \infty} x(t) = c\mathbf{1}$ without the knowledge of c, we can define the error state e_{c_i} as, for some chosen $i \in \{1, \cdots, n\}$,

$$e_{c_i} = \left[(x_1 - x_i) \cdots (x_{i-1} - x_i)(x_{i+1} - x_i) \cdots (x_n - x_i) \right]^T \in \Re^{n-1}. \quad (5.3)$$

It follows that, given the ∞-norm preserving property, System 5.1 is asymptotically cooperatively stable if and only if $\lim_{t \to \infty} e_{c_i}(t) = 0$. The following lemma is useful to find the dynamic equation of error state e_{c_i}.

Lemma 5.2. *Consider matrix set $\mathcal{M}_1(c_0)$ defined in Section 4.2. Let $W_i \in \Re^{(n-1) \times n}$ be the resulting matrix after inserting -1 as the ith column into $I_{(n-1) \times (n-1)}$, and let $G_i \in \Re^{n \times (n-1)}$ denote the resulting matrix after eliminating as the ith column from $I_{n \times n}$, that is,*

$$W_1 \triangleq \left[-\mathbf{1} \; I_{(n-1) \times (n-1)} \right], \quad G_1 \triangleq \begin{bmatrix} 0 \\ I_{(n-1) \times (n-1)} \end{bmatrix}.$$

Then, $W_i G_i = I_{(n-1) \times (n-1)}$, $H = HG_i W_i$ for any matrix $H \in \mathcal{M}_1(0)$, and $W_i D = W_i D G_i W_i$ for matrix $D \in \mathcal{M}_1(c_0)$ with $c_0 \in \Re$.

Proof: The first equation is obvious. To show $H = HG_i W_i$, note that $G_i W_i$ is the resulting matrix after inserting 0 as the ith row into W_i. Thus, all the columns of $HG_i W_i$ are the same as those of H except for the ith column which is the negative sum of the rest columns of H and hence is the same as the ith column of H since $H \in \mathcal{M}_1(0)$.

It follows from $D \in \mathcal{M}_1(c_0)$ that $[-cI + D] \in \mathcal{M}_1(0)$ and, by the preceding discussion, $[-cI + D] = [-cI + D]G_i W_i$. Pre-multiplying W_i on both sides yields

$$
\begin{aligned}
W_i D &= W_i[-cI + D] + cW_i \\
&= W_i[-cI + D]G_i W_i + cW_i \\
&= W_i D G_i W_i - cW_i G_i W_i + cW_i \\
&= W_i D G_i W_i,
\end{aligned}
$$

which is the last equation. $\qquad\square$

It follows from Lemma 5.2 that, under transformation $e_{c_i} \overset{\triangle}{=} W_i x$, System 5.1 is mapped into

$$
\begin{aligned}
\dot{e}_{c_i} &= W_i[-I + D(t)]x \\
&= W_i[-I + D(t)]G_i W_i x \\
&= W_i[-I + D(t)]G_i e_{c_i} \\
&= [-I + W_i D(t)G_i]e_{c_i}, \qquad (5.4)
\end{aligned}
$$

where $e_{c_i} \in \Re^{n-1}$ with $e_{c_i}(0) = W_i x(0)$. Hence, asymptotic cooperative stability of System 5.1 is converted into the standard stability problem of System 5.4 which is also linear and piecewise-constant.

Conceptually, there are three methods to analyze asymptotic cooperative stability for System 5.1. The first method is to extend further the analysis in Section 5.1.1, and it will be presented in Chapter 6 as it applies to both linear and non-linear systems. The second method is to convert the asymptotic cooperative stability problem into a standard asymptotic stability problem as shown above in (5.4) but, since the matrix product $W_i D(t)G_i$ is no longer a non-negative matrix, this method is studied later in Section 5.6. The third method is to utilize explicitly all the properties of $D(t)$ and then apply Theorem 4.53, which leads to the following results. Specifically, it follows from (5.1) that

$$
x(t_{k+1}) = P_k x(t_k), \quad x_0 \text{ given}, \quad k \geq 0, \qquad (5.5)
$$

where $P_k = e^{[-I + D(t_k)](t_{k+1} - t_k)}$ is the state transient matrix, Lemma 5.3 summarizes the properties of P_k, and hence Theorem 5.4 is the consequence of applying Lemma 5.3 and Theorem 4.53.

Lemma 5.3. *Consider matrix* $P = e^{(-I+D)\tau}$ *where* $D \in \Re_{+}^{n \times n}$ *is row-stochastic and constant. Then, for any finite* $\tau > 0$, *matrix* P *is diagonally positive and also row-stochastic, P is positive if and only if D is irreducible, and P is lower triangularly positive if and only if D is lower triangularly complete.*

Proof: It follows that

$$P = e^{(-I+D)\tau} = e^{-2\tau} e^{(I+D)\tau} = e^{-2\tau} \sum_{i=0}^{\infty} \frac{1}{i!} (I+D)^i \tau^i,$$

from which diagonal positiveness is obvious. It follows from $(I+D)^i \mathbf{1} = 2^i \mathbf{1}$ that $P\mathbf{1} = \mathbf{1}$. It follows from Corollary 4.4 that $(I+E)^i > 0$ for $i \geq (n-1)$ and hence $P > 0$ if and only if D is irreducible. If D is lower triangularly complete, so are $(I+E)^i$ and hence P, and *vice versa*. By Lemma 4.44, P is also lower triangularly positive. □

Theorem 5.4. *System 5.1 or its piecewise solution in (5.5) is uniformly asymptotically cooperatively stable if and only if* $\{D(t_k) : k \in \aleph\}$ *is uniformly sequentially complete.*

Theorem 5.4 provides the conditions on both stability and cooperative controllability, and the conditions are in terms of topologies changes as well as the properties of matrix $D(t)$. While topology changes may not be predicted, cooperative control design is to render a networked control system in the form of (5.1) and to ensure the properties for matrix $D(t)$.

5.1.3 A Simple Cooperative System

Arguably, the simplest cooperative system is the *alignment* problem shown in Fig. 5.1 in which particles move at the same constant absolute velocity and their directions of motion are adjusted according to the local information available. Through experimentation, Vicsek [47, 263] discovered that alignment can be achieved by the so-called neighboring rule: motion direction of any given particle is adjusted regularly to the average direction of motion of the particles in its neighborhood of radius r. To ensure alignment for those particles far from the rest, random perturbation should be added.

Mathematically, the neighboring rule can be expressed as

$$\theta_i(k+1) = \frac{1}{1+n_i(k)} \left(\theta_i(k) + \sum_{j \in N_i(k)} \theta_j(k) \right), \qquad (5.6)$$

where $N_i(k)$ is the index set of those particles in the neighborhood of the ith particle, $n_i(k)$ is the number of elements in $N_i(k)$, and both of them are time-varying. The continuous-time counterpart of (5.6) is given by

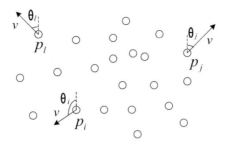

Fig. 5.1. Motion alignment and cooperative behavior of self-driven particles

$$\dot{\theta}_i = -\theta_i(t) + \frac{1}{1+n_i(t)} \left(\theta_i(t) + \sum_{j \in N_i(t)} \theta_j(t) \right), \qquad (5.7)$$

where $N_i(t)$ and $n_i(t)$ correspond to $N_i(k)$ and $n_i(k)$, respectively. Since Neighboring Rule 5.7 is already in the form of (5.1), Theorem 5.4 provides the theoretical guarantee that this remarkably simple model can render the cooperative behavior of $\theta_i(t) \to c$ for all i. Later, (5.6) will be studied (as a special case of (5.20)), and its convergence will be claimed (as an application of Theorem 5.12).

It should be noted that Neighboring Rule 5.6 is intuitive and reflects the group moving behavior. Indeed, this simple rule has been used in cellular automata [275], reaction-diffusion models [160, 224], flow of granular materials [88], animal aggregation [217, 260], *etc.*

5.2 Linear Cooperative Control Design

In this section, the cooperative control problem is studied for a set of heterogeneous systems whose dynamics can be transformed into the following form: for $i = 1, \cdots, q$,

$$\dot{x}_i = (-I + A_i)x_i + B_i u_i, \quad y_i = C_i x_i, \quad \dot{\varphi}_i = \psi_i(t, \varphi_i, x_i), \qquad (5.8)$$

where $y_i \in \Re^m$ is the output, $u_i \in \Re^m$ is the cooperative control to be designed, integer $l_i \geq 1$ represents the relative degree of the ith system, $x_i \in \Re^{l_i m}$ is the linear state of the ith system, $K_c \in \Re^{m \times m}$ is a non-negative design matrix chosen to be row-stochastic and irreducible,

$$
A_i = \begin{bmatrix} 0 & I_{m\times m} & 0 & \cdots & 0 & 0 \\ 0 & 0 & I_{m\times m} & \ddots & 0 & 0 \\ \vdots & \ddots & \ddots & \ddots & \ddots & \vdots \\ 0 & 0 & \cdots & 0 & I_{m\times m} & 0 \\ 0 & 0 & 0 & \cdots & 0 & I_{m\times m} \\ K_c & 0 & 0 & \cdots & 0 & 0 \end{bmatrix} = \begin{bmatrix} 0 & I_{(l_i-1)\times(l_i-1)} \otimes I_{m\times m} \\ K_c & 0 \end{bmatrix},
$$

$$
B_i = \begin{bmatrix} 0 \\ I_{m\times m} \end{bmatrix} \in \Re^{(l_i m)\times m}, \quad C_i = \begin{bmatrix} I_{m\times m} & 0 \end{bmatrix} \in \Re^{m\times(l_i m)},
$$

$\varphi_i \in \Re^{n_i - l_i m}$ is the vector of state variables associated with internal dynamics, and the internal dynamics of $\dot{\varphi}_i = \psi_i(t, \varphi_i, x_i)$ are assumed to be input-to-state stable. If $l_i - 1 = 0$, the corresponding rows and columns of $I_{(l_i-1)\times(l_i-1)} \otimes I_{m\times m}$ are empty, i.e., they are removed from matrix $A_i \in \Re^{(l_i m)\times(l_i m)}$.

Equation 5.8 is selected to be the canonical form for heterogenous systems because it has the following general yet useful properties:

(a) Except for the presence of K_c, matrix pair $\{A_i, B_i\}$ is in the controllable canonical form, matrix pair $\{A_i, C_i\}$ is in the observable canonical form; and hence pairs $\{-I + A_i, B_i\}$ and $\{-I + A_i, C_i\}$ are controllable and observable, respectively. Indeed, by defining a proper output vector, any stabilizable linear time-invariant system can be transformed (under a state transformation and a self-feedback control) into (5.8), and so can an input-output feedback linearizable non-linear (or linear) system with input-to-state stable internal dynamics.

(b) It is obvious that matrix A_i has the same properties of being row-stochastic and irreducible as matrix K_c. Thus, by Theorem 5.4, the ith system in (5.8) and with $u_i = 0$ is asymptotically cooperatively stable by itself.

(c) It is straightforward to show that, under the output-feedback set-point control $u_i = K_c(c_0 \mathbf{1} - y_i) = c_0 \mathbf{1} - K_c y_i$ for any fixed $c_0 \in \Re$, the linear portion in (5.8) is by itself asymptotically cooperatively stable as $x_i \to c_0 \mathbf{1}$.

The linear cooperative control problem to be addressed is to ensure that the collection of all the systems becomes asymptotically cooperatively stable. To solve this problem, cooperative controls need to be designed according to the information available, and the corresponding feedbacks are typically selective and intermittent, which represents one of main differences between cooperative control and standard control problems.

5.2.1 Matrix of Sensing and Communication Network

Unless mentioned otherwise, we consider the general case that the systems may operate by themselves for most of the time and that exchange of output information among the systems occurs only intermittently and selectively (according to their bandwidth, physical postures, distances, *etc.*). To focus upon the analysis and design of cooperative control, we choose to admit any sensing/communication protocol that enables networked information exchange.

Without loss of any generality, let infinite sequence $\{t_k : k \in \aleph\}$ denote the time instants at which the topology of sensing/communication network changes. Then, the following binary and piecewise-constant matrix $S(t)$, called *sensing/communication matrix*, captures the instantaneous topology of information exchange: $S(t) = S(k)$ for all $t \in [t_k, t_{k+1})$ and

$$S(t) = \begin{bmatrix} s_{11} & s_{12}(t) & \cdots & s_{1q}(t) \\ s_{21}(t) & s_{22} & \cdots & s_{2q}(t) \\ \vdots & \vdots & \vdots & \vdots \\ s_{q1}(t) & s_{q2}(t) & \cdots & s_{qq} \end{bmatrix} \tag{5.9}$$

where $s_{ii} \equiv 1$, $s_{ij}(t) = 1$ if the output of the jth dynamical system is known to the ith system at time t, and $s_{ij} = 0$ if otherwise. As the default setting, it is assumed that $0 < \underline{c}_t \leq t_{k+1}^s - t_k^s \leq \overline{c}_t < \infty$, where $\underline{c}_t$ and $\overline{c}_t$ are constant bounds. If $S(t)$ is constant or becomes so after some finite time, time sequence $\{t_k : k \in \aleph\}$ can always be chosen to be infinitely long with incremental upper bound $\overline{c}_t$ by letting $S(t_k)$ be identical. In the case that some topology changes occurred in certain time intervals of infinitesimal or zero length, these changes could not be detected or reacted by the physical systems, and hence they should be excluded from control design, and the subsequent analysis and design hold for the rest of topological changes.

Analysis and design of cooperative control do not require that time sequence $\{t_k : k \in \aleph\}$ or the corresponding changes of $S(t)$ be predictable or prescribed or known *a priori* or modeled in any way. Instead, the time sequence and the corresponding changes of $s_{ij}(t)$ (where $j = 1, \cdots, q$) in the ith row of matrix $S(t)$ are detectable instantaneously by (and only by) the ith system, and the cooperative control u_i reacts to the changes by taking the available information into its calculation. Specifically, the cooperative control should be of general form

$$u_i = U_i(s_{i1}(t)(y_1 - y_i), \cdots, s_{iq}(t)(y_q - y_i)), \tag{5.10}$$

which requires only a selective set of feedback on relative output measurements. In addition to topology changes, the sensing/comminucation network may also have latency, which is the subject of Section 6.6.1.

5.2.2 Linear Cooperative Control

A general class of linear relative-output-feedback cooperative controls are of form: for $i = 1, \cdots, q$,

$$u_i = \frac{1}{\sum\limits_{\eta=1}^{q} w_{i\eta}(t)s_{i\eta}(t)} \sum\limits_{j=1}^{q} w_{ij}(t)K_c[s_{ij}(t)(y_j - y_i)], \qquad (5.11)$$

where $w_{ij}(t) > 0$ are generally time-varying weights, $s_{ij}(t)$ are piecewise-constant as defined in (5.9), and K_c is the matrix contained in A_i. It follows that, if $s_{ij}(t) = 0$, u_i does not depend on y_j and hence is always implementable with all and only the available information. To facilitate the derivation of closed-loop dynamics, we rewrite Control 5.11 as

$$u_i = \sum\limits_{j=1}^{q} G_{ij}(t)y_j - K_c y_i \triangleq G_i(t)y - K_c y_i, \qquad (5.12)$$

where $n_i = l_i m$, $l = \sum_{i=1}^{q} l_i$, $n = \sum_{i=1}^{q} n_i = lm$, $y = [y_1^T \ \cdots \ y_q^T]^T \in \Re^{qm}$ is the overall output, $G_i(t) \triangleq [G_{i1}(t) \cdots G_{iq}(t)]$ is the feedback gain matrix, and its elements are defined by

$$G_{ij}(t) \triangleq \frac{w_{ij}(t)s_{ij}(t)}{\sum\limits_{\eta=1}^{q} w_{i\eta}(t)s_{i\eta}(t)} K_c.$$

Combining (5.8) and (5.11) for all i and exploring the special structure of matrices $\{A_i, B_i, C_i\}$ yields the following closed-loop and overall system:

$$\dot{x} = [-I + \overline{A} + BG(t)C]x = [-I + D(t)]x, \qquad (5.13)$$

where

$$\overline{A} = \mathrm{diag}\{(A_1 - B_1 K_c), \cdots, (A_q - B_q K_c)\} \in \Re^{n \times n},$$

$$B = \mathrm{diag}\{B_1, \cdots, B_q\} \in \Re^{n \times (mq)}, \quad C = \mathrm{diag}\{C_1, \cdots, C_q\} \in \Re^{(mq) \times n},$$

$$G(t) = [G_1^T(t) \cdots G_q^T(t)]^T \in \Re^{(mq) \times (mq)}, \text{ and } D(t) \in \Re^{n \times n} \text{ is given by}$$

$$D(t) = \begin{bmatrix} D_{11}(t) & \cdots & D_{1q}(t) \\ \vdots & \vdots & \vdots \\ D_{q1}(t) & \cdots & D_{qq}(t) \end{bmatrix}, \quad \begin{cases} D_{ii}(t) = \begin{bmatrix} 0 & I_{(l_i-1)\times(l_i-1)} \otimes I_{m\times m} \\ G_{ii}(t) & 0 \end{bmatrix}, \\ D_{ij}(t) = \begin{bmatrix} 0 & 0 \\ G_{ij}(t) & 0 \end{bmatrix}, \quad \text{if } i \neq j. \end{cases}$$

$$(5.14)$$

It is obvious that matrix $D(t)$ is piecewise-constant (as matrix $S(t)$) and also row-stochastic. The following lemma establishes the equivalence of structural properties among matrices $S(t)$, $G(t)$ and $D(t)$. Combining the lemma with Theorem 5.4 yields Theorem 5.6.

Lemma 5.5. *The following statements are true:*

(a) Consider matrix $E \in \Re_+^{(qm) \times (qm)}$ with sub-blocks $E_{ij} \in \Re_+^{m \times m}$. Then, matrix

$$\overline{E} = \begin{bmatrix} 0 & I_{m \times m} & 0 & \cdots & 0 \\ E_{11} & 0 & E_{12} & \cdots & E_{1q} \\ \vdots & \vdots & \vdots & & \vdots \\ E_{q1} & 0 & E_{q2} & \cdots & E_{qq} \end{bmatrix}$$

is irreducible (or lower triangularly complete in its canonical form) if and only if E is irreducible (or lower triangularly complete in its canonical form).

(b) Given matrices $S \in \Re_+^{q \times q}$ and $F \in \Re_+^{m \times m}$ with F being irreducible and row-stochastic, matrix $S \otimes F$ is irreducible (or lower triangularly complete in its canonical form) if and only if S is irreducible (or lower triangularly complete in its canonical form).

(c) Matrix $D(t)$ in (5.14) is irreducible (or lower triangularly complete) if and only if $S(t)$ is irreducible (or lower triangularly complete).

Proof:

(a) To show that $\overline{E}$ is irreducible if E is irreducible, consider vectors $z_1, \alpha \in \Re_+^m$, $z_2 \in \Re_+^{(q-1)m}$, $z = \begin{bmatrix} z_1^T & z_2^T \end{bmatrix}^T$, $\overline{z} = \begin{bmatrix} z_1^T & \alpha^T & z_2^T \end{bmatrix}^T$, and $z' = \begin{bmatrix} \alpha^T & z_2^T \end{bmatrix}^T$. It follows that, for any $\gamma \geq 1$, inequality $\gamma \overline{z} \geq \overline{E} \overline{z}$ is equivalent to

$$\gamma z_1 \geq \alpha, \quad \gamma z' \geq Ez, \tag{5.15}$$

which implies that

$$\gamma^2 z \geq \gamma z' \geq Ez.$$

If E is irreducible, we know from Corollary 4.3 and from the above inequality that $z > 0$ and $Ez > 0$. It follows from (5.15) that $\overline{z} > 0$ and, by Corollary 4.3, $\overline{E}$ is irreducible.

On the other hand, if E is reducible, there is non-zero and non-positive vector z such that $\gamma z \geq Ez$ for some $\gamma \geq 1$. Inequality $\gamma z \geq Ez$ implies $\gamma \overline{z}' \geq \overline{E} \overline{z}'$, where vector $\overline{z}' = \begin{bmatrix} z_1^T & z_1^T & z_2^T \end{bmatrix}^T$ is also non-zero and non-positive. Hence, by Corollary 4.3, $\overline{E}$ must be reducible.

Matrix $\overline{E}$ can be permutated into

$$\overline{E}' = \begin{bmatrix} 0 & E_{11} & E_{12} & \cdots & E_{1q} \\ I_{m \times m} & 0 & 0 & \cdots & 0 \\ 0 & E_{21} & E_{22} & \cdots & E_{2q} \\ \vdots & \vdots & \vdots & & \vdots \\ 0 & E_{q1} & E_{q2} & \cdots & E_{qq} \end{bmatrix}.$$

It is apparent that, if $\overline{E}$ is permutated to be lower triangular (either complete or incomplete), $\overline{E}'$ has the same property.

(b) The proof is parallel to that of (a) but uses the facts that, for any $\bar{z} \in \Re_+^{mq}$, inequality $z' \otimes \mathbf{1}_m \leq \bar{z} \leq z'' \otimes \mathbf{1}_m$ holds for some $z', z'' \in \Re_+^q$ and that, for any $z \in \Re_+^q$, $\gamma(z \otimes \mathbf{1}_m) \geq (S \otimes F)(z \otimes \mathbf{1}_m)$ holds if and only if $\gamma z \geq S z$.

(c) Matrix $G(t)$ in (5.13) and (5.12) has the property that

$$\mathcal{B}(G(t)) = \mathcal{B}(S(t) \otimes K_c),$$

where $\mathcal{B}(\cdot)$ is defined in (4.22). The conclusion can be made by invoking (b) and then by using (a) inductively with respect to relative degrees of l_i and by applying appropriate permutations.

$\square$

Theorem 5.6. *System 5.8 is uniformly asymptotically cooperatively stable under Cooperative Control 5.11 if and only if the sensing/communication matrix $S(t)$ in (5.9) has the property that $\{S(t_k) : k \in \aleph\}$ is uniformly sequentially complete.*

5.2.3 Conditions of Cooperative Controllability

The analysis leading to Theorem 5.6 illustrates that, for a group of heterogeneous linear systems, their cooperative controllability calls for two conditions: (a) the individualized property that each and every system in the group is input-output dynamically feedback linearizable and has input-to-state stable internal dynamics, and (b) the network property that $\{S(t_k) : k \in \aleph\}$ is sequentially complete.

The requirement of individual systems being dynamic feedback linearizable will be relaxed in Chapter 6. In many applications, outputs of the systems are physically determined, and a cooperative behavior of system outputs is desired. In these cases, the following output cooperative stability should be pursued.

Definition 5.7. *Systems of (5.16) are said to be asymptotically output cooperatively stable if all the states are Lyapunov stable and if $\lim_{t \to \infty} y(t) = c\mathbf{1}$.*

To show sufficiency of feedback linearization and to demonstrate the flexibility in choosing self-feedback controls for achieving output cooperative stability, consider the ith system in its original state equation as

$$\dot{z}_i = F_i(z_i) + B_i(z_i) w_i, \quad y_i = h(z_i), \tag{5.16}$$

where $z_i \in \Re^{n_i}$ and $y_i, w_i \in \Re^m$. Since the system is input-output feedback linearizable, there exists a transformation $w_i = \alpha_i(z_i) + \beta_i(z_i) u_i$ such that

$$y_i^{(l_i)} = -K_i \left[y_i^T \cdots \left(y_i^{(l_i-1)} \right)^T \right]^T + u_i.$$

Gain matrix K_i can be chosen to match the above equation with the canonical form of (5.8). For output cooperative stability, there are many choices of K_i and u_i, and they are summarized into the following theorem.

Theorem 5.8. *Systems in the form of (5.16) are uniformly asymptotically output cooperatively stable under Cooperative Control 5.11 if the following conditions hold:*

(a) There exists a mapping $w_i = \alpha_i(z_i) + \beta_i(z_i)u_i$ such that the dynamics from output y_i to input u_i are linear and positive and that, under the control $u_i = K_c(\mathbf{c1} - y_i)$, y_i converges to the steady state of $\mathbf{c1}$.

(b) Sensing/communication matrix $\{S(t_k) : k \in \aleph\}$ is uniformly sequentially complete.

As shown in Section 4.4.4, the network condition of $\{S(t_k) : k \in \aleph\}$ being uniformly sequentially complete is both sufficient and necessary for achieving uniform asymptotic cooperative stability. A simple way of monitoring uniform sequential completeness is to compute the binary products of sensor/communication matrices sequence $S(t_k)$ over certain consecutive time intervals. That is, for any sub-sequence $\{k_\eta : \eta \in \aleph\}$, the cumulative effect of network topology changes is determined by the cumulative exchange of information, and the latter over time interval $[t_{k_\eta}, t_{k_{\eta+1}})$ is described by the composite matrix

$$S_\Lambda(\eta) \triangleq S(t_{k_{\eta+1}-1}) \bigwedge S(t_{k_{\eta+1}-2}) \bigwedge \cdots \bigwedge S(t_{k_\eta}), \qquad (5.17)$$

where $\bigwedge$ denotes the operation of generating a binary product of two binary matrices. Hence, sequence $S(t_k)$ is sequentially complete if $S_\Lambda(\eta)$ is lower triangularly complete for all $\eta \in \aleph$, that is, there exists a permutation matrix $\mathcal{T}(\eta)$ under which

$$\mathcal{T}^T(\eta)S_\Lambda(\eta)\mathcal{T}(\eta) = \begin{bmatrix} S'_{\Lambda,11}(\eta) & 0 & \cdots & 0 \\ S'_{\Lambda,21}(\eta) & S'_{\Lambda,22}(\eta) & \cdots & 0 \\ \vdots & \vdots & \ddots & \vdots \\ S'_{\Lambda,p1}(\eta) & S'_{\Lambda,p2}(\eta) & \cdots & S'_{\Lambda,pp}(\eta) \end{bmatrix}, \qquad (5.18)$$

where $p(\eta) > 0$ is an integer, $S'_{\Lambda,ii}(\eta)$ are square and irreducible, and $S'_{\Lambda,ij}(\eta) \neq 0$ for some $j < i$. Once k_η is known, $k_{\eta+1}$ should inductively be found such that lower triangular completeness of (5.18) holds. If the composite matrix in (5.18) would be observable real-time, sensing/communication protocols could be devised and implemented to allocate resources, to improve information exchange, and to achieve the lower triangular completeness. Of course, such a monitoring and optimization scheme would require more than a truly decentralized, local-information-sharing, and memoryless network. The following examples provide and illustrate sequentially complete sensor/communication sequences.

Example 5.9. Consider a sensor/communication sequence $\{S(t_k), k \in \aleph\}$ defined by $S(t_k) = S_1$ for $k = 4\eta + 3$, $S(t_k) = S_2$ for $k = 4\eta + 2$, $S(t_k) = S_3$ for $k = 4\eta + 1$, and $S(t_k) = S_4$ for $k = 4\eta$, where $\eta \in \aleph$,

$$S_1 = \begin{bmatrix} 1 & 0 & 0 \\ 1 & 1 & 0 \\ 0 & 0 & 1 \end{bmatrix}, \quad S_2 = \begin{bmatrix} 1 & 1 & 0 \\ 0 & 1 & 0 \\ 0 & 0 & 1 \end{bmatrix},$$

$$S_3 = \begin{bmatrix} 1 & 0 & 0 \\ 0 & 1 & 0 \\ 1 & 0 & 1 \end{bmatrix}, \quad \text{and} \quad S_4 = \begin{bmatrix} 1 & 0 & 0 \\ 0 & 1 & 0 \\ 0 & 0 & 1 \end{bmatrix}. \tag{5.19}$$

It follows that

$$S_\Lambda(\eta) \triangleq S_1 \bigwedge S_2 \bigwedge S_3 \bigwedge S_4 = \begin{bmatrix} 1 & 1 & 0 \\ 1 & 1 & 0 \\ 1 & 0 & 1 \end{bmatrix} \triangleq \begin{bmatrix} S'_{\Lambda,11} & \emptyset \\ S'_{\Lambda,21} & 1 \end{bmatrix},$$

from which sequential completeness of sequence $\{S(t_k), k \in \aleph\}$ becomes apparent.

Similarly, it can be shown that every sequence generated from patterns S_i is sequentially complete as long as the sequence contains infinite entries of all the patterns of S_1, S_2, S_3 and S_4. Thus, for the purpose of simulations and verification, one can generate a sequentially complete sequence by randomly selecting the next time instant t_k and then by randomly choosing $S(t_k)$ from S_1 up to S_4. Unless mentioned otherwise, the following choices are made in all the subsequent simulation studies: $t_{k+1} - t_k = 1$ is set, and a random integer is generated under uniform distribution to select $S(t_k)$ from S_1 up to S_4. That is, the sensing/communication sequences generated in the simulations have probability 1 to be uniformly sequentially complete. ◇

Example 5.10. A lower triangularly complete composite sensor/communication matrix $S_\Lambda(\cdot)$ may arise from lower triangularly incomplete matrices $S(t_k)$ of different block sizes. For example, consider sequence $\{S(t_k) : k \in \aleph\}$ defined by $S(t_k) = S_1$ if $k = 2\eta$ and $S(t_k) = S_2$ if $k = 2\eta + 1$, where $\eta \in \aleph$,

$$S_1 = \begin{bmatrix} 1 & 0 & 0 & 0 & 0 \\ 1 & 1 & 1 & 0 & 0 \\ 0 & 1 & 1 & 0 & 0 \\ 0 & 0 & 0 & 1 & 0 \\ 0 & 0 & 0 & 0 & 1 \end{bmatrix}, \quad \text{and} \quad S_2 = \begin{bmatrix} 1 & 0 & 0 & 0 & 0 \\ 0 & 1 & 1 & 0 & 0 \\ 0 & 0 & 1 & 1 & 0 \\ 0 & 0 & 0 & 1 & 1 \\ 0 & 0 & 1 & 0 & 1 \end{bmatrix}.$$

It follows that

$$S_\Lambda(\eta) \triangleq S_2 \bigwedge S_1 = \begin{bmatrix} 1 & 0 & 0 & 0 & 0 \\ 1 & 1 & 1 & 0 & 0 \\ 0 & 1 & 1 & 1 & 0 \\ 0 & 0 & 0 & 1 & 1 \\ 0 & 1 & 1 & 0 & 1 \end{bmatrix},$$

which implies $\{S_\Lambda(\eta) : \eta \in \aleph\}$ is sequentially lower-triangularly complete. ◇

Example 5.11. A lower triangularly complete composite sensor/communication matrix $S_\Lambda(\cdot)$ may also arise from lower triangularly incomplete matrices $S(t_k)$

whose canonical forms require different permutation matrices. For instance, consider sequence $\{S(t_k) : k \in \aleph\}$ defined by $S(t_k) = S_1$ if $k = 3\eta$, $S(t_k) = S_2$ if $k = 3\eta + 1$, and $S(t_k) = S_3$ if $k = 3\eta + 2$, where $\eta \in \aleph$,

$$S_1 = \begin{bmatrix} 1 & 0 & 0 & 0 & 0 \\ 0 & 1 & 0 & 0 & 0 \\ 0 & 0 & 1 & 0 & 0 \\ 0 & 1 & 1 & 1 & 0 \\ 0 & 0 & 0 & 0 & 1 \end{bmatrix}, \quad S_2 = \begin{bmatrix} 1 & 0 & 1 & 0 & 0 \\ 0 & 1 & 0 & 1 & 0 \\ 1 & 0 & 1 & 0 & 0 \\ 0 & 1 & 0 & 1 & 0 \\ 0 & 0 & 0 & 0 & 1 \end{bmatrix},$$

and

$$S_3 = \begin{bmatrix} 1 & 0 & 0 & 0 & 0 \\ 0 & 1 & 0 & 0 & 0 \\ 0 & 0 & 1 & 0 & 0 \\ 0 & 0 & 0 & 1 & 0 \\ 0 & 0 & 1 & 0 & 1 \end{bmatrix}.$$

It is apparent that matrices S_1, S_2 and S_3 are all reducible and lower triangularly incomplete. Their canonical forms are of different sizes on the diagonal and require different permutation matrices: $T_s(t_{3\eta}) = T_s(t_{3\eta+2}) = I$, and

$$T_s(t_{3\eta+1}) \overset{\triangle}{=} \begin{bmatrix} 0 & 1 & 0 & 0 & 0 \\ 0 & 0 & 0 & 1 & 0 \\ 1 & 0 & 0 & 0 & 0 \\ 0 & 0 & 1 & 0 & 0 \\ 0 & 0 & 0 & 0 & 1 \end{bmatrix}.$$

Nonetheless, sequence $\{S(k), k \in \aleph\}$ is sequentially complete because an infinite lower-triangularly complete sub-sequence $\{S_\Lambda(\eta) : \eta \in \aleph\}$ can be constructed as $S_\Lambda(\eta) \overset{\triangle}{=} S(t_{3\eta+2}) \bigwedge S(t_{3\eta+1}) \bigwedge S(t_{3\eta}) = S_3 \bigwedge S_2 \bigwedge S_1$ and its canonical lower triangular form is

$$T_\Lambda^T S_\Lambda(\eta) T_\Lambda = \begin{bmatrix} 1 & 1 & 0 & 0 & 0 \\ 1 & 1 & 0 & 0 & 0 \\ 1 & 0 & 1 & 1 & 0 \\ 1 & 0 & 1 & 1 & 0 \\ 1 & 1 & 0 & 0 & 1 \end{bmatrix},$$

where $T_\Lambda = T_s(t_{3\eta+1})$. $\diamond$

5.2.4 Discrete Cooperative System

Parallel to the study of Continuous System 5.1, discrete-time cooperative system

$$x(k+1) = P_k x(k), \quad x_0 \text{ given}, \quad k \geq 0, \tag{5.20}$$

can be investigated, the following theorem is a restatement of Theorem 4.53, and Theorem 5.12 corresponds to Theorem 5.4.

Theorem 5.12. *System 5.20 is uniformly asymptotically cooperatively stable if and only if $\{P_k : k \in \aleph\}$ is uniformly sequentially complete and uniformly diagonally positive.*

For discrete-time heterogeneous systems, their canonical form is given by

$$x_i(k+1) = [c_d I + (1 - c_d)A_i]x_i(k) + (1 - c_d)B_i u_i(k), \quad y_i(k) = C_i x_i(k),$$
$$\varphi_i(k+1) = g_i(k, \varphi_i(k), x_i(k)),$$

$$(5.21)$$

where A_i, B_i, C_i are the same as those in (5.8), and $0 < c_d < 1$ is a design constant. Then, under the same cooperative control $u_i(k)$ given by (5.11) except that t is replaced by k everywhere therein, the overall closed-loop system becomes (5.20), where $P_k = c_d I + (1 - c_d)D(k)$ and matrix $D(k)$ is structurally the same as $D(t)$ in (5.13). Thus, applying Lemma 5.5 and Theorem 5.12 yields Theorem 5.13 (which corresponds to Theorem 5.6).

Theorem 5.13. *System 5.21 is uniformly asymptotically cooperatively stable under Cooperative Control 5.11 (after replacing t with k) if and only if the corresponding sensing/communication matrix $\{S(k) : k \in \aleph\}$ is uniformly sequentially complete.*

5.3 Applications of Cooperative Control

In this section, Cooperative Control 5.11 is applied to several typical problems and its performance is illustrated by simulation results. Unless stated otherwise, topology changes in the simulations are chosen according to Example 5.9, *i.e.*, the sensing/communication matrix is randomly switched among the seed matrices in (5.19). It is assumed that, by utilizing the hierarchical control structure described in Section 1.4, individual systems have been mapped into the canonical form of (5.8) for the purpose of cooperative control design.

5.3.1 Consensus Problem

The *consensus* problem is to ensure that $x(t) \to c\mathbf{1}$ for some $c \in \Re$, it is the final outcome of asymptotic cooperative stability, and hence Cooperative Control 5.11 can directly be applied. As an extension of the simple alignment problem in Section 5.1.3, consider that dynamics of three agents are described by (5.8) with $m = 2$ and $l_i = 1$. Then, the consensus problem is to align both velocity components along the two primary axes in the two-dimensional space. Cooperative Control 5.11 is simulated with the choices of $w_{ij} = 1$ and

$$K_c = \begin{bmatrix} 0 & 1 \\ 1 & 0 \end{bmatrix},$$

which is irreducible and row-stochastic. Consensus of the velocity components is shown in Fig. 5.2, where the initial velocities of the three agents are set to be $[0.2 \ -0.3]^T$, $[0.6 \ 0.2]^T$, and $[-0.3 \ 0.5]^T$, respectively.

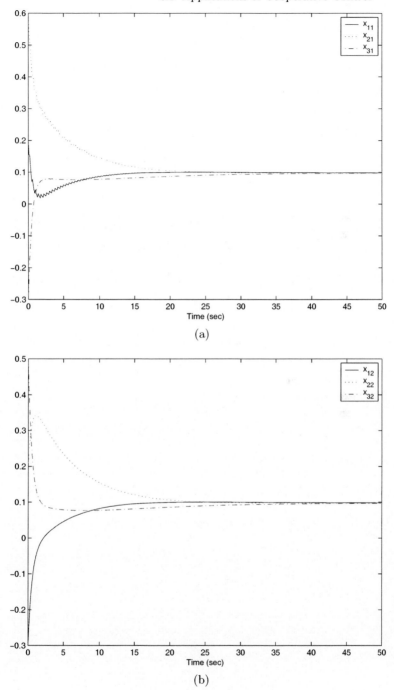

Fig. 5.2. Velocity consensus

5.3.2 Rendezvous Problem and Vector Consensus

Vector consensus is to ensure that $x \to \mathbf{1} \otimes c$ for some vector $c \in \Re^m$. The simplest approach to achieve the behavior of vector consensus is to decouple the output channels of the systems by setting $K_c = I_{m \times m}$ in which case, by Theorem 5.6, asymptotic cooperative stability is concluded per output channel. For example, let us relax the velocity consensus problem in Section 5.3.1 so that velocities v_{x_i} along the x axis reach a consensus for all i while v_{y_i} along the y axis also have a (different) consensus. This relaxation gives us back the alignment problem (as their orientation angles $\theta_i = v_{x_i}/v_{y_i}$ have a consensus) but without assuming a constant linear absolute velocity. In this case, we can keep all the settings as before except for $K_c = I_{2 \times 2}$, and the simulation results are shown in Fig. 5.3.

The *rendezvous* problem is to ensure that all the agents converge to the same point in either 2-D or 3-D space. This goal can be achieved using either consensus or vector consensus. For instance, we can apply the same cooperative controls (used in the aforementioned problems of velocity consensus and alignment) to the position alignment problem, and the simulation results are shown in Fig. 5.4 for initial positions of $[8 \ 1]^T$, $[1 \ 6]^T$, and $[4 \ -1]^T$, respectively. Specifically, Fig. 5.4(a) shows consensus of all the coordinates under the single-consensus cooperative control, while Fig. 5.4(b) shows the general case of rendezvous under the vector-consensus cooperative control.

5.3.3 Hands-off Operator and Virtual Leader

A cooperative system described by (5.1) is autonomous as a group (or groups). To enable the interaction between a hands-off operator and the cooperative system, we can model the operator as a virtual leader. Communication between the virtual leader and the systems in the cooperative system can also be intermittent and local, depending upon availability of the operator and the means of communication. Thus, we can introduce the following augmented sensor/communication matrix and its associated time sequence $\{\bar{t}_k : k \in \aleph\}$ as: $\overline{S}(t) = \overline{S}(\bar{t}_k)$ for all $t \in [\bar{t}_k, \bar{t}_{k+1})$ and

$$
\overline{S}(t) = \begin{bmatrix} 1 & s_{01}(t) \cdots s_{0q}(t) \\ s_{10}(t) & \\ \vdots & S(t) \\ s_{q0}(t) & \end{bmatrix} \in \Re^{(q+1) \times (q+1)}, \tag{5.22}
$$

where $s_{0i}(t) = 1$ if x_i is known to the operator at time t and $s_{0i}(t) = 0$ if otherwise, and $s_{i0}(t) = 1$ if command $y_0(t)$ is received by the ith system at time t and $s_{i0}(t) = 0$ if otherwise. Accordingly, cooperative control is modified from (5.11) to the following version: for $i = 1, \cdots, q$,

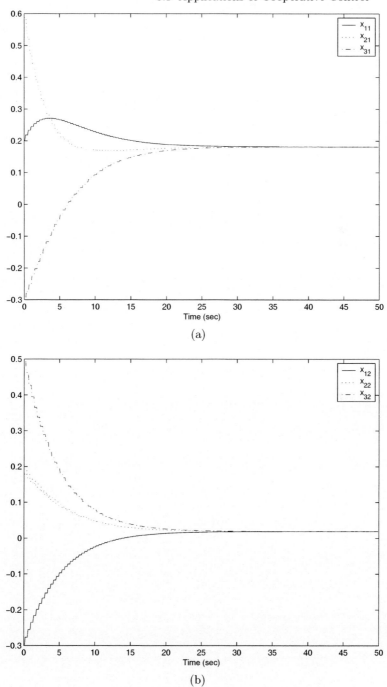

Fig. 5.3. Vector consensus: convergence per channel

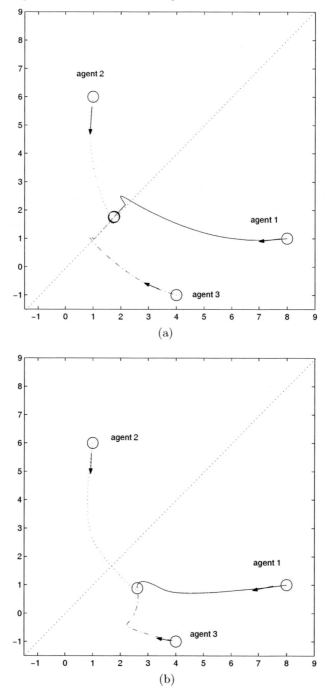

Fig. 5.4. Rendezvous under cooperative controls

$$u_i(t) = \frac{1}{\sum\limits_{\eta=0}^{q} w_{i\eta} s_{i\eta}(t)} \sum_{j=0}^{q} w_{ij} K_c [s_{ij}(t)(y_j - y_i)]. \tag{5.23}$$

Evolution of operator's command can be modeled as

$$\dot{y}_0 = \left(1 - \sum_{i=1}^{q} \epsilon_i s_{0i}(t)\right)(u_0 - y_0) + \sum_{i=1}^{q} \epsilon_i s_{0i}(t)(y_i - y_0), \tag{5.24}$$

where $y_0(t) \in \Re^m$, $\epsilon_i \geq 0$ are constants with $\sum_{i=1}^{q} \epsilon_i \leq 1$, and u_0 is a constant (or piecewise-constant) vector specifying the *desired cooperative behavior*. It will be shown in Section 5.8 that, under sequential completeness of $\{\overline{S}(\bar{t}_k)\}$, a single consensus behavior can be achieved under the choices of $u_0 = c_0^d \mathbf{1}$ for some $c_0^d \in \Re$ and K_c being irreducible, and vector consensus behaviors can be ensured under $u_0 = c \in \Re^m$ and $K_c = I$. In general, both u_0 and initial condition of $y_0(t_0)$ affect the cooperative behavior(s).

Virtual Leader 5.25 generates filtered terms of y_i, and hence Cooperative Control 5.23 is a combination of Cooperative Control 5.11 and its filtered version. To see analogous derivations in standard control theory, consider scalar system

$$\dot{z} = a + v,$$

where a is an unknown constant, and its adaptive control with a leakage adaptation law [172] is

$$v = -z + \hat{a}, \quad \dot{\hat{a}} = -\hat{a} + z.$$

Hence, Cooperative Control 5.23 can be viewed as an *adaptive cooperative control* or *integral cooperative control* based on (5.11).

In the special case of $\epsilon_i = 0$ for all i and $y_0(t_0) = u_0$, Dynamics 5.24 of the virtual leader reduce to

$$\dot{y}_0 = 0, \tag{5.25}$$

with which the overall Sensing/communication Matrix 5.22 becomes

$$\overline{S}(t) = \begin{bmatrix} 1 & 0 \cdots 0 \\ s_{10}(t) & \\ \vdots & S(t) \\ s_{q0}(t) & \end{bmatrix}. \tag{5.26}$$

In this case, we know from Theorem 5.6 that, if $\{\overline{S}(\bar{t}_k)\}$ is sequentially completed, the desired cooperative behavior of $y_i \to u_0$ is achieved. Without the virtual leader, the cooperative behavior of $y_i = y_j$ can be achieved but not necessarily the desired behavior of $y_i = u_0$. While $y_i \to u_0$ can be cast as a standard control problem (as $(y_i - u_0) \to 0$), the cooperative control design

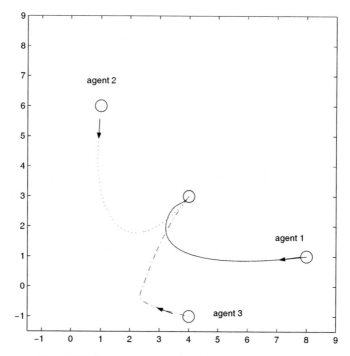

Fig. 5.5. Rendezvous to the desired target position

framework makes it possible to handle both intermittent delivery of command (or the reference input) and intermittent feedback.

As an example, reconsider the rendezvous problem in Section 5.3.2 and assume that the desired target position is $[4 \quad 3]^T$ and that the first system receives the target information from the virtual leader once a while. The corresponding sensor/communiction matrix is given by (5.26) where $S(t)$ is the same as before, s_{10} is a binary value randomly assigned over time, and $s_{20} = s_{30} = 0$. Under the same initial conditions, Cooperative Control 5.23 with $w_{ij} = 0.9$ for $i \neq j$ and $w_{ii} = 0.1$ is simulated together with Virtual Leader 5.25, and the result in Fig. 5.5 demonstrates convergence to the desired target position.

In addition to commanding the desired cooperative behavior and ensuring its convergence, the virtual leader can also play critical roles in enhancing robustness and maintaining network connectivity. These topics will be discussed in Sections 5.7 and 5.4, respectively.

5.3.4 Formation Control

To illustrate the options in designing cooperative formation control, consider without loss of any generality the second-order dynamics: $i = 1, \cdots, q,$

$$\ddot{\psi}_i = v_i,$$

where $\psi_i \in \Re^m$ is the output, and v_i is the control. Suppose that the formation is presented by the following vertices in the moving frame $\mathcal{F}^d(t)$ that moves along a contour[2] described by time function $\psi^d(t)$:

$$p_i^d(t) = \sum_{j=1}^{m} \alpha_{ij} e_j(t), \quad i = 1, \cdots, q,$$

where basis vectors e_j are calculated from $\psi^d(t)$ and according to the discussions in Section 3.4.1. As illustrated in Section 1.4, we can define individual behavior $y_i = \psi_i - \psi^d(t) - p_i^d$ and choose vehicle-level control to be

$$v_i = u_i + \ddot{\psi}^d(t) + \ddot{p}_i^d - \ddot{y}_i = u_i - 2\dot{\psi}_i - 2\dot{\psi}^d(t) + \ddot{\psi}^d(t) + \ddot{p}_i^d - 2\dot{p}_i^d, \quad (5.27)$$

under which the systems are mapped into the canonical form of (5.8) with $l_i = 2$ and $K_c = 1$:

$$\dot{y}_i = -y_i + z_i, \quad \dot{z}_i = -z_i + y_i + u_i.$$

In this case, the cooperative control to achieve and maintain the formation is given by (5.11), that is,

$$u_i = \frac{1}{s_{i1}(t) + \cdots + s_{iq}(t)} \sum_{l=1}^{q} s_{il}(t)(y_l - y_i)$$

$$= -\psi_i + \frac{1}{s_{i1}(t) + \cdots + s_{iq}(t)} \sum_{l=1}^{q} s_{il}(t) \left(\psi_l + \sum_{j=1}^{m} (\alpha_{ij} - \alpha_{lj}) e_j(t) \right) \quad (5.28)$$

$$\triangleq -\psi_i + \hat{\psi}_i^d,$$

in which $\hat{\psi}_i^d$ is in the form of (3.99). Nonetheless, compared to the tracking formation control design in Section 3.4.1, the above cooperative formation control design of v_i does not requires time derivatives of $\hat{\psi}_i^d$, and hence it only involves dynamics of the ith system but not dynamics of any other systems.

As an example, consider the case of three agents whose formation contour is given by $\dot{\psi}^d(t) = [0.2 \quad 0.2]^T$ and whose formation vertices in the moving frame are $p_1^d = [\frac{1}{\sqrt{2}} \quad \frac{1}{\sqrt{2}}]^T$, $p_2^d = [-\frac{1}{\sqrt{2}} \quad \frac{1}{\sqrt{2}}]^T$, and $p_3^d = [\frac{1}{\sqrt{2}} \quad -\frac{1}{\sqrt{2}}]^T$. Simulation is done under the randomly-switched sensing/communication sequence specified earlier in this section and for the initial positions $\psi_1(0) = [4 \quad 2.5]^T$, $\psi_2(0) = [5 \quad 2]^T$ and $\psi_3(0) = [3 \quad 1]^T$. The simulation results in Fig. 5.6 verify performance of cooperative formation control.

---

[2] The formation is supposed to move along the contour or in the phase portrait of $\psi^d(t)$, but it may not necessarily track $\psi^d(t)$ closely. If $\psi^d(t)$ is to be tracked, a virtual leader can be introduced as described in Section 5.3.3.

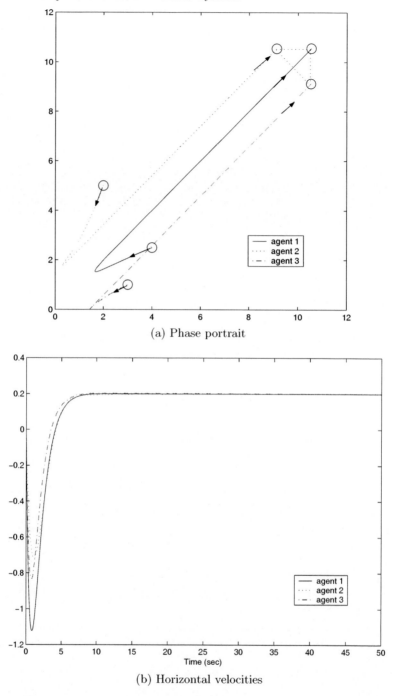

(a) Phase portrait

(b) Horizontal velocities

Fig. 5.6. Performance under Cooperative Formation Control 5.28

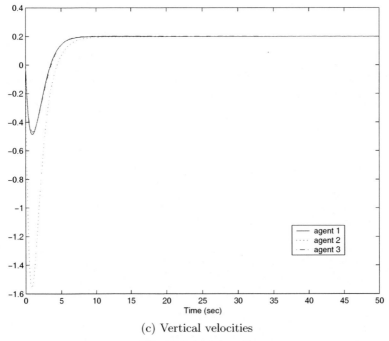

(c) Vertical velocities

Fig. 5.6 (continued)

To reduce further the information required for formation control, consider the case that $p_i^d(t)$ is piecewise-constant[3] and that $\psi^d(t)$ is not available to all the vehicles. In this case, the formation can be decomposed into geometrical relationship with respect to any neighboring vehicle l (with the virtual leader as vehicle 0), $\psi^d(t)$ is dropped explicitly from the individual behavior (but instead is injected through the sensing/communication network by the virtual leader), and hence the vehicle-level control in (5.27) reduces to

$$v_i = u_i - 2\dot{\psi}_i, \quad p_i^d = \sum_{j=1}^{m} \alpha_{ij} e_j(t_k) \approx \sum_{j=1}^{m} \alpha'_{ilj} e'_{ilj}(t_k),$$

where e'_{ilj} are the orthonormal basis vectors of moving frame $\mathcal{F}_l(t)$ at the lth vehicle, α'_{ilj} are the coordinates of the desired location for the ith vehicle with respect to the desired formation frame $\mathcal{F}_l^d(t)$. That is, whenever $s_{il}(t) = 1$, the ith system can observe (or receive information from) the lth system about its position $\psi_l(t)$ (or simply $[\psi_l(t) - \psi_i(t)]$) and its current heading basis vectors $\hat{e}'_{ilj}(t_k)$ (which can also be estimated if not directly observed). Hence, Cooperative Formation Control 5.28 can be modified to be

---

[3] This will be relaxed in Section 5.8 by using the design of dynamical cooperative control.

$$u_i = \frac{1}{s_{i0}(t) + s_{i1}(t) + \cdots + s_{iq}(t)} \sum_{l=0}^{q} s_{il}(t) \left(\psi_l - \psi_i + \sum_{j=1}^{m} \alpha'_{ilj} \hat{e}'_{ilj}(t) \right),$$

(5.29)

which is of form (3.102).

To illustrate the above estimation-based formation control, let us consider again the case of three entities whose formation motion contour is given by $\psi^d(t) = [0.2 \ 0.2]^T$ with formation vertices in the moving frame with respect to $\psi^d(t)$ being $[1 \ 0]^T$, $[0 \ 1]^T$, and $[0 \ -1]^T$, respectively. The desired formation can be decomposed into desired geometric relationship between any pair of two entities. For instance, consider the neighboring structure specified by the following augmented sensing/communication matrix

$$\overline{S} = \begin{bmatrix} 1 & 0 & 0 & 0 \\ 1 & 1 & 0 & 0 \\ 0 & 1 & 1 & 0 \\ 0 & 1 & 0 & 1 \end{bmatrix}.$$

It follows that relative coordinates of α'_{ilj} and current heading basis vectors of the neighbors are

$$\begin{cases} \alpha'_{101} = 1, \ \alpha'_{102} = 0 \\ \hat{e}'_{101} = e'_{101} = \begin{bmatrix} \frac{1}{\sqrt{2}} \\ \frac{1}{\sqrt{2}} \end{bmatrix} \\ \hat{e}'_{102} = e'_{102} = \begin{bmatrix} -\frac{1}{\sqrt{2}} \\ \frac{1}{\sqrt{2}} \end{bmatrix} \end{cases}, \qquad \begin{cases} \alpha'_{211} = -1, \ \alpha'_{212} = 1 \\ \hat{e}'_{211} = \begin{bmatrix} \frac{\dot\psi_{11}}{\sqrt{\dot\psi_{11}^2 + \dot\psi_{12}^2}} \\ \frac{\dot\psi_{12}}{\sqrt{\dot\psi_{11}^2 + \dot\psi_{12}^2}} \end{bmatrix} \\ \hat{e}'_{212} = \begin{bmatrix} -\frac{\dot\psi_{12}}{\sqrt{\dot\psi_{11}^2 + \dot\psi_{12}^2}} \\ \frac{\dot\psi_{11}}{\sqrt{\dot\psi_{11}^2 + \dot\psi_{12}^2}} \end{bmatrix} \end{cases},$$

and

$$\begin{cases} \alpha'_{311} = -1, \ \alpha'_{312} = -1 \\ \hat{e}'_{311} = \begin{bmatrix} \frac{\dot\psi_{11}}{\sqrt{\dot\psi_{11}^2 + \dot\psi_{12}^2}} \\ \frac{\dot\psi_{12}}{\sqrt{\dot\psi_{11}^2 + \dot\psi_{12}^2}} \end{bmatrix} \\ \hat{e}'_{312} = \begin{bmatrix} -\frac{\dot\psi_{12}}{\sqrt{\dot\psi_{11}^2 + \dot\psi_{12}^2}} \\ \frac{\dot\psi_{11}}{\sqrt{\dot\psi_{11}^2 + \dot\psi_{12}^2}} \end{bmatrix} \end{cases},$$

where $\dot\psi_{11}$ and $\dot\psi_{12}$ are the two elements of vector $\dot\psi_1$ (and they can also be estimated online). The corresponding simulation results are provided in Fig. 5.7, and they have similar performance as those in Fig. 5.6. For the general case of changing topologies, it is straightforward to derive the expressions of α'_{ilj} and $\hat{e}'_{ilj}(t)$ for all the pairs of (i, l), based on which Cooperative Formation Control 5.29 can be implemented.

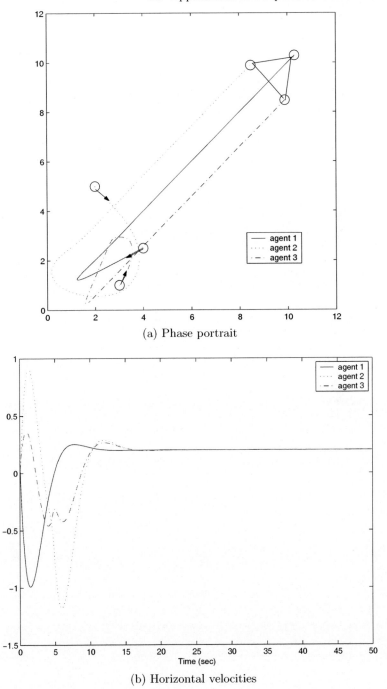

(a) Phase portrait

(b) Horizontal velocities

Fig. 5.7. Performance under Cooperative Formation Control 5.29

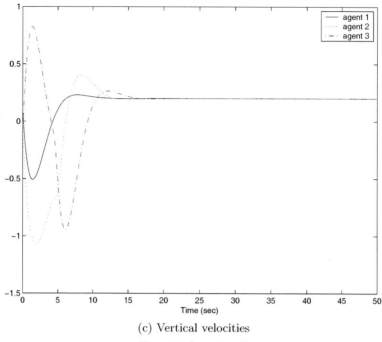

(c) Vertical velocities

Fig. 5.7 (continued)

5.3.5 Synchronization and Stabilization of Dynamical Systems

Closely related to vector consensus is the so-called synchronization problem. Consider the following linear systems of identical dynamics: for $i = 1, \cdots, q$,

$$\dot{x}_i = Ax_i + Bu_i, \tag{5.30}$$

where $x_i \in \Re^n$ and $u_i \in \Re^m$. Then, the systems are said to be *synchronized* if $\|x_i(t) - x_j(t)\| \to 0$ as $t \to \infty$.

A time-invariant synchronizing control is

$$u_i = K \sum_{j=1}^{q} g_{ij}(x_j - x_i),$$

where K is a gain matrix to be chosen, and $G = [g_{ij}] \geq 0$ is the coupling matrix. It follows that the overall closed-loop system is

$$\dot{x} = (I_{q \times q} \otimes A)x + [(-H + G) \otimes (BK)]x,$$

where $x = \begin{bmatrix} x_1^T & \cdots & x_q^T \end{bmatrix}^T$ and $H = \text{diag}\{\sum_{j=1}^{q} g_{ij}\}$. If G is irreducible, matrix $(-H + G)$ with zero row sum has its eigenvalues at λ_i where $Re(\lambda_1) \leq Re(\lambda_2) \leq \cdots < \lambda_q = 0$. Therefore, there exists a matrix T (whose last column is $\mathbf{1}$) such that

$$T^{-1}(-H + G)T = J,$$

where J is the Jordan canonical form with diagonal elements of λ_i. Applying state transformation $z = (T \otimes I_{n \times n})x$ yields

$$\dot{z} = (I_{q \times q} \otimes A)z + [J \otimes (BK)]z.$$

If K is chosen such that $A + \lambda_i BK$ is Hurwitz for $i = 1, \cdots, (q-1)$, the above system has the limit that $z_i \to 0$ for all i except that $z_q(t) = e^{At} z_q(0)$. Hence,

$$x(t) = Tz(t) \to \mathbf{1} \otimes z_q(t),$$

which establishes synchronization. Compared to cooperative control with constant network topology, synchronization under constant coupling does not require that matrix A be Metzler.

In the case that the synchronization control is chosen to be time-varying as

$$u_i = K \sum_{j=1}^{q} g_{ij}(t)(x_j - x_i),$$

the above eigenvalue analysis is no longer applicable. Should matrix $G(t)$ be known, one can attempt to analyze synchronization by applying Lyapunov direct method and derive the Riccati-equation-like conditions [276, 277] since finding the Lyapunov function is generally very difficult. In the general case that the synchronization control is implemented through a network, matrix $G(t)$ is not known a priori but nonetheless synchronization can be concluded using Theorem 5.8 by first designing an individual control law for every system such that matrix A becomes Metzler.

If matrix A is also Hurwitz (or simply stabilizable), the above synchronization problem reduces to the stabilization problem of interconnected systems. General conditions on heterogeneous systems and their connectivity topology can be derived.

5.4 Ensuring Network Connectivity

The preceding analysis of cooperative systems requires that topological changes of their sensing/communication network is uniformly sequentially complete over time. This requirement of network connectivity is both necessary and sufficient and, based on this requirement, cooperative control is designed for cooperative motion behaviors. Depending upon the network which acquires and transmits the information, motion behaviors could have significant impact on network connectivity since efficiency and accuracy of sensing and communication are dependent on distance and direction.

If feedback information needed to implement cooperative control is transmitted (for instance, by wireless communication), a proper communication protocol should be implemented to manage power and bandwidth in such a way that network connectivity is ensured over time. Should it become necessary, one way to maintain and enhance network connectivity is for the hands-off operator to act as one virtual leader. Typically, the command center has the best equipment and communication capacity so that the virtual leader can seek out each and every system in the group and its dynamics in (5.24) play the role of implicit information relay. Specifically, we know that, by broadcasting (*i.e.*, $s_{i0}(t) = 1$) individually to and demanding response (*i.e.*, $s_{0i}(t) = 1$) from every system repeatedly over time, $\{\overline{S}(\overline{t}_k)\}$ in (5.22) would become sequentially complete even if $\{S(t_k)\}$ would never be sequentially complete. In the case that the virtual leader only broadcasts to some of the systems but receives nothing, sensing/communication matrix is prescribed by (5.26), and then $\{\overline{S}(\overline{t}_k)\}$ is sequentially complete if and only if $\{S(t_k)\}$ is sequentially complete. This also implies that, if any of the systems acts maliciously as a fake leader (*i.e.*, by only broadcasting but not reacting), it can cause the whole group of systems to oscillate unless corrective measures can be taken.

If all the feedback information needed to implement cooperative control is acquired by the sensors onboard any of the individual vehicles, cooperative control must be designed to generate motions that enable the sensors to continue data acquisition and hence maintain network connectivity. In this case, motion and network connectivity cannot be separated, and network connectivity needs to be maintained by properly choosing weights in cooperative control law, that is, $w_{ij}(t) > 0$ in (5.11). To illustrate the process, consider the one-dimensional six-agent rendezvous problem in which all the particles have the same sensing radius of $R_s > 0$ and their dynamics are given by

$$z_i(k+1) = u_i(k), \quad i = 1, \cdots, 6,$$

where cooperative control $u(k)$ is chosen to be in the form of (5.11) as

$$u_i(k) = \frac{1}{\sum_{l=1}^{6} w_{il}(k)s_{il}(k)} \sum_{j=1}^{6} w_{ij}(k)s_{ij}(k)z_j(k). \tag{5.31}$$

Assume that $z_1(0) = z_2(0) = 0$, $z_3(0) = R_s$, $z_4(0) = 2R_s$, and $z_5(0) = z_6(0) = 3R_s$. Clearly, the sensing network is connected at $k = 0$ but could become disconnected at $k = 1$ and beyond. Indeed, once the sensing network becomes disconnected, it may never recover in this case that only passive sensors of fixed range are employed. Hence, maintaining network connectivity is the key to ensure cooperative behavior of rendezvous. In what follows, two design cases of Cooperative Control 5.31 are elaborated and compared.

The first case is that all the weights are set to be identical and hence $w_{ij}(k) = 1$. It follows that

$$u_1(0) = u_2(0) = \frac{1}{3}R_s, \ u_3(0) = \frac{3}{4}R_s, \ u_4(0) = 2\frac{1}{4}R_s, \ u_5(0) = u_6(0) = 2\frac{2}{3}R_s,$$

which yields

$$z_1(1) = z_2(1) = \frac{1}{3}R_s, \; z_3(1) = \frac{3}{4}R_s, \; z_4(1) = 2\frac{1}{4}R_s, \; z_5(1) = z_6(1) = 2\frac{2}{3}R_s.$$

Hence, at $k = 1$, the sensing network is no longer connected since $z_4(1) - z_3(1) = 1.5R_s$. As k increases, the network stays disconnected and, for all $k \geq 2$,

$$z_1(k) = z_2(k) = z_3(k) = \frac{17}{36}R_s, \quad z_4(k) = z_5(k) = z_6(k) = 2\frac{19}{36}R_s.$$

Intuitively, breakup of the sensing network is primarily due to the movement of agents 3 and 4. At $k = 0$, each of these two agents has two other agents barely within and at both ends of their sending ranges and, since either knows how its neighbors will move, both of agents 3 and 4 should not move at $k = 1$ as any movement could break up network connectivity.

The second case is to choose the weights according to the so-called *circumcenter algorithm*: for each agent, find its neighbors in the sensing range, calculate the circumcenter location of its neighbors (*i.e.*, the center of the smallest interval that includes all its neighbors), and move itself to the circumcenter. Obviously, the algorithm reflects the intuition aforementioned. For the six-agent rendezvous problem, let the circumcenter location for the ith agent at step k be denoted by $\xi_i(k)$. Applying the algorithm, we know that

$$\xi_1(0) = \xi_2(0) = R_s, \quad \xi_3(0) = R_s, \quad \xi_4(0) = 2R_s, \quad \xi_5(0) = \xi_6(0) = 2R_s,$$

and

$$\xi_1(1) = \xi_2(1) = \xi_3(1) = \xi_4(1) = \xi_5(1) = \xi_6(1) = 1.5R_s.$$

It follows that, at $k = 0$, the corresponding non-zero weights for implementing Cooperative Control 5.31 are

$$\begin{cases} w_{11}(0) = w_{12}(0) = \epsilon, \; w_{13}(0) = 1 - 2\epsilon; \\ w_{21}(0) = w_{22}(0) = \epsilon, \; w_{23}(0) = 1 - 2\epsilon; \\ w_{31}(0) = w_{32}(0) = \epsilon, \; w_{33}(0) = 1 - 3\epsilon, \; w_{34}(0) = \epsilon; \\ w_{43}(0) = \epsilon, \; w_{44}(0) = 1 - 3\epsilon, \; w_{45}(0) = w_{46}(0) = \epsilon; \\ w_{54}(0) = 1 - 2\epsilon, \; w_{55}(0) = w_{56}(0) = \epsilon; \\ w_{64}(0) = 1 - 2\epsilon, \; w_{65}(0) = w_{66}(0) = \epsilon; \end{cases}$$

and

$$w_{ij}(k) = 1, \quad k \geq 2,$$

where $\epsilon > 0$ is a small constant. It can be easily verified that, as long as ϵ is small, Cooperative Control 5.31 can maintain network connectivity. As ϵ approaches zero, Cooperative Control 5.31 becomes the circumcenter algorithm which provides the best convergence speed and is, strictly speaking, a non-linear cooperative control. Nonetheless, the circumcenter algorithm provides useful guidance in selecting the weights of Control 5.31 for the systems equipped with a bidirectional sensing/communication network. Non-linear cooperative systems will be studied in Chapter 6, and the circumcenter algorithm will be analyzed further in Section 6.4.

The circumcenter algorithm or its linear approximation for Control 5.31 can be extended to arbitrary dimensions if each agent has the knowledge of a common frame and if its sensing range is rectangular (or contains a non-trivial rectangular sub-set). However, typical sensing ranges in the 2-D and 3-D spaces are a limited sector and a limited cone, respectively. In those cases, maintaining network connectivity is not difficult if a simple chain of leader-follower motion pattern can be imposed.

5.5 Average System and Its Properties

It is shown by Theorems 5.6 and 5.13 as well as by the discussions in Section 5.2.3 that asymptotic cooperative stability of System 5.1 is determined by cumulative information exchange over time. Another way to quantify explicitly cumulative effects of topological changes is to develop the so-called average system (introduced in Section 2.4.2) for Cooperative System 5.1. Specifically, Lemma 5.14 provides the properties on an average cooperative system, and it together with Theorem 5.6 yields Theorem 5.15.

Lemma 5.14. *Exponential functions of non-negative matrices have the following properties:*

(a) Define

$$\xi(\alpha) = [e^{A\alpha}e^{B\tau}]^{\frac{\alpha}{\alpha+\tau}},$$

where $\alpha \geq 0$ is the scalar argument, $A, B \in \Re_+^{n \times n}$ are constant matrices, and $\tau > 0$ is a constant. Then, $[\xi(\alpha) - I]$ is a non-negative matrix for any sufficiently small values of $\alpha > 0$.

(b) Consider a pair of non-negative and row-stochastic matrices D_i and D_j. If $\mathcal{B}[D_i + D_j] = \mathcal{B}[(D_i + D_j)^2]$ where $\mathcal{B}(\cdot)$ is defined in (4.22), there is a non-negative and row-stochastic matrix $D_{a,ij}$ such that, for any $\tau_i, \tau_j > 0$

$$e^{D_{a,ij}(\tau_i+\tau_j)} = e^{D_i\tau_i}e^{D_j\tau_j}.$$

(c) Given any row-stochastic matrix $D_a = \mathbf{1}c^T$ with $c \in \Re_+^n$, $D_a = D_a^2$ and $e^{-(I+D_a)t} \to D_a$ as $t \to \infty$.

Proof:

(a) It follows that $\xi(0) = I$ and that, since $\ln C$ is well defined for any invertible matrix C,

$$\frac{d\xi(\alpha)}{d\alpha} = \frac{\partial e^{A\alpha}e^{B\tau}}{\partial \alpha} \cdot [e^{A\alpha}e^{B\tau}]^{\frac{\alpha}{\alpha+\tau}-1}$$

$$+ \frac{\partial \frac{\alpha}{\alpha+\tau}}{\partial \alpha} \cdot \ln[e^{A\alpha}e^{B\tau}] \cdot [e^{A\alpha}e^{B\tau}]^{\frac{\alpha}{\alpha+\tau}}$$

$$= \left\{ A + \frac{\tau}{(\alpha+\tau)^2} \ln[e^{A\alpha}e^{B\tau}] \right\} \xi(\alpha).$$

Therefore, the Taylor series expansion of $\xi(\alpha)$ for sufficiently small and positive values of α is

$$\xi(\alpha) = I + [A + B]\alpha + h.o.t., \tag{5.32}$$

where $h.o.t.$ stands for high-order terms. By repeatedly taking higher-order derivatives, one can show that their values at $\alpha = 0$ introduce new non-negative matrix product terms while, except for scalar multipliers, non-positive matrix terms are the same as those non-negative matrix terms already appeared in the lower-order derivatives. For example, it follows that

$$\frac{d^2\xi(\alpha)}{d\alpha^2} = \left\{ A + \frac{\tau}{(\alpha + \tau)^2} \ln[e^{A\alpha}e^{B\tau}] \right\}^2 \xi(\alpha)$$

$$+ \left\{ -\frac{2\tau}{(\alpha + \tau)^3} \ln[e^{A\alpha}e^{B\tau}] + \frac{\tau}{(\alpha + \tau)^2} A \right\} \xi(\alpha),$$

which implies that the second-order term in Taylor Expansion 5.32 is

$$\left. \frac{d^2\xi(\alpha)}{d\alpha^2} \right|_{\alpha=0} \alpha^2 = \left\{ [A + B]^2 - \frac{2}{\tau}B + \frac{1}{\tau}A \right\} \alpha^2.$$

In the above expression, non-negative terms in terms of products AB and BA appear, the only non-positive term is $-2\alpha^2 B/\tau$ and this term is dominated by αB which is one of the non-negative terms already appeared in the first-order Taylor expansion. By induction, one can claim that, in the Taylor series expansion in (5.32), all non-positive terms appeared in $h.o.t.$ are dominated by the non-negative terms, which proves (a).

(b) It follows from Baker-Campbell-Hausdorff formula [213] that

$$D_{a,ij}(\tau_i + \tau_j) = D_i\tau_i + D_j\tau_j + \frac{1}{2}ad_{D_i\tau_i}(D_j\tau_j) + \frac{1}{12}\{ad_{D_i\tau_i}(ad_{D_i\tau_i}D_j\tau_j)$$

$$+ ad_{D_j\tau_j}(ad_{D_i\tau_i}D_j\tau_j)\} + \cdots,$$

where $ad_A B = AB - BA$ is the Lie bracket. Since D_i and D_j are row-stochastic, $[ad_{D_i\tau_i}(D_j\tau_j)]\mathbf{1} = 0$ and so do the higher-order Lie brackets. Hence, matrix $D_{a,ij}$ is row-stochastic. To show that $D_{a,ij}$ is non-negative, let $\alpha_i = \tau_i/\beta$ for any chosen positive integer $\beta > 1$ and note that

$$e^{D_{a,ij}(\tau_i+\tau_j)} = \left[\prod_{k=1}^{\beta} e^{D_i\alpha_i} \right] e^{D_j\tau_j} = \left[\prod_{k=1}^{\beta-1} e^{D_i\alpha_i} \right] e^{H(\alpha_i+\tau_j)},$$

where $e^{H(\alpha_i+\tau_j)} \triangleq e^{D_i\alpha_i}e^{D_j\tau_j}$. On the other hand, we know from Taylor expansion

$$\left[e^{H(\alpha_i+\tau_j)} \right]^{\frac{\alpha_i}{\alpha_i+\tau_i}} = e^{H\alpha_i} = I + H\alpha_i + \frac{1}{2}H^2\alpha_i^2 + h.o.t.$$

and from (a) that, if $\mathcal{B}(H) = \mathcal{B}(H^2)$, H is non-negative for all sufficiently small $\alpha_i > 0$ and by induction matrix $D_{a,ij}$ is non-negative. Comparing the above expansion with (5.32) yields the conclusion of (b).

(c) Since $D_a = \mathbf{1}c^T$, $D_a^2 = D_a$ and hence

$$\lim_{t\to\infty} e^{-(I+D_a)t} = \lim_{t\to\infty} e^{-t} \sum_{j=0}^{\infty} \frac{1}{j!} D_a^j t^j = \lim_{t\to\infty} e^{-t} \left[I + D_a(e^t - 1)\right] = D_a,$$

which completes the proof.

□

Theorem 5.15. *The following statements are equivalent:*

(a) System 5.1 is uniformly asymptotically cooperatively stable.

(b) For all $\eta \in \aleph$, composite sensing/communication matrices $S_\Lambda(\eta)$ defined in (5.17) are lower triangularly complete, and the lengths $(t_{k_\eta} - t_{k_{\eta-1}})$ of their composite time intervals are uniformly bounded.

(c) For any $k \in \aleph$, there exists $\kappa > k$ such that System 5.1 has the following average system:

$$\dot{x}_a = [-I + D_a(\kappa : k)]x_a, \quad t \in [t_k, t_\kappa), \tag{5.33}$$

where terminal conditions are $x_a(t_\kappa) = x(t_\kappa)$ and $x_a(t_k) = x(t_k)$, and $D_a(\kappa : k)$ is a non-negative and row-stochastic matrix satisfying

$$e^{[-I+D_a(\kappa:k)](t_\kappa - t_k)} = e^{[-I+D(t_{\kappa-1})](t_\kappa - t_{\kappa-1})} e^{[-I+D(t_{k_\kappa-2})](t_{\kappa-1} - t_{\kappa-2})}$$
$$\cdots e^{[-I+D(t_k)](t_{k+1} - t_k)}. \tag{5.34}$$

Statement (b) of Lemma 5.14 spells out the condition under which an average system can be found for System 5.1 over any two consecutive intervals. In general, the average system always exists over a composite interval of sufficient length, which is stated by Theorem 5.15 and is due to (c) of Lemma 5.14. The following example illustrates some of the details.

Example 5.16. Control 5.11 with $m = l_i = w_{ij} = 1$ renders the following matrices for the Sensing/communication Sequence 5.19 in Example 5.9: $D(t_k) = D_1$ for $k = 4\eta + 3$, $D(t_k) = D_2$ for $k = 4\eta + 2$, $D(t_k) = D_3$ for $k = 4\eta + 1$, and $D(t_k) = D_4$ for $k = 4\eta$, where $\eta \in \aleph$,

$$D_1 = \begin{bmatrix} 1 & 0 & 0 \\ 0.5 & 0.5 & 0 \\ 0 & 0 & 1 \end{bmatrix}, \quad D_2 = \begin{bmatrix} 0.5 & 0.5 & 0 \\ 0 & 1 & 0 \\ 0 & 0 & 1 \end{bmatrix}, \quad D_3 = \begin{bmatrix} 1 & 0 & 0 \\ 0 & 1 & 0 \\ 0.5 & 0 & 0.5 \end{bmatrix}, \quad D_4 = \begin{bmatrix} 1 & 0 & 0 \\ 0 & 1 & 0 \\ 0 & 0 & 1 \end{bmatrix}.$$

Assume that $t_k - t_{k-1} = 1$ for all k. It follows that

$$D_a(3:2) = \begin{bmatrix} 0.6887 & 0.3113 & 0 \\ 0.1888 & 0.8112 & 0 \\ 0 & 0 & 1.0000 \end{bmatrix}$$

is the average matrix of D_2 and D_1 as

$$e^{2[-I+D_a(3:2)]} = e^{-I+D_1}e^{-I+D_2} = \begin{bmatrix} 0.6065 & 0.3935 & 0 \\ 0.2387 & 0.7613 & 0 \\ 0 & 0 & 1 \end{bmatrix}.$$

However, non-negative matrix $D_a(3:0)$ does not exists to be the average matrix of D_4 up to D_1 since $\mathcal{B}[(D_a(3:2) + D_3)^2] \neq \mathcal{B}[D_a(3:2) + D_3]$, while

$$e^{-I+D_1}e^{-I+D_2}e^{-I+D_3}e^{-I+D_4} = e^{-I+D_a(3:2)}e^{-I+D_3}$$

$$= \begin{bmatrix} 0.6065 & 0.3935 & 0 \\ 0.2387 & 0.7613 & 0 \\ 0.3935 & 0 & 0.6065 \end{bmatrix}.$$

Nonetheless, for any $\alpha \in \aleph$, matrix

$$D_a((4\alpha + 4 * 32) : (4\alpha)) = \begin{bmatrix} 0.3775 & 0.6225 & 0 \\ 0.3775 & 0.6225 & 0 \\ 0.5 & 0 & 0.5 \end{bmatrix}$$

is the matrix of the average system over interval $[4\alpha, \ 4\alpha + 128)$ (or extended interval $[4\alpha, \ t)$ for any $t > 4\alpha + 128$) as

$$e^{32[-I+D_a((4\alpha+4*32):(4\alpha))]} = \left[e^{-I+D_1}e^{-I+D_2}e^{-I+D_3}e^{-I+D_4} \right]^{32}$$

$$= \begin{bmatrix} 0.3775 & 0.6225 & 0 \\ 0.3775 & 0.6225 & 0 \\ 0.3775 & 0.6225 & 0 \end{bmatrix}.$$

Clearly, matrix $D_a(128:0)$ corresponds $S_\Lambda(\eta)$ defined by (5.17) with $t_{k_\eta} = k_\eta$ and $k_\eta = 128\eta$, where $\eta \in \aleph$. $\diamond$

5.6 Cooperative Control Lyapunov Function

The previous analysis of cooperative stability is based on the solution of piecewise-constant linear networked systems and convergence of the resulting multiplicative matrix sequence. However, an exact solution cannot be analytically found for networked systems of non-linear dynamics. In order to analyze and design non-linear networked systems, we need to apply non-linear tools that do not require explicit solution. One prominent candidate is the Lyapunov direct method which is universally applicable to linear and non-linear systems. In this section, Lyapunov function is sought for Cooperative System 5.8, and the results build the foundation for non-linear analysis in Chapter 6.

Should System 5.8 be uniformly asymptotically cooperatively stable, the corresponding Relative-error System 5.4 must be uniformly asymptotically stable, and hence existence of a Lyapunov function is guaranteed by Lyapunov converse theorem. The subsequent discussion provides a Lyapunov function that is in the form for concluding cooperative stability and is also strictly decreasing over time.

Definition 5.17. $V_c(x(t), t)$ *is said to be a* cooperative control Lyapunov function (CCLF) *for system* $\dot{x} = \mathcal{F}(x, t)$ *if* $V_c(x, t)$ *is uniformly bounded with respect to* t, *if* $V_c(x, t) = V(e_{c_i}, t)$ *is positive definite with respect to* e_{c_i} *for some or all* $i \in \{1, \cdots, n\}$ *where* e_{c_i} *is defined in (5.3), and if* $V_c(x(t'), t') > V_c(x(t), t)$ *along the solution of the system and for all* $t > t' \geq t_0$ *unless* $x(t') = c\mathbf{1}$.

In the sequel, cooperative control Lyapunov function is found first for a linear cooperative system of fixed topology and then for one of varying topologies.

5.6.1 Fixed Topology

Consider the following time-invariant system

$$\dot{x} = -(I - D)x \overset{\triangle}{=} -Ax, \tag{5.35}$$

where $x \in \Re^n$, $D \geq 0$ is row-stochastic and constant, and hence A is a singular M-matrix. It has been shown in Section 4.3.5 that, if A is irreducible, a Lyapunov function of the following simple form can be found:

$$V(x) = x^T P x = \sum_{i=1}^{n} \lambda_i x_i^2, \quad P = \mathrm{diag}\{\lambda_1, \cdots, \lambda_n\}, \tag{5.36}$$

where $\lambda = \begin{bmatrix} \lambda_1 \cdots \lambda_n \end{bmatrix}^T$ is the unity left eigenvector of D and associated with eigenvalue $\rho(D) = 1$, that is,

$$D^T \lambda = \lambda, \quad \lambda^T \mathbf{1} = 1. \tag{5.37}$$

Nonetheless, Lyapunov Function 5.36 is not a control Lyapunov function for System 5.35 because V is positive definite with respect to x but $\dot{V}$ is only negative semi-definite and because, as shown in Example 4.35, such a Lyapunov function does not exist if A is reducible. Hence, we need to find a cooperative control Lyapunov function first for any singular irreducible M-matrix and then for a singular reducible M-matrix. The following lemma and theorem provide a constructive way of finding such a cooperative control Lyapunov function.

Lemma 5.18. *For any constants* λ_1 *up to* λ_n *and along the trajectory of System 5.35, the time derivative of*

$$V_c = \sum_{i=1}^{n} \sum_{j=1}^{n} \lambda_i \lambda_j (x_j - x_i)^2 = \sum_{i=1}^{n} \lambda_i e_{c_i}^T G_i^T P G_i e_{c_i}, \tag{5.38}$$

can be expressed as

$$\dot{V}_c = -2 \sum_{i=1}^{n} \lambda_i e_{c_i}^T Q_{c_i} e_{c_i}, \tag{5.39}$$

where P is that in (5.36), e_{c_i} is defined by (5.3), $G_i \in \Re^{n \times (n-1)}$ is the resulting matrix after eliminating the ith column from $I_{n \times n}$, and

$$Q_{c_i} \triangleq G_i^T [P(I - D) + (I - D)^T P] G_i$$
$$= (G_i^T P G_i)(I - G_i^T D G_i) + (I - G_i^T D G_i)^T (G_i^T P G_i). \qquad (5.40)$$

Proof: To verify the expression of $\dot{V}_c$ for function V_c in (5.38), consider first the time derivative of term $(x_i - x_j)^2$. It follows that, for any pair of $i, j \in \{1, \cdots, n\}$,

$$\frac{1}{2} \frac{d(x_i - x_j)^2}{dt}$$
$$= (x_i - x_j)[\dot{x}_i - \dot{x}_j]$$
$$= (x_i - x_j)\left[-x_i + \sum_{k=1}^{n} d_{ik} x_k + x_j - \sum_{k=1}^{n} d_{jk} x_k \right]$$
$$= (x_i - x_j)\left\{ \left[-(1 - d_{ii})x_i + \sum_{k \neq j,\, k \neq i,\, k=1}^{n} d_{ik} x_k + (1 - d_{jj})x_j \right.\right.$$
$$\left.\left. - \sum_{k \neq i,\, k \neq i,\, k=1}^{n} d_{jk} x_k \right] + d_{ij} x_j - d_{ji} x_i \right\}. \qquad (5.41)$$

Since D is row-stochastic, we have

$$\begin{cases} d_{ij} = 1 - d_{ii} - \displaystyle\sum_{k \neq i,\, k \neq j,\, k=1}^{n} d_{ik} \\ d_{ji} = 1 - d_{jj} - \displaystyle\sum_{k \neq j,\, k \neq j,\, k=1}^{n} d_{jk} \end{cases}.$$

Substituting the above two expressions of d_{ij} and d_{ji} into (5.41) yields

$$\frac{1}{2} \frac{d(x_i - x_j)^2}{dt}$$
$$= (x_i - x_j)\left[-(2 - d_{ii} - d_{jj})x_i + \sum_{k \neq j,\, k \neq i,\, k=1}^{n} d_{ik}(x_k - x_j) + (2 - d_{ii} - d_{jj})x_j \right.$$
$$\left. + \sum_{k \neq i,\, k \neq j,\, k=1}^{n} d_{jk}(x_i - x_k) \right]$$
$$= -(1 - d_{ii})(x_i - x_j)^2 - (1 - d_{jj})(x_j - x_i)^2 + \sum_{k \neq i,\, k=1}^{n} d_{ik}(x_i - x_j)(x_k - x_j)$$

$$+ \sum_{k \neq j,\, k=1}^{n} d_{jk}(x_i - x_j)(x_i - x_k). \tag{5.42}$$

It follows from (5.38) and (5.42) that

$$
\frac{1}{2}\dot{V}_c = -\sum_{i=1}^{n}\sum_{j=1}^{n}\lambda_i\lambda_j(1-d_{ii})(x_i-x_j)^2 - \sum_{i=1}^{n}\sum_{j=1}^{n}\lambda_i\lambda_j(1-d_{jj})(x_j-x_i)^2
$$

$$
+\sum_{i=1}^{n}\sum_{j=1}^{n}\left[\sum_{k \neq i,\, k=1}^{n}\lambda_i\lambda_j d_{ik}(x_i-x_j)(x_k-x_j)\right]
$$

$$
+\sum_{i=1}^{n}\sum_{j=1}^{n}\left[\sum_{k \neq j,\, k=1}^{n}\lambda_i\lambda_j d_{jk}(x_i-x_j)(x_i-x_k)\right]
$$

$$
= -2\sum_{i=1}^{n}\sum_{j=1}^{n}\lambda_i\lambda_j(1-d_{ii})(x_i-x_j)^2
$$

$$
+2\sum_{i=1}^{n}\sum_{j=1}^{n}\left[\sum_{k \neq j,\, k=1}^{n}\lambda_i\lambda_j d_{jk}(x_i-x_j)(x_i-x_k)\right]. \tag{5.43}
$$

On the other hand, it follows from (5.40) and P that Q_{c_i} is the resulting square matrix after removing the ith row and column from matrix

$$
Q = P(I-D) + (I-D)^T P
$$
$$
= \begin{bmatrix}
2\lambda_1(1-d_{11}) & -\lambda_1 d_{12} - \lambda_2 d_{21} & \cdots & -\lambda_1 d_{1n} - \lambda_n d_{n1} \\
-\lambda_2 d_{21} - \lambda_1 d_{12} & 2\lambda_2(1-d_{22}) & \cdots & -\lambda_2 d_{2n} - \lambda_n d_{n2} \\
\vdots & \vdots & \vdots & \vdots \\
-\lambda_n d_{n1} - \lambda_1 d_{1n} & -\lambda_n d_{n2} - \lambda_2 d_{2n} & \cdots & 2\lambda_n(1-d_{nn})
\end{bmatrix}.
$$

Hence, we have

$$
-\sum_{i=1}^{n}\lambda_i x_{c_i}^T Q_{c_i} x_{c_i}
$$

$$
= -\sum_{i=1}^{n}\lambda_i \left\{ \sum_{j \neq i,\, j=1}^{n} 2\lambda_j(1-d_{jj})(x_j-x_i)^2 \right.
$$

$$
\left. - \sum_{j \neq i,\, j=1}^{n}\left[\sum_{k \neq i,\, k \neq j,\, k=1}^{n}(\lambda_j d_{jk}+\lambda_k d_{kj})(x_j-x_i)(x_k-x_i)\right]\right\}
$$

$$
= -\sum_{i=1}^{n}\sum_{j=1}^{n} 2\lambda_i\lambda_j(1-d_{jj})(x_j-x_i)^2
$$

$$+\sum_{i=1}^{n}\sum_{j=1}^{n}\left[\sum_{k\neq j,\ k=1}^{n}\lambda_i(\lambda_j d_{jk}+\lambda_k d_{kj})(x_j-x_i)(x_k-x_i)\right]$$

$$=-2\sum_{i=1}^{n}\sum_{j=1}^{n}\lambda_i\lambda_j(1-d_{jj})(x_j-x_i)^2$$

$$+\sum_{i=1}^{n}\sum_{j=1}^{n}\left[\sum_{k\neq j,\ k=1}^{n}\lambda_i\lambda_j d_{jk}(x_j-x_i)(x_k-x_i)\right]$$

$$+\sum_{i=1}^{n}\sum_{j=1}^{n}\left[\sum_{k\neq j,\ k=1}^{n}\lambda_i\lambda_k d_{kj}(x_j-x_i)(x_k-x_i)\right]$$

$$=-2\sum_{i=1}^{n}\sum_{j=1}^{n}\lambda_i\lambda_j(1-d_{jj})(x_j-x_i)^2$$

$$+2\sum_{i=1}^{n}\sum_{j=1}^{n}\left[\sum_{k\neq j,\ k=1}^{n}\lambda_i\lambda_j d_{jk}(x_j-x_i)(x_k-x_i)\right],$$

which is the same as (5.43). Hence, Matrix Expression 5.39 is established. $\square$

Theorem 5.19. *If matrix D is irreducible, function V_c in (5.38) is a cooperative control Lyapunov function for System 5.35 by choosing λ according to (5.37).*

Proof: It follows from Theorem 4.8 that $\lambda > 0$ and hence V_c is positive definite with respect to e_{c_i}. To show Q_{c_i} is positive definite and hence $\dot{V}_c$ is negative definite, recall that $Q_{c_i} = G_i^T Q G_i$, where $Q = P(I-D)+(I-D)^T P$. According to (d) of Theorem 4.27, Q is positive semi-definite and hence so is Q_{c_i}. Furthermore, since D is irreducible, PD and in turn $(PD+D^T P)$ as well as $2(\max\lambda_i I - P)+(PD+D^T P)$ are irreducible and, by (a) of Theorem 4.27, matrix $Q = 2P-(PD+D^T P) = 2\max\lambda_i I - [2(\max\lambda_i I - P)+(PD+D^T P)]$ is of rank $n-1$. Thus, $x^T Q x > 0$ for all $x \neq c\mathbf{1}$ with $c \in \Re$, and $\mathbf{1}^T Q x = x^T Q \mathbf{1} = 0$. In particular, $x = x_i\mathbf{1} + G_i e_{c_i}$, $x \neq c\mathbf{1}$ if and only if $e_{c_i} = 0$, and hence

$$x^T Q x = e_{c_i}^T G_i^T Q G_i e_{c_i}$$

is positive definite with respect to e_{c_i}. $\square$

The above proof also suggests a close relationship between Lyapunov Function 5.36 and Cooperative Control Lyapunov Function 5.38. Specifically, it follows that

$$\dot{V} = -\frac{1}{n}\sum_{i=1}^{n}x^T[P(I-D)+(I-D)^T P]x$$

$$= -\frac{1}{n}\sum_{i=1}^{n}[(G_i e_{c_i} + x_i \mathbf{1})^T[P(I-D)+(I-D)^T P](G_i e_{c_i} + x_i \mathbf{1})]$$

$$= -\frac{1}{n}\sum_{i=1}^{n} e_{c_i}^T Q_{c_i} e_{c_i},$$

which is negative definite with respect to e_{c_i}. On the other hand, it follows that

$$V = x^T P x$$
$$= (G_i e_{c_i} + x_i \mathbf{1})^T P(G_i e_{c_i} + x_i \mathbf{1})$$
$$= e_{c_i}^T (G_i^T P G_i) e_{c_i} + 2 x_i \mathbf{1}^T P G_i e_{c_i} + x_i^2$$
$$= e_{c_i}^T (G_i^T P G_i) e_{c_i} + 2 x_i \mathcal{O} - x_i^2,$$

where $\mathcal{O}$ is the stationary center of mass studied in (5.2). Clearly, V is not positive definite with respect to e_{c_i}, but function V_c is and it can be rewritten as

$$V_c \triangleq \sum_{i=1}^{n} \lambda_i e_{c_i}^T (G_i^T P G_i) e_{c_i}$$

$$= \sum_{i=1}^{n} \lambda_i V - 2\mathcal{O}\sum_{i=1}^{n} \lambda_i x_i + \sum_{i=1}^{n} \lambda_i x_i^2$$

$$= 2V - 2\mathcal{O}^2.$$

Taking the time derivative on both sides of the above expression, invoking the expression of $\dot{V}$ and recalling the property of $\mathcal{O}$ yield

$$\dot{V}_c = 2\dot{V} = -\frac{2}{n}\sum_{i=1}^{n} e_{c_i}^T Q_{c_i} e_{c_i},$$

which is an alternative expression equivalent to (5.39).

If matrix A is reducible, Cooperative Control Lyapunov Function 5.38 can still be found but Lyapunov function in the form of (5.36) may not exist. Theorem 5.20 provides a general solution, while Corollary 5.21 gives a simpler version (and its proof is a special case of $V_c = V_2$ where V_2 is defined by (5.47)).

Theorem 5.20. *Consider a non-negative, reducible and row-stochastic matrix D, and let $F_\searrow$ be its lower-triangular canonical form in (4.1). Then, System 5.35 has a cooperative control Lyapunov function if and only if $F_\searrow$ is lower triangularly complete, and the Lyapunov function is of form*

$$V_c = \sum_{i=1}^{n}\sum_{k=1}^{n} p_{\mu,k}(x_\mu - x_k)^2, \tag{5.44}$$

where $p_{\mu,k}$ are positive constants. If $F_{\searrow}$ is not lower triangularly complete, $p_{\mu,k} > 0$ exist such that $\dot{V}_c \leq 0$ and $\dot{V}_c$ is negative definite with respect to $(x_\mu - x_k)$ for all μ, k corresponding to the same lower triangularly complete block.

Proof: To show sufficiency, assume without loss of any generality that $p = 2$ (if otherwise, the subsequent proof can be extended to $p > 2$ by induction). Since $p = 2$, we have that

$$\dot{x} = \left(-I + F_{\searrow}\right) x, \quad x = \begin{bmatrix} y \\ z \end{bmatrix}, \quad F_{\searrow} = \begin{bmatrix} F_{11} & 0 \\ F_{21} & F_{22} \end{bmatrix}, \qquad (5.45)$$

where $F_{11} \in \Re^{r_1 \times r_1}$ and $F_{22} \in \Re^{r_2 \times r_2}$ are irreducible, and $r_1 + r_2 = n$.

It follows from Theorem 5.19 that there exist positive constants λ_1 up to λ_{r_1} such that, along the trajectory of $\dot{y} = (-I + F_{11})y$, the time derivative of

$$V_1 = \sum_{i=1}^{r_1} \sum_{j=1}^{r_1} \lambda_i \lambda_j (y_j - y_i)^2 \qquad (5.46)$$

is

$$\dot{V}_1 = -2 \sum_{i=1}^{r_1} \lambda_i y_{c_i}^T Q_{1,c_i} y_{c_i},$$

where $y_{c_i} = G_{1,i}^T(y - y_i \mathbf{1})$, $G_{1,i}$ is the same as G_i in Lemma 5.18 except for its dimension, and Q_{1,c_i} are all positive definite. On the other hand, let

$$x_{c_i} = \begin{bmatrix} y_{c_i} \\ z_{c_i} \end{bmatrix}, \quad \begin{cases} y_{c_i} = G_{1,i}^T(y - y_i \mathbf{1}), \quad z_{c_i} = z - y_i \mathbf{1}, \\ \qquad \text{if } i \in \{1, \cdots, r_1\}; \\ y_{c_i} = y - z_{i-r_1} \mathbf{1}, \quad z_{c_i} = G_{2,(i-r_1)}^T(z - z_{i-r_1} \mathbf{1}), \\ \qquad \text{if } i \in \{r_1 + 1, \cdots, r_1 + r_2\}; \end{cases}$$

where $G_{2,j}$ is the same as G_j in Lemma 5.18 except for its dimension. Then, for any $i \in \{1, \cdots, r_1\}$, dynamics of z can be expressed as

$$\dot{z}_{c_i} = [F_{21}G_i + \mathbf{1}e_i^T(-I + F_{11})G_i]y_{c_i} + (-I + F_{22})z_{c_i},$$

where $e_i \in \Re^{r_1}$ is the vector with zero as its elements except that its ith element is 1. Since $F_{21} \neq 0$, it follows from Lemma 4.38 and Theorem 4.31 that there exists a positive definite and diagonal matrix P_2 of diagonal elements λ_{r_1+1} up to λ_n such that

$$Q_2 \triangleq P_2(I - F_{22}) + (I - F_{22})^T P_2$$

is positive definite. Now, consider Lyapunov function

$$V_2 = \alpha_1 V_1 + \sum_{i=1}^{r_1} z_{c_i}^T P_2 z_{c_i}, \qquad (5.47)$$

where $\alpha_1 > 0$ is a positive constant to be determined. It follows that

$$\dot{V}_2 = -2\alpha_1 \sum_{i=1}^{r_1} \lambda_i y_{c_i}^T Q_{1,c_i} y_{c_i} - \sum_{j=1}^{r_1} z_{c_i}^T Q_2 z_{c_i}$$

$$+2\sum_{i=1}^{r_1} z_{c_i}^T P_2 [F_{21}G_i + \mathbf{1}e_i^T(-I + F_{11})G_i] y_{c_i},$$

which is negative definite by simply choosing

$$\alpha_1 > \max_{1 \le i \le r_1} \frac{\lambda_{max}^2(P_2)\|F_{21}G_i + \mathbf{1}e_i^T(-I + F_{11})G_i\|^2}{2\lambda_i \lambda_{min}(Q_{1,c_i})\lambda_{min}(Q_2)}.$$

Choose the following candidate of cooperative control Lyapunov function:

$$V_c = V_3 + \alpha_2 V_1 + \alpha_3 V_2,$$

where $\alpha_2 > 0$ and $\alpha_3 \ge 0$ are to be chosen shortly, V_1 is defined by (5.46), V_2 is given by (5.47), and

$$V_3 = \sum_{i=1}^{n} \sum_{j=1}^{n} \lambda_i \lambda_j (x_j - x_i)^2, \quad \lambda_i \ge 0.$$

It follows from Lemma 5.18 that

$$\dot{V}_3 = -2\sum_{i=1}^{n} \lambda_i x_{c_i}^T Q_{c_i}' x_{c_i},$$

where $x_{c_i} = G_i^T(x - x_i\mathbf{1})$, $P = \mathrm{diag}\{\lambda_1, \cdots, \lambda_n\}$, and $Q_{c_i}' = G_i^T[P(I - F_\searrow) + (I - F_\searrow)^T P]G_i$. Therefore, we know that

$$\dot{V}_c = -2\sum_{i=1}^{n} \lambda_i x_{c_i}^T Q_{c_i}' x_{c_i} + \alpha_3 \dot{V}_2,$$

where $Q_{2,c_i} = G_{2,(i-r_1)}^T Q_2 G_{2,(i-r_1)}$ for $r_1 < i \le r_1 + r_2$,

$$Q_{c_i}' = \begin{cases} \begin{bmatrix} (1+\alpha_2)Q_{1,c_i} & -G_{1,i}^T F_{12} \\ -F_{12}^T G_{1,i} & Q_2 \end{bmatrix} & \text{if } 1 \le i \le r_1, \\[2em] \begin{bmatrix} Q_1 & -F_{12}G_{2,(i-r_1)} \\ -G_{2,(i-r_1)}^T F_{12}^T & Q_{2,c_i} \end{bmatrix} & \text{if } r_1 + 1 \le i \le r_1 + r_2. \end{cases}$$

Since Q_{1,c_i} and Q_2 are positive definite, matrices Q_{c_i}' with $1 \le i \le r_1$ can be made positive definite by simply increasing α_2. Since Q_1 is only positive semi-definite, matrices Q_{c_i}' with $r_1 + 1 \le i \le r_1 + r_2$ cannot be guaranteed to be positive definite. Nonetheless, all the indefinite terms are from the product of $2z_{c_i}^T G_{2,(i-r_1)}^T F_{12}^T(y - z_i \mathbf{1})$, and they all are of form $(x_j - x_i)(x_k - x_i)$ for $r_1 + 1 \leq i, j \leq r_1 + r_2$ and $1 \leq k \leq r_1$. Noting that matrices Q_{2,c_i} are positive definite and they are associated with $(x_j - x_i)$ and that $\dot{V}_2$ has been shown above to be negative definite with respect to $(x_k - x_i)$, we can conclude that $\dot{V}_c$ can be made negative definite by increasing both α_2 and α_3.

To show necessity, note that matrix $F_{\searrow}$ must be block diagonal if $F_{\searrow}$ is not lower triangularly complete and that both sub-matrices $[-I + F_{11}]$ and $[-I + F_{22}]$ have one of their eigenvalues at 0. As a result, their block-systems are decoupled and hence they together cannot be asymptotically cooperatively stable. In other words, $\dot{V}_c$ cannot be negative definite, which shows necessity.

In the case that $F_{21} = 0$, it follows from the expression of matrix Q'_{c_i} that it is positive semi-definite. Thus, $\dot{V}_c$ is negative semi-definite by setting $V_c = V_3$. □

Corollary 5.21. *Consider a non-negative, reducible and row-stochastic matrix D, and let $F_{\searrow}$ be its lower-triangular canonical form in (4.1). Then,*

$$V_c(x) = \sum_{i=1}^{n} \sum_{k \in \Omega_1} p_{\mu,k}(x_\mu - x_k)^2 \tag{5.48}$$

is a cooperative control Lyapunov function for System 5.35 if and only if $F_{\searrow}$ is lower triangularly complete, where $p_{\mu,k}$ are positive constants, and Ω_1 is the set of the indices corresponding to those indices of F_{11} in (4.1).

5.6.2 Varying Topologies

In the presence of topology changes, the cooperative control Lyapunov function can be sought as before for Piecewise-Constant System 5.1 within each and every interval. Over time, a cooperative control Lyapunov function can also be found for its average system. Combining Theorems 5.15, 5.19 and 5.20 as well as Lemmas 5.14 and 5.22 yields the following theorem.

Lemma 5.22. *Cooperative control Lyapunov function for system $\dot{x} = [-I + \mathbf{1}c^T]x$ with $c \in \Re_+^n$ can be chosen to be V_c in (5.44) in which $p_{\mu,k} > 0$ are arbitrary.*

Proof: It follows from $\dot{x} = [-I + \mathbf{1}c^T]x$ that $\dot{x}_i - \dot{x}_j = -(x_i - x_j)$, which completes the proof. □

Theorem 5.23. *System 5.1 is uniformly asymptotically cooperatively stable if and only if its average system over any sufficiently-long time interval has a Cooperative Control Lyapunov Function in the form of (5.44).*

Example 5.24. Consider the following sensing/communication matrices S_i and their corresponding cooperative matrices D_i: over two consecutive intervals $[t_k, t_{k+1}]$ and $[t_{k+1}, t_{k+2})$,

$$S(t_k) = \begin{bmatrix} 1 & 0 & 0 \\ 1 & 1 & 0 \\ 0 & 0 & 1 \end{bmatrix}, \quad D(t_k) = \begin{bmatrix} 1 & 0 & 0 \\ 0.5 & 0.5 & 0 \\ 0 & 0 & 1 \end{bmatrix},$$

$$S(t_{k+1}) = \begin{bmatrix} 1 & 0 & 0 \\ 0 & 1 & 0 \\ 1 & 0 & 1 \end{bmatrix}, \quad \text{and} \quad D(t_{k+1}) = \begin{bmatrix} 1 & 0 & 0 \\ 0 & 1 & 0 \\ 0.5 & 0 & 0.5 \end{bmatrix}.$$

Assume that $t_{k+2} - t_{k+1} = t_{k+1} - t_k$ for simplicity. It follows from $D(t_k)$ and $D(t_{k+1})$ being commutative that the average system over interval $[t_k, t_{k+2})$ is defined by matrix

$$D_a((k+2) : k) = \begin{bmatrix} 1 & 0 & 0 \\ 0.25 & 0.75 & 0 \\ 0.25 & 0 & 0.75 \end{bmatrix}.$$

Since D_a is lower triangularly complete, Cooperative Control Lyapunov Function 5.44 exists for the average system, and it can be found by either following the proof of Theorem 5.20 or through simple computation shown below.

It follows from

$$V_c = p_1(x_1 - x_2)^2 + p_2(x_1 - x_3)^2 + p_3(x_2 - x_3)^2$$

that, along the trajectory of $\dot{x}_a = (-I + D_a)x_a$,

$$\dot{V}_c = -0.25p_1(x_1 - x_2)^2 - 0.25p_2(x_1 - x_3)^2 - 0.25p_3(x_2 - x_3)^2.$$

Hence, all choices of V_c with $p_i > 0$ are cooperative control Lyapunov functions over the composite interval $[t_k, t_{k+2})$.   ◇

It should be noted that, in the presence of topological changes, common (cooperative control) Lyapunov function may not exist. As such, the average system and its (cooperative control) Lyapunov function are used to describe cumulative characteristics of a linear (cooperative) system over time.

5.7 Robustness of Cooperative Systems

To analyze robustness, consider first the ideal cooperative system

$$z(k + 1) = P(k)z(k), \tag{5.49}$$

which represents either the continuous-time version of (5.5) or the discrete-time version of (5.20). It is assumed that, if all the feedbacks are accurate, System 5.49 is (uniformly) asymptotically cooperatively stable; that is, P_k is non-negative and row-stochastic, and the system has the limit that, for some constant vector $c \in \mathfrak{R}^n$,

$$\lim_{k \to \infty} \Phi(k + 1, 0) = \mathbf{1}c^T$$

where

$$\Phi(k+1,j) = \prod_{\eta=j}^{k} P(\eta) = P(k)P(k-1)\cdots P(j). \tag{5.50}$$

In practice, the information available for feedback is often inaccurate due to measurement errors, communication noises, and disturbances. Unless relative measurements are sufficient, a universal coordinate is needed to interpret absolute measurements, which may also induce errors for different entities. To account for these errors, Cooperative Model 5.49 has to be changed into

$$z(k+1) = P(k)z(k) + \begin{bmatrix} \sum_{j=1}^{n} P_{1j}(k)w'_{1j}(k) \\ \vdots \\ \sum_{j=1}^{n} P_{nj}(k)w'_{nj}(k) \end{bmatrix}, \tag{5.51}$$

where $w'_{ij}(k)$ is the feedback error of the ith entity about the jth entity at time k. Since matrix $P(k)$ is row-stochastic, (5.51) can be simplified as

$$z(k+1) = P(k)z(k) + \begin{bmatrix} w_1(k) \\ \vdots \\ w_n(k) \end{bmatrix} \stackrel{\triangle}{=} P(k)z(k) + w(k), \tag{5.52}$$

where $w_i(k)$ is the lumped error at the ith entity and has the property that

$$|w_i(k)| \le \max_j |w'_{ij}(k)|. \tag{5.53}$$

The solution to System 5.52 is

$$z(k+1) = \Phi(k+1,0)z(0) + \sum_{i=0}^{k-1} \Phi(k+1,i+1)w(i) + w(k). \tag{5.54}$$

Property 5.53 implies that, no matter how many entities a cooperative system contains, the lumped error at one of the entities is no worse than the worst of its individual errors of acquiring information only from one of its neighboring entities. While Property 5.53 is nice, the presence of $w(k)$ may degrade or even destroy cooperativeness (cooperative stability). In what follows, impacts of three different types of errors are investigated, and strategies are developed to enhance robustness of cooperative systems.

Sector-Bounded Errors

The first case is that lumped error $w(k)$ satisfies the linear growth condition, that is, $w(k)$ is proportional to $z(k)$ as

$$w(k) = W(k)z(k), \tag{5.55}$$

where $W(k)$ is a matrix of bounded values. Substituting (5.55) into (5.52) yields

$$z(k+1) = [P(k) + W(k)]z(k) \triangleq P_d(k)y(k). \tag{5.56}$$

In general, the presence of $W(k)$ prevents matrix $P_d(k)$ from keeping the properties of $P(k)$. That is, matrix $P_d(k)$ may no longer be row-stochastic, it may not even be non-negative, and its structural property may be altered. The following example shows that, in either of these three cases, cooperative stability becomes void.

Example 5.25. Consider System 5.56 with

$$W(k) = \begin{bmatrix} w_{11} & 0 & 0 \\ w_{21} & w_{22} & 0 \\ w_{31} & w_{32} & w_{33} \end{bmatrix}, \quad \text{and} \quad P(k) = \begin{bmatrix} 1 & 0 & 0 \\ 0.5 & 0.5 & 0 \\ 0 & 0.1 & 0.9 \end{bmatrix}.$$

If $w_{ij} = 0.01$ in the above matrix $W(k)$, matrix $P_d(k)$ is non-negative but not row-stochastic. It is straightforward to verify that $z_i(k) \to \infty$ for all i and $|z_i(k) - z_j(k)| \to \infty$ for all $i \neq j$ as $k \to \infty$.

If $w_{ij} = 0$ except $w_{31} = -0.01$ in the above matrix $W(k)$, matrix $P_d(k)$ is not non-negative or row-stochastic. It follows from

$$\lim_{k\to\infty} [P(k) + W(k)]^k = \begin{bmatrix} 1 & 0 & 0 \\ 1 & 0 & 0 \\ 0.9 & 0 & 0 \end{bmatrix}$$

that System 5.56 is not asymptotically cooperatively stable.

If $w_{ij} = 0$ except $w_{32} = -0.1$ and $w_{33} = 0.1$ in the above matrix $W(k)$, matrix $P_d(k)$ is still non-negative and row-stochastic. However, unlike $P(k)$, matrix $P_d(k)$ is no longer lower triangularly complete. Hence,

$$\lim_{k\to\infty} [P(k) + W(k)]^k = \begin{bmatrix} 1 & 0 & 0 \\ 1 & 0 & 0 \\ 0 & 0 & 1 \end{bmatrix},$$

which implies the loss of asymptotic cooperative stability for System 5.56. ◇

Example 5.25 shows that, in the presence of sector-bounded errors, asymptotic cooperative stability cannot be maintained unless matrix $W(k)$ has the properties that $W(k)\mathbf{1} = 0$, $|w_{ij}(k)| = 0$ if $p_{ij}(k) = 0$, and $|w_{ij}(k)| < p_{ij}(k)$ if $p_{ij}(k) > 0$. These properties ensure that matrix $P_d(k)$ is non-negative, row-stochastic and structurally the same as matrix $P(k)$. However, for unknown error matrix $W(k)$, these properties are too restrictive to be satisfied.

Uniformly Bounded Errors

Consider the case that lumped error $w(k)$ is uniformly bounded as $|w_i(k)| \leq c_w$ for all k and for some constant c_w. In this case, approximate cooperativeness can be maintained but Lyapunov stability is typically lost. To quantify robustness better, the following definitions are introduced, and Lemma 5.27 is the basic result.

Definition 5.26. *System 5.52 is said to be* robust cooperative *in the presence of $w(k)$ with $|w_i(k)| \leq c_w$ if*

$$\lim_{\eta \to \infty} \sup_{k \geq \eta} \max_{i,j} |y_i(\eta) - y_j(\eta)| \leq \gamma_1(c_w),$$

where $\gamma_1(\cdot)$ is a strictly monotone increasing function of its argument.

Lemma 5.27. *Suppose that System 5.49 is asymptotically cooperatively stable and that, for some integer $\kappa > 0$, its state transition matrix $\Phi(k+1, i+1)$ is scrambling for all $(k - i) \geq \kappa$ and uniformly with respect to k, i. Then, in the presence of uniformly bounded error $w(k)$, System 5.52 is robust cooperative.*

Proof: It follows from $|w_i(k)| \leq c_w$, from the definition of $\delta(\cdot)$ in (4.15), and from Solution 5.54 that

$$\max_{i,j} |z_i(k+1) - z_j(k+1)|$$

$$\leq \delta(\Phi(k+1, 0)) \max_i \{|z_i(0)|\} + 2\delta \left(\sum_{i=0}^{k-1} \Phi(k+1, i+1) \right) c_w + 2c_w$$

$$\leq \delta(\Phi(k+1, 0)) \max_i \{|z_i(0)|\} + 2 \sum_{i=0}^{k-1} \delta \left(\Phi(k+1, i+1) \right) c_w + 2c_w. \quad (5.57)$$

On the other hand, $\Phi(k+1, i+1)$ being uniformly scrambling implies $\lambda(\Phi(k+1, i+1)) \leq 1 - \epsilon_\phi$, where $\lambda(\cdot)$ is defined by (4.15) and $\epsilon_\phi > 0$. Hence, it follows from Inequality 4.18 that, for any positive integer l and for any k satisfying $(l-1)\kappa + 1 \leq k < l\kappa + 1$,

$$\sum_{i=0}^{k-1} \delta \left(\Phi(k+1, i+1) \right) = \sum_{i=(l-1)\kappa+1}^{k-1} \delta \left(\Phi(k+1, i+1) \right) + \sum_{i=0}^{(l-1)\kappa} \delta \left(\Phi(k+1, i+1) \right)$$

$$\leq \kappa + \sum_{i=0}^{(l-1)\kappa} \delta \left(\Phi((l-1)\kappa + 2, i+1) \right)$$

$$\leq \kappa + \sum_{j=1}^{l-1} \sum_{i=(j-1)\kappa}^{j\kappa} \delta \left(\Phi((l-1)\kappa + 2, i+1) \right)$$

$$\leq \kappa + \kappa \sum_{j=1}^{l-1} \delta \left(\Phi((l-1)\kappa + 2, j\kappa + 1) \right)$$

$$\leq \kappa + 1 + \kappa \sum_{j=1}^{l-2} \prod_{i=j}^{l-2} \delta \left(\Phi((i+1)\kappa + 1, i\kappa + 1) \right)$$

$$\leq \kappa + 1 + \kappa \sum_{j=1}^{l-2}(1 - \epsilon_\phi)^{l-1-j} \quad \rightarrow \quad \frac{\kappa + \epsilon_\phi}{\epsilon_\phi} \text{ as } k, l \to \infty.$$

The conclusion can be drawn by substituting the above result into (5.57) and by noting that $\delta(\Phi(k+1,0)) \to 0$ as $k \to \infty$. $\qquad\square$

Recall that $\Phi(k+1, i+1)$ being scrambling uniformly for all $(k-i) \geq \kappa$ implies exponential convergence of System 5.49 to its consensus. Lemma 5.27 remains to be the same without requiring the exponential convergence since its proof can be generalized to the case that $\Phi(k+1, i+1)$ becomes scrambling for sufficient large $(k-i)$. The following result shows that, if the errors are vanishing, robust cooperativeness as well as asymptotic cooperative convergence can be maintained.

Definition 5.28. *System 5.52 is said to be* robust cooperatively convergent *in the presence of $w(k)$ with $|w_i(k)| \leq c_w$ if*

$$\lim_{k\to\infty} \max_{i,j} |z_i(\eta) - z_j(k)| = 0.$$

Lemma 5.29. *Suppose that System 5.49 is asymptotically cooperatively stable. Then, if cumulative error magnitude series $\sum_{i=0}^{k} \max_j |w_j(i)|$ are Cauchy sequences as $\lim_{k,l\to\infty} \sum_{i=l}^{k} \max_j |w_j(i)| = 0$, System 5.52 is robust cooperatively convergent. Furthermore, if error series $w_j(k)$ are absolutely convergent series as $\sum_{k=0}^{\infty} \max_j |w_j(k)| < \infty$, System 5.52 is robust cooperatively convergent and uniformly bounded.*

Proof: Since System 5.49 is asymptotically cooperatively stable, $\lim_{k\to\infty} \Phi(k+1,0)z(0) = c_1 \mathbf{1}$ for some $c_1 \in \Re$. Hence, it follows from Solution 5.54 that System 5.52 is robust cooperatively convergent if and only if, for some $c_2 \in \Re$,

$$\lim_{k\to\infty}\left[\sum_{i=0}^{k-1} \Phi(k+1, i+1)w(i) - c_2(k)\mathbf{1}\right] = 0.$$

For robust cooperative convergence and uniform boundedness, c_1 is a constant (for any given series of $w(i)$).

Let $l = k/2$ if k is even or $l = (k+1)/2$ if otherwise. Then,

$$\sum_{i=0}^{k-1} \Phi(k+1, i+1)w(i) = \sum_{i=l}^{k-1} \Phi(k+1, i+1)w(i) + \sum_{i=0}^{l} \Phi(k+1, i+1)w(i).$$

It follows from $\lim_{k,l\to\infty} \sum_{i=l}^{k} \max_j |w_j(i)| = 0$ and from $\Phi(k+1, i+1)$ being row-stochastic that

$$\lim_{k\to\infty} \left| \sum_{i=l}^{k-1} \Phi(k+1,i+1)w(i) \right| \le \lim_{k\to\infty} \sum_{i=l}^{k-1} \Phi(k+1,i+1)\,|w(i)|$$

$$\le \lim_{k\to\infty} \sum_{i=l}^{k-1} \max_j |w_j(i)| \mathbf{1}$$

$$= 0,$$

where $|\cdot|$ is the operation defined in Section 4.1. On the other hand, it follows from System 5.49 being asymptotically cooperatively stable that, for some $\eta \in \Re_+^n$ with $\eta^T \mathbf{1} = 1$,

$$\lim_{k\to\infty} \Phi(k+1,l+2) = \mathbf{1}\eta^T,$$

and hence

$$\lim_{k\to\infty} \sum_{i=0}^{l} \Phi(k+1,i+1)w(i) = \lim_{k\to\infty} \Phi(k+1,l+2)\sum_{i=0}^{l} \Phi(l+2,i+1)w(i)$$

$$= \mathbf{1} \lim_{l\to\infty} \sum_{i=0}^{l} \eta^T \Phi(l+2,i+1)w(i)$$

$$\overset{\triangle}{=} \mathbf{1} \lim_{k\to\infty} c_2(k),$$

from which robust cooperative convergence is concluded.

If $\sum_{k=0}^{\infty} \max_j |w_j(k)| < \infty$, we have

$$\left| \lim_{l\to\infty} \sum_{i=0}^{l} \eta^T \Phi(l+2,i+1)w(i) \right| \le \lim_{l\to\infty} \sum_{i=0}^{l} \max_j |w_j(i)| < \infty,$$

from which $c_2(k)$ is uniformly bounded and convergent. Therefore, robust cooperative convergence and uniform boundedness are concluded. $\qquad\square$

Robust cooperative convergence requires only $[z_i(k) - z_j(k)] \to 0$ and, unless $z_i(k)$ converges to a finite consensus, it does not imply that the system is uniformly bounded. Due to the presence of uniformly bounded errors, the state likely becomes unbounded because the unperturbed System 5.49 is merely Lyapunov stable (*i.e.*, not asymptotically stable or exponentially stable) and Lyapunov stability is not robust in general, and robust cooperativeness is the best achievable. The following example shows a few simple but representative cases.

Example 5.30. Consider System 5.52 with

$$P(k) = \begin{bmatrix} 1 & 0 & 0 \\ 0.5 & 0.5 & 0 \\ 0 & 0.5 & 0.5 \end{bmatrix}.$$

It follows that, for any $k > i$,

$$
\Phi(k,i) = \begin{bmatrix} 1 & 0 & 0 \\ 1 - 0.5^{k-i} & 0.5^{k-i} & 0 \\ 1 - (k - i + 1)0.5^{k-i} & (k - i)0.5^{k-i} & 0.5^{k-i} \end{bmatrix}.
$$

Case (1): $w(k) = \begin{bmatrix} w_1 & w_2 & w_3 \end{bmatrix}^T$ where $w_1 \neq 0$. In this case, the system is subject to constant errors, its state $z(t)$ is not uniformly bounded, but it is robust cooperative because

$$
\lim_{k\to\infty} \left[\sum_{i=0}^{k-1} \Phi(k+1, i+1) - k\mathbf{1}\begin{bmatrix} 1 & 0 & 0 \end{bmatrix} \right] = \begin{bmatrix} 0 & 0 & 0 \\ -1 & 1 & 0 \\ 2 & 2 & 1 \end{bmatrix}.
$$

Case (2): $w(k) = \begin{bmatrix} 0 & w_2 & w_3 \end{bmatrix}^T$ where $w_2 w_3 \neq 0$. In this case, System 5.52 is also subject to constant errors; nonetheless its state $z(t)$ is both uniformly bounded and robust cooperative because $z_1(k)$ is identical to that of System 5.13.

Case (3): $w(k) = 1/(k+1)$. In this case, System 5.52 is subject to vanishing errors whose series are Cauchy. It can be verified that $\lim_{k\to\infty}[z(k) - \mathcal{H}(k)\mathbf{1} - z_1(0)\mathbf{1}] = 0$, where $\mathcal{H}(k) \triangleq \sum_{i=1}^{k} \frac{1}{i}$ is the so-called harmonic number. Hence, the system is robust cooperatively convergent but not uniformly bounded.

Case (4): $w(k) = \mathbf{1}\epsilon_w^k$ for $0 \leq \epsilon_w < 1$. In this case, System 5.52 is subject to exponentially vanishing errors (whose series are convergent). It can be verified that $\lim_{k\to\infty} z(k) = \frac{1}{1-\epsilon_w}\mathbf{1} + z_1(0)\mathbf{1}$. Hence, the system is both robust cooperatively convergent and uniformly bounded. $\diamond$

Robustification *via* Virtual Leader

Case (2) in Example 5.30 tells us that uniform boundedness of the state can be ensured if the system topology has certain special property. The following lemma provides such a result, and its proof can be completed by simply invoking Theorem 5.13 and Lemmas 5.27 and 5.29. Recall from Section 5.3.3 that the system structural condition of (5.58) can be achieved by introducing a virtual leader and its corresponding Sensing/communication Matrix 5.26.

Lemma 5.31. *Suppose that System 5.49 is asymptotically cooperatively stable and that System 5.52 has the structure that, for some integer $n_1 > 0$,*

$$
P(k) = \begin{bmatrix} P_{11}(k) & 0 \\ P_{21}(k) & P_{22}(k) \end{bmatrix}, \quad P_{11}(k) \in \Re^{n_1 \times n_1}, \quad w_1(k) = \cdots = w_{n_1}(k) = 0.
$$
(5.58)

Then, System 5.52 is robust cooperative and uniformly bounded if $w(k)$ is uniformly bounded, and it is robust cooperatively convergent if $\sum_{i=0}^{k} |w_j(i)|$ are Cauchy sequences as $\lim_{k,l\to\infty} \sum_{i=l}^{k} |w_j(i)| = 0$ for all j.

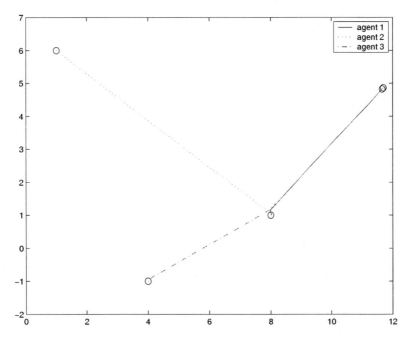

Fig. 5.8. Rendezvous in the presence of measurement bias

As an illustration of Lemma 5.31, consider the rendezvous problem studied in Sections 5.3.2 and 5.3.3. Here, we assume that $w(t) = 0.01$ is a small constant bias incurred in measuring neighbors' positions. In the presence of this bias, the system is robust cooperative but not robust cooperatively convergent; the response is shown in Fig. 5.8 for time interval $[0, 50]$ and afterwards the state variables continue their motion in the upper right direction toward infinity. In comparison, the virtual leader introduced to make the state converge to a specific location also keeps the state from going to infinity, which is shown in Fig. 5.9(a). Robustfication is obvious from comparing Fig. 5.9(a) and the corresponding bias-free response in Fig. 5.9(b).

Robustification can also be achieved by designing a non-linear cooperative control, which will be discussed in Section 6.3.1.

Robustification Using Deadzone on Relative Measurement and Relative Motion

In the absence of disturbance, a cooperative system can also be made robust against bounded measurement errors by only taking relative measurement and by only making relative movement when the feedback exceeds certain error level. To motivate this, reconsider Ideal System 5.49. It follows from $P(k)$ being row-stochastic that, for any $i = 1, \cdots, n$,

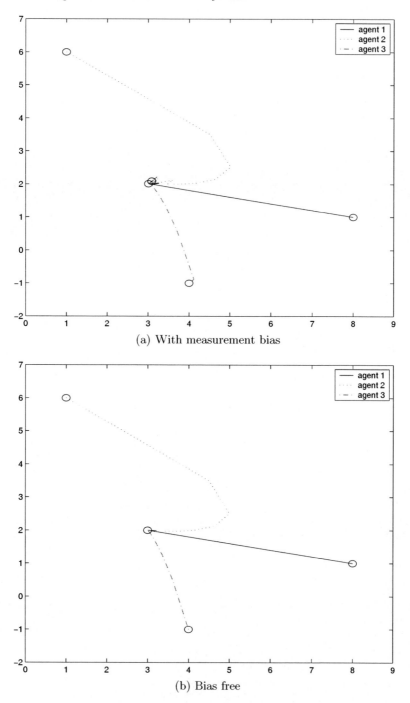

(a) With measurement bias

(b) Bias free

Fig. 5.9. Robustification *via* a virtual leader

$$z_i(k+1) - z_i(k) = \sum_{j=1}^n P_{ij}(k)z_j(k) - z_i(k) = \sum_{j=1,j\neq i}^n P_{ij}(k)\Delta_{ij}(k),$$

in which relative positions $\Delta_{ij}(k) \triangleq [z_j(k) - z_i(k)]$ are measured whenever $s_{ij}(k) = 1$ (i.e., $P_{ij}(k) > 0$), and $z_i(k+1) - z_i(k)$ are the relative motions executed at the kth step. Since only the measurement errors are considered in this case, the corresponding system reduces to

$$z_i(k+1) - z_i(k) = \sum_{j=1,j\neq i}^n P_{1j}(k)\hat{\Delta}_{ij}(k)$$

$$= \sum_{j=1,j\neq i}^n P_{1j}(k)\Delta_{ij}(k) + \sum_{j=1,j\neq i}^n P_{ij}(k)w'_{ij}(k), \quad (5.59)$$

where $\hat{\Delta}_{ij}(k)$ is the measurement of $\Delta_{ij}(k)$, and $w'_{ij}(k)$ is the measurement error uniformly bounded as $|w'_{ij}(k)| \leq c'_w$. Note that System 5.59 is in the same form as (5.51) and (5.52) except that the lumped error $w_i(k)$ is now bounded as

$$|w_i(k)| \leq \epsilon c'_w,$$

where $\epsilon = 1 - \min_j P_{jj}(k)$ and hence $\epsilon \in [0,1)$. According to Example 5.30, Motion Protocol 5.59 may render unbounded state $z_i(k)$. In order to ensure robustness of the state being bounded, we can simply introduce the deadzone function

$$\mathcal{D}_{c'_w}[\eta] = \begin{cases} 0 & \text{if } |\eta| \leq c'_w \\ \eta & \text{if } |\eta| > c'_w \end{cases}$$

into (5.59) and obtain the following result.

Lemma 5.32. *Suppose that System 5.49 is asymptotically cooperatively stable. Then, the relative motion protocol*

$$z_i(k+1) - z_i(k) = \sum_{j=1,j\neq i}^n P_{1j}(k)\mathcal{D}_{c'_w}[\hat{\Delta}_{ij}(k)]$$

$$= \sum_{j=1,j\neq i}^n P_{1j}(k)\mathcal{D}_{c'_w}[\Delta_{ij}(k) + w'_{ij}(k)], \quad i = 1,\cdots,n, \quad (5.60)$$

ensures that the state $z(k)$ is both uniformly bounded and robust cooperative.

Proof: Whenever $|\Delta_{ij}(k)| > c'_w$, System 5.60 is identical to (5.59) and (5.52). Hence, robust cooperativeness is readily concluded by invoking Lemma 5.27.

To show that $z(k)$ is uniformly bounded, it is sufficient to show that, with respect to k, $\max_i z_i(k)$ is non-increasing and $\min_i z_i(k)$ is non-decreasing. Suppose that, at time instant k, $z_{i^*} = \max_i z_i(k)$. Then, we have

$$\Delta_{i^*j}(k) \triangleq z_j(k) - z_{i^*}(k) \leq 0$$

for all j, which implies that

$$\mathcal{D}_{c'_w}[\Delta_{i^*j}(k) + w'_{ij}(k)] \leq 0.$$

Substituting the above result into (5.60) yields

$$z_{i^*}(k+1) - z_{i^*}(k) \leq 0,$$

which shows that $\max_i z_i(k)$ is non-increasing. Similarly, it can be shown that $\min_i z_i(k)$ is non-decreasing. This concludes the proof.   □

Note that System 5.60 is non-linear but nonetheless Lemma 5.32 is established. Non-linear cooperative systems will be systematically studied in Chapter 6.

5.8 Integral Cooperative Control Design

The virtual leader introduced in Sections 5.3.3 and 5.3.4 allows the hands-off operator(s) to adjust the group behavior of consensus or formation, but the discussion is limited to the case that the operator's output is constant. In Section 5.7, the presence of a virtual leader is also shown to increase robustness of the group behavior. To account for possible changes in the operator's output, a multi-layer cooperative control design is presented below, and the design renders an integral-type cooperative control law.

Consider first the simple case that the sub-systems are given by

$$\dot{z}_i = u_i, \quad i = 1, \cdots, q, \tag{5.61}$$

and that the virtual leader is given by

$$\dot{z}_0 = u_0, \tag{5.62}$$

where $u_i \in \Re$ is the control input to the ith sub-system, and $u_0 \in \Re$ is the operator control input. Assuming that $u_0 = v_0$ for a (piecewise-) constant velocity command v_0, we have

$$\dot{z}_0 = v_0, \quad \dot{v}_0 = 0.$$

In this case, cooperative control v_i can be chosen to be of the integral form

$$u_i = u_{p_i} + v_i, \quad \dot{v}_i = u_{v_i}, \tag{5.63}$$

where u_{p_i} is a relative-position feedback cooperative control law, and u_{v_i} is a relative-velocity feedback cooperative control law. For instance, in the spirit of Cooperative Control 5.11, u_{p_i} and u_{v_i} can be chosen to be

$$u_{p_i} = \frac{1}{\sum\limits_{\eta=0}^{q} s_{i\eta}(t)} \sum_{j=0}^{q} s_{ij}(t)(z_j - z_i) \quad \text{and} \quad u_{v_i} = \frac{1}{\sum\limits_{\eta=0}^{q} s_{i\eta}(t)} \sum_{j=0}^{q} s_{ij}(t)(v_j - v_i),$$

respectively. It follows that the overall closed-loop system can be written as

$$\dot{z} = [-I + \overline{D}(t)]z + v, \quad \dot{v} = [-I + \overline{D}(t)]v, \tag{5.64}$$

where

$$z = \begin{bmatrix} z_0 \\ z_1 \\ \vdots \\ z_q \end{bmatrix}, \quad \overline{S}(t) = \begin{bmatrix} 1 & 0 \cdots 0 \\ s_{10}(t) & \\ \vdots & S(t) \\ s_{q0}(t) & \end{bmatrix}, \quad \overline{D}(t) = \begin{bmatrix} \dfrac{\overline{S}_{ij}(t)}{\sum_{\eta=0}^{q} s_{i\eta}(t)} \end{bmatrix}, \quad v = \begin{bmatrix} v_0 \\ v_1 \\ \vdots \\ v_q \end{bmatrix}.$$

Closed-loop System 5.64 consists of two cascaded position and velocity sub-systems. It follows from Theorem 5.6 that, under uniform sequential completeness of $\overline{S}(t)$, $v \to c_v \mathbf{1}$ for some constant c_v. Defining $e_z = z - c_v t\mathbf{1}$, we can rewrite the position sub-system in (5.64) as

$$\dot{e}_z = [-I + \overline{D}(t)]e_z + (v - c_v\mathbf{1}),$$

whose convergence properties can be concluded using Lemmas 5.27 and 5.29. Specifically, if $\overline{S}(t)$ is uniformly sequentially complete, convergence of $v \to c_v \mathbf{1}$ is exponential, and hence convergence of $z \to (z_0(t_0) + c_v t)\mathbf{1}$ is in turn established.

As an illustration on the performance under Integral Control 5.63, consider the consensus problem of three entities defined by (5.61) with $q = 3$, let the virtual leader be that in (5.62) and with $u_0 = v_0 = 1$, and set the augmented sensor/communication matrix to be

$$\overline{S} = \begin{bmatrix} 1 & 0 & 0 & 0 \\ 1 & 1 & 1 & 0 \\ 0 & 0 & 1 & 1 \\ 0 & 1 & 0 & 1 \end{bmatrix}.$$

For the initial conditions of $[z_0(0)\ z_1(0)\ z_2(0)\ z_3(0)] = [0\ 6\ 2\ 4]$, the simulation results are given in Fig. 5.10 .

Obviously, System 5.61 can be extended to those in the form of (5.8), and the above design process remains. On the other hand, Virtual Leader 5.61 can also be extended to be that of (piecewise-) constant acceleration command, in which case the overall system under the integral cooperative control design becomes

$$\dot{z} = [-I + \overline{D}(t)]z + v, \quad \dot{v} = [-I + \overline{D}(t)]v + a, \quad \dot{a} = [-I + \overline{D}(t)]a,$$

for which cooperative stability can be concluded in the same fashion. More generalized operations by virtual leader, such as Models 5.24 and 5.22, can also be analyzed and designed.

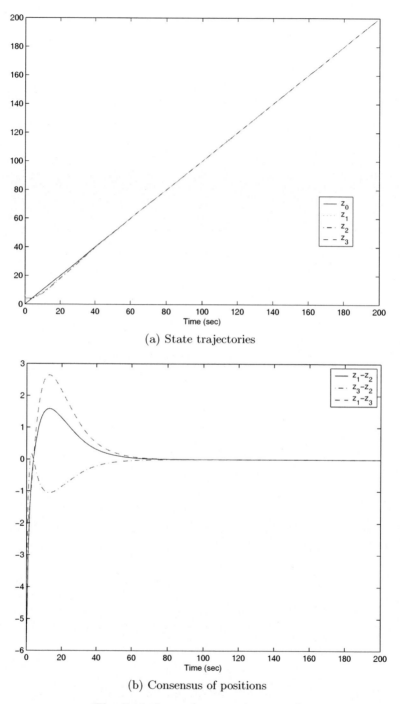

(a) State trajectories

(b) Consensus of positions

Fig. 5.10. Integral cooperative control

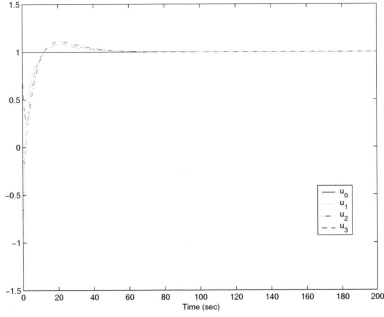

(c) Control inputs and consensus of velocities

Fig. 5.10 (continued)

5.9 Notes and Summary

Combined into the hierarchical control structure in Section 1.4, analysis of linear cooperative systems in this chapter makes it possible to design linear cooperative controls for autonomous vehicles operating in a dynamically changing and uncertain environment and to exhibit not only certain group behavior but also their individual behaviors.

Early work on distributed robotics is based on artificial intelligence methods [167], behavior-based control paradigm [7], and probabilistic exploration [69]. For example, decentralized architecture, task allocation, mapping building, coordination and control algorithms are pursued for multi-vehicle teams [5, 17, 71, 184]; rule-based formation behaviors are defined and evaluated through simulations [13]; heuristic rules are used to make mobile robots to form certain geometric patterns [248] or to converge to a point [4, 252]. Indeed, by observing animal behaviors and experimenting through computer animation, biologically inspired rules are obtained, for instance, cooperative rules [115], the rules of cohesion, separation and alignment in a flock [217], and avoidance, aggregation and dispersion [149].

The alignment problem [263] represents a truly local-feedback cooperative control problem, and the nearest neighbor rule is the solution initially derived from experimentation. Using the graph theoretical approach, convergence of the nearest neighbor rule is analyzed for a connected and undirected graph [98], for a directed graph with a spanning tree [214], for a directed graph with a globally reachable node [135]. Many of these analyses are done for agents of relative degree one; an extension to second-order dynamics is possible but involved [255, 256]. A recent account on convergence of multi-agent coordination, consensus and flocking can be found in [24].

Complementary to the graph theoretical approach, the matrix theoretical method is adopted in this chapter. This approach is standard in systems and control theory, and the lower triangular form and lower triangular completeness arise naturally due to their physical meanings [196, 198, 199, 201, 200]. As shown in Section 4.2, all the results presented have their graphical explanations and can be visually as explicit as a graph. In terms of analysis and design, the matrix theoretical approach has the advantage that heterogeneous systems of high relative-degree can be handled [202], control Lyapunov function can be found [205], and robustness can be analyzed [209].

In the simpler case that the topology of sensing and communication network is time-invariant, analysis and control of multi-vehicle systems can be carried out using such standard approaches in control theory as non-linear control, sliding mode control, and so on [50, 105, 128, 179, 182, 253, 269]. Early work on asymptotic stability analysis based on connective topology can be found in [228, 229]. Closely related is the synchronization problem of electric circuits [41, 277], and the Lyapunov direct method is used to derive Riccati-equation-like condition [276]. It is shown in Section 5.3 that the cooperative control framework can be used to solve the problems of consensus, rendezvous, tracking of a virtual leader, formation control, synchronization, and stabilization of interconnected systems. For autonomous vehicles that only use onboard sensors to acquire neighbors' information, maintaining network connectivity is crucial, and the circumcenter algorithm based on proximity graph [4, 43, 134, 147] or one of its variations [146] should be either directly implemented or used for choosing the weights in the cooperative control law.

6

Cooperative Control of Non-linear Systems

In this chapter, cooperative stability of the following networked-connected non-linear system is investigated:

$$\dot{x} = F^c(x, D(t)), \tag{6.1}$$

or equivalently element-by-element,

$$\dot{x}_i = F_i^c(x, D_i(t)) = F_i^c(d_{i1}(t)x_1, d_{i2}(t)x_2, \cdots, d_{in}(t)x_n),$$

where $x \in \Re^n$ is the state, $D(t) = [d_{ij}(t)] \in \Re_+^{n \times n}$ is a non-negative and piecewise-constant matrix (but, unlike in the preceding analysis, $D(t)$ is not required in general to be row-stochastic), $D_i(t)$ is the ith row of $D(t)$, and $F_i^c(\cdot)$ is the closed-loop dynamics of the ith state variable. In the case that x_i and x_j are state variables of the same physical sub-system in the network, the values of $d_{ij}(t)$ and $d_{ji}(t)$ characterize the structural property of dynamics coupling. If x_i and x_j are state variables of different physical sub-systems, $d_{ij}(t)$ and $d_{ji}(t)$ capture connectivity property of the sensing/communication network between the two sub-systems. While the structural property of sub-systems are typically fixed, the network topology often changes. Hence, matrix $D(t)$ is generally time-varying, and changing values of its entries are not assumed to be known *a priori* in either stability analysis or cooperative control design in order to consider arbitrary topology changes.

Cooperative stability of linear systems has been studied in Chapter 5 and based on convergence of multiplicative row-stochastic matrix sequences. However, non-linear systems do not have a piecewise-constant matrix solution. The standard method to overcome this difficulty and handle non-linear systems is to use the Lyapunov direct method. It will be shown in Section 6.1 that, if the network has either bidirectional or balanced irreducible topologies, a common and known cooperative control Lyapunov function is available to derive cooperative stability conditions. If the network has sequentially complete but otherwise arbitrary topology changes, common Lyapunov function

does not exist in general and, as shown in Section 5.6, a successful cooperative control Lyapunov function has to be solved based upon the topological
changes to be experienced by the network and hence cannot be found *a priori*. This marks the major distinction between Lyapunov stability analysis of
non-linear systems and cooperative stability analysis of networked non-linear
systems, and it is also the main difficulty in analyzing non-linear cooperative
systems. To overcome the difficulty, it is observed that cooperative control
Lyapunov functions found in Section 5.6 all have the same and known components. In Section 6.2, a multi-variable topology-based comparison theorem
and its extensions are introduced so that the knowledge of cooperative control Lyapunov function is no longer required and that asymptotic cooperative
stability can instead be analyzed using the known components of cooperative
control Lyapunov function. In Section 6.3, cooperative control is designed for
several classes of systems including cascaded systems, systems in the feedback
form, affine systems, *etc.* Discrete non-linear cooperative systems, cooperative
control of non-holonomic systems, and cooperative and pliable behaviors are
subsequently discussed in Sections 6.4, 6.5, and 6.6, respectively.

6.1 Networked Systems with Balanced Topologies

The Lyapunov direct method can be readily applied to conclude asymptotic
cooperative stability if a common cooperative control Lyapunov function can
be found for all the topologies under consideration. One such case is that the
network topology is restricted to be *balanced* in the sense that both $D(t)$ and
$D^T(t)$ are row-stochastic. As shown in Chapter 5, $D(t)$ being row-stochastic
can be satisfied. Should the network be bidirectional, $D(t) = D^T(t)$, and the
network is always balanced. The following lemma summarizes the conditions
on both system dynamics and network topology.

Lemma 6.1. *Consider Non-linear Networked System 6.1 and suppose that the
network changes are always balanced. Then, System 6.1 is uniformly asymptotically cooperatively stable if $D(t)$ is row-stochastic, piecewise-constant and
uniformly sequentially complete and if the following inequality holds along its
trajectory and for all i and k:*

$$\frac{d}{dt}(x_i - x_k)^2 \leq -2(x_i - x_k)^2 + 2\sum_{j=1}^{n}(x_i - x_k)[d_{ij}(t) - d_{kj}(t)]x_j. \quad (6.2)$$

Proof: System 6.1 includes as its special case the linear networked system
$\dot{x} = [-I + D(t)]x$ for which the time derivative of $(x_i - x_k)^2$ is

$$\frac{d}{dt}(x_i - x_k)^2 = -2(x_i - x_k)^2 + 2\sum_{j=1}^{n}(x_i - x_k)[d_{ij}(t) - d_{kj}(t)]x_j. \quad (6.3)$$

Consider the following candidate of common cooperative control Lyapunov function:

$$V_c(x) = \sum_{i=1}^{n} \sum_{k=1}^{n} (x_i - x_k)^2,$$

which is the special case of Cooperative Control Lyapunov Function 5.38 with $\lambda_i = 1$. It follows from $D(t)$ being balanced that $D^T \mathbf{1} = \mathbf{1}$ and hence Lemma 5.18 holds with $\lambda_i = 1$ and $P = I$. Hence, comparing the two expressions of (6.3) and (6.2) and then following the proof of Lemma 5.18, we conclude that the time derivative of $V_c(x)$ along any trajectory of System 6.1 is

$$\dot{V}_c(x) \leq -2 \sum_{i=1}^{n} e_{c_i}^T Q_{c_i}(t) e_{c_i}, \qquad (6.4)$$

where G_i is the matrix defined in Lemma 5.2, $e_{c_i} = G_i^T(x - x_i \mathbf{1})$, $Q_{c_i}(t) = G_i^T Q(t) G_i$, and

$$Q(t) = 2I - D(t) - D^T(t).$$

It follows from $Q(t)\mathbf{1} = 0$, from (d) of Theorem 4.27 and from $Q(t)$ being symmetric that $Q(t)$ is a singular M-matrix and also positive semi-definite. Thus, we know from (6.4) and (6.2) that $(x_i - x_k)^2$ are both uniformly bounded and uniformly continuous.

Assume without loss of any generality that, at some time t, non-negative $D(t)$ has the following lower triangular canonical form: for some permutation matrix $T(t)$ and for some integer $p(t) \geq 1$,

$$T^T(t)D(t)T(t) = \begin{bmatrix} E_{11} & 0 & \cdots & 0 \\ E_{21} & E_{22} & \cdots & 0 \\ \vdots & \vdots & \ddots & \vdots \\ E_{p1} & E_{p2} & \cdots & E_{pp} \end{bmatrix} \triangleq E_{\searrow}(t).$$

It follows from Lemma 4.44 that, if $D(t)$ is lower triangularly complete, $[2I + D(t) + D^T(t)]^k \geq [I + D(t)]^k + [I + D^T(t)]^k > 0$ for sufficiently large integer k. Hence, $[D(t) + D^T(t)]$ as well as $Q(t)$ are irreducible. By (a) of Theorem 4.27, matrix $Q(t)$ is of rank $(n-1)$ and, since $Q(t)\mathbf{1} = 0$, $z^T Q(t)z > 0$ for all $z \neq c\mathbf{1}$ and therefore matrix $Q_{c_i}(t)$ is positive definite. In this case, asymptotic cooperative stability can be concluded from (6.4).

In general, $D(t)$ may not be lower triangularly complete at any time t, but the cumulative topology changes over a finite period of time are lower triangularly complete, and it follows from (6.4) that

$$\int_{t_0}^{\infty} \left[\sum_{i=1}^{n} e_{c_i}^T Q_{c_i}(t) e_{c_i} \right] dt < \infty. \qquad (6.5)$$

In this case, the preceding analysis applies to lower-triangularly complete sub-blocks of $D(t)$ at every instant of time. That is, if k_1, k_2 are two indices corresponding to any lower-triangularly complete sub-matrix of $E_{\setminus}(t)$ given above and at any time t, there is a positive definite term of form $(x_{k_1} - x_{k_2})^2$ in the left hand of Inequality 6.5. Since the cumulative topology changes over a finite period of time are lower triangularly complete, all combinations of quadratic terms $(x_{k_1} - x_{k_2})^2$ appear in the left hand of Inequality 6.5. Since the sensing/communication sequence is uniformly sequentially complete, every term of $(x_{k_1} - x_{k_2})^2$ in (6.5) is integrated over an infinite sub-sequence of finite time intervals and, by Lemma 2.6, $(x_{k_1} - x_{k_2}) \to 0$ uniformly for all k_1 and k_2. That is, $x \to c\mathbf{1}$ for some $c \in \Re$, which completes the proof.   □

Lemma 6.1 is limited to networked systems with balanced topologies. Without any centralized coordination, balanced topologies can only be achieved by requiring that the network be bidirectional. Clearly, Lemma 6.1 needs to be generalized to overcome this restriction. Indeed, in many applications such as those governed by a leader-follower teaming strategy, the sensing and communication network is inherently unbalanced. In those cases, analysis and control of cooperative systems with arbitrary topologies should be carried out, which is the subject of the next section.

6.2 Networked Systems of Arbitrary Topologies

To analyze non-linear cooperative systems with arbitrary topologies, it is natural to develop a Lyapunov-based argument since the Lyapunov direct method does not require any knowledge of system solution and is a universal approach for analyzing non-linear systems. For linear networked cooperative systems, cooperative control Lyapunov function is explicitly found as complete-square function V_c in (5.44). Indeed, dynamics of linear cooperative systems are simpler in the sense that all the sub-systems can be expressed in a canonical form and, since there is no time-varying or non-linear dynamics except for piecewise-constant network changes, their Lyapunov functions are determined primarily by the topology changes. Unlike Lyapunov functions in typical control problems [108, 116, 192], cooperative control Lyapunov function V_c changes over time and is usually unknown and not differentiable because the network topology changes are instantaneous and not known *a priori*. For non-linear cooperative systems, finding a cooperative Lyapunov function becomes more challenging. In this section, we present a comparative Lyapunov argument by requiring the knowledge of not the cooperative control Lyapunov function itself but its components. The Lyapunov comparison theorem is then used to analyze and design several classes of non-linear cooperative systems.

6.2.1 A Topology-based Comparison Theorem

To motivate the basic concept of cooperative control Lyapunov function components and their comparative Lyapunov argument, consider first Linear Cooperative System 5.1, its Cooperative Control Lyapunov Function 5.44, and the closely related Lyapunov Function 5.36. It follows that, as topology changes, cooperative control Lyapunov function V_c and Lyapunov function V change accordingly. Nonetheless, V_c and V have the same components of x_i^2 and $(x_\mu - x_k)^2$ for any $i, \mu, k \in \{1, \cdots, n\}$, and hence these components are called *components of (cooperative control) Lyapunov functions*. It follows from (5.1) that

$$\frac{d}{dt}\left[\frac{1}{2}x_i^2\right] = -x_i^2 + \sum_{l=1}^{n} d_{il}(t)x_i x_l, \qquad (6.6)$$

and that

$$\frac{d}{dt}\left[\frac{1}{2}(x_\mu - x_k)^2\right] = -(x_\mu - x_k)^2 + \sum_{l=1}^{n}(x_\mu - x_k)[d_{\mu l}(t) - d_{kl}(t)]x_l. \quad (6.7)$$

Clearly, Linear Cooperative System 5.1 being uniformly bounded is equivalent to x_i^2 being uniformly bounded for all i, and Linear Cooperative System 5.1 being asymptotically cooperatively stable implies and is implied by that $(x_\mu - x_k)^2$ converges to zero for all μ, k. The first question arising is whether Lyapunov stability and asymptotic cooperative stability can be directly concluded from Equalities 6.6 and 6.7. Given Theorem 5.20, the answer to the first question should be affirmative. The second question arising is whether Equalities 6.6 and 6.7 can be generalized so that Lyapunov stability and asymptotic cooperative stability of non-linear cooperative systems can be concluded. In this regard, Lemma 6.1 provides an affirmative answer for the special case of balanced topologies. In what follows, a Lyapunov argument is provided for systems of arbitrary topologies. To this end, Equalities 6.6 and 6.7 are generalized into the following two conditions.

Definition 6.2. *A time function $\alpha(t)$ is said to be* strictly increasing *(or* strictly decreasing*) over a time interval $[t_1, t_2]$ if $\alpha(t_1) < \alpha(t_2)$ (or $\alpha(t_1) > \alpha(t_2)$) and if, for any $[t_1', t_2'] \subset [t_1, t_2]$, $\alpha(t_1') \le \alpha(t_2')$ (or $\alpha(t_1') \ge \alpha(t_2')$).*

Condition 6.3 *System 6.1 is said to be* amplitude dominant on the diagonal *if, for all i, differential inequality*

$$\frac{d}{dt}V_i(x_i) \le -\xi_i(|x_i|) + \eta_i(x_i)\sum_{l=1}^{n}d_{il}(t)[\beta_{i,l}(x_l) - \beta_{i,l}(x_i)], \qquad (6.8)$$

or

$$\frac{d}{dt}V_i(x_i) \le -\xi_i(|x_i|) + \eta_i(x_i)\sum_{l=1}^{n}d_{il}(t)\beta_{i,l}(x_l - x_i), \qquad (6.9)$$

holds for some positive definite, radially unbounded and differentiable function $V_i(\cdot)$, piecewise-constant entries $d_{il}(t)$ of non-negative matrix $D(t)$, non-negative function $\xi_i(\cdot)$, and strictly monotone increasing functions $\eta(\cdot)$ and $\beta(\cdot)$ with $\eta(0) = \beta(0) = 0$.

Condition 6.4 *System 6.1 is said to be* relative amplitude dominant on the diagonal *if, for any index pair* $\{\mu, k\}$, *the following differential inequality holds:*

$$\frac{d}{dt}L_{\mu,k}(x_\mu - x_k) \leq [\eta'_{\mu,k}(x_\mu) - \eta'_{\mu,k}(x_k)] \sum_{l=1}^{n} d_{\mu l}(t)[\beta'_{\mu,k,l}(x_l) - \beta'_{\mu,k,l}(x_\mu)]$$

$$- [\eta'_{\mu,k}(x_\mu) - \eta'_{\mu,k}(x_k)] \sum_{l=1}^{n} d_{kl}(t)[\beta''_{\mu,k,l}(x_l) - \beta''_{\mu,k,l}(x_k)]$$

$$- \xi'_{\mu,k}(|x_\mu - x_k|), \tag{6.10}$$

or

$$\frac{d}{dt}L_{\mu,k}(x_\mu - x_k) \leq \eta'_{\mu,k}(x_\mu - x_k) \sum_{l=1}^{n} [d_{\mu l}(t)\beta'_{\mu,k,l}(x_l - x_\mu)$$

$$- d_{kl}(t)\beta''_{\mu,k,l}(x_l - x_k)] - \xi'_{\mu,k}(|x_\mu - x_k|), \tag{6.11}$$

where $L_{\mu,k}(\cdot)$ *is a positive definite, radially unbounded and differentiable function, piecewise-constant time functions* $d_{il}(t)$ *are the entries of non-negative matrix* $D(t)$, *scalar function* $\xi'_{\mu,k}(\cdot)$ *is non-negative, and scalar functions* $\eta'_{\mu,k}(\cdot)$, $\beta'_{\mu,k,l}(\cdot)$, *and* $\beta''_{\mu,k,l}(\cdot)$ *are strictly monotone increasing functions and pass through the origin.*

Both Conditions 6.3 and 6.4 contain two inequalities, and the two inequalities have the same implications but are not the same unless the functions on the right hand sides are linear. In the conditions, matrix $D(t)$ is not required to be row-stochastic. Should $V_i(s) = 0.5s^2$, $\beta_{i,l}(s) = \eta_i(s) = s$, and matrix $D(t)$ be row-stochastic (*i.e.*, $\sum_l d_{il}(t) = 1$), Inequality 6.8 becomes

$$\frac{d}{dt}V_i(x_i) \leq -\xi_i(|x_i|) + x_i \sum_{l=1}^{n} d_{il}(t)[x_l - x_i],$$

$$= -\xi_i(|x_i|) - x_i^2 + \sum_{l=1}^{n} d_{il}(t)x_i x_l,$$

$$\leq -x_i^2 + \sum_{l=1}^{n} d_{il}(t)x_i x_l, \tag{6.12}$$

which includes (6.6) as a special case. Similarly, Inequality 6.10 includes (6.7) as a special case. Based on Conditions 6.3 and 6.4, the following theorem provides a straightforward Lyapunov argument on Lyapunov stability and asymptotic cooperative stability without requiring the knowledge of any specific Lyapunov function or cooperative control Lyapunov function.

Theorem 6.5. *Consider Networked System 6.1 satisfying the following conditions:*

(a) Matrix sequence of $D(t)$ over time is uniformly sequentially complete (which, by the structural property of each sub-system and through control design, is true if and only if the sensor/communnication sequence of $S(t)$ is uniformly sequentially complete over time). Furthermore, whenever its element $d_{ij}(t) \neq 0$, it is uniformly bounded from below by a positive constant.
(b) Conditions 6.3 and 6.4 hold.

Then System 6.1 is both uniformly Lyapunov stable and uniformly asymptotically cooperatively stable.

Proof: Let $\Omega = \{1, \cdots, n\}$ be the set of indices on state variables. Then, the following three index sub-sets of Ω are introduced to describe instantaneous values of the state variables: at any instant time t,

$$\Omega_{max}(t) = \{i \in \Omega : x_i = x_{max}\}, \quad \Omega_{mid}(t) = \{i \in \Omega : x_{min} < x_i < x_{max}\},$$

and

$$\Omega_{min}(t) = \{i \in \Omega : x_i = x_{min}\}.$$

where

$$x_{max}(t) = \max_{j \in \Omega} x_j(t), \quad \text{and} \quad x_{min}(t) = \min_{j \in \Omega} x_j(t).$$

It is apparent that, unless $x_i = x_j$ for all i and j, $x_{min} < x_{max}$ and set Ω is partitioned into the three mutually disjoint sub-sets of Ω_{max}, Ω_{mid} and Ω_{min}. In addition, the index set of maximum magnitude is defined as

$$\Omega_{mag}(t) = \{i \in \Omega : |x_i| = x_{mag}\}, \quad x_{mag}(t) = \max_{j \in \Omega} |x_j(t)|.$$

Thus, if $i \in \Omega_{mag}(t)$, either $i \in \Omega_{max}(t)$ or $i \in \Omega_{min}(t)$ but not both unless x_j are identical for all $j \in \Omega$. On the other hand, for each state variable x_i, we can define the set of its neighbors as

$$\Theta_i(t) = \{j \in \Omega : j \neq i \text{ and } d_{ij} > 0\}.$$

In addition, the set of its neighbors with distinct values is characterized by

$$\Theta_i'(t) = \{j \in \Omega : j \neq i, \quad d_{ij} > 0, \quad \text{and} \quad x_j \neq x_i\}.$$

In particular, the set of the neighbors for those variables of maximum magnitude is denoted by

$$\Theta_{i,mag}'(t) = \{i \in \Omega_{mag}, j \in \Omega : j \neq i, \quad d_{ij} > 0, \quad \text{and} \quad j \notin \Omega_{mag}\}.$$

Among all the state variables, the maximum relative distance is defined as

$$\delta_{max}(t) = \max_{\mu,k\in\Omega} |x_\mu(t) - x_k(t)|.$$

It is obvious that $\delta_{max}(t) = x_{max}(t) - x_{min}(t)$.

The proof is completed in six steps. The first step is regarding Lyapunov stability, and the rest are about convergence and asymptotic cooperative asymptotic stability. It is worth noting that, if $x_{min} = x_{max}$ at some instant of time t, $\Omega = \Omega_{max} = \Omega_{min}$ while Ω_{mid} is empty and that, by Step 3, the system is already asymptotically cooperatively stable. Thus, in the analysis after Step 3, $x_{min} < x_{max}$ is assumed without loss of any generality.

Step 1: Lyapunov Stability: to show Lyapunov stability, it is sufficient to demonstrate that the maximum magnitude of the state variables does not increase over time. Suppose without loss of any generality that, at time t, $i^* \in \Omega_{mag}(t)$. It follows from (6.8) or (6.9), the definition of $\Omega_{mag}(t)$, and $D(t)$ being non-negative that

$$\frac{d}{dt}V_{i^*}(|x_{i^*}|^2) \le -\xi_{i^*}(|x_{i^*}|)$$

$$-\begin{cases} \eta_{i^*}(|x_{i^*}|) \sum_{l=1}^{n} d_{i^*l}(t)[\beta_{i^*,l}(|x_{i^*}|) - \beta_{i^*,l}(x_l)\mathrm{sign}(x_{i^*})] \\ \eta_{i^*}(|x_{i^*}|) \sum_{l=1}^{n} d_{i^*l}(t)\beta_{i^*,l}(|x_{i^*}| - |x_l|) \end{cases}$$

$$\le -\xi_{i^*}(|x_{i^*}|) \le 0, \tag{6.13}$$

from which $x_{mag}(t)$ is known to be non-increasing over time.

Step 2: Strictly decreasing magnitude of $x_{mag}(t)$ over any time interval over which, for all $i \in \Omega_{mag}(t)$, the corresponding index sets $\Theta'_{i,mag}(t)$ do not remain empty: this conclusion is established by showing that, if $|x_i(t)| = x_{mag}(t)$ and $\Theta'_{i,mag}(t)$ is non-empty, $|x_i(t)|$ is strictly decreasing. By definition, we know from $\Theta'_{i,mag}(t)$ being non-empty that there exists some $j \ne i$ such that $d_{ij} > 0$ and $j \notin \Omega_{mag}$. In this case, Inequality 6.13 is strict and hence $|x_i(t)|$ is strictly decreasing. Furthermore, $x_{mag}(t)$ is strictly monotone decreasing if $\Theta'_{i,mag}(t)$ are non-empty for all t and for all $i \in \Omega_{mag}$. In addition, since $d_{ij}(t)$ is uniformly bounded away from zero whenever $d_{ij}(t) \ne 0$, it follows from the strict version of (6.13) that, if $x_{max}(t)$ is decreasing, the decreasing is uniformly with respect to time.

Step 3: Maximum distance $\delta_{max}(t)$ being non-increasing: recall that $\eta'_{\mu,k}(\cdot)$, $\beta'_{\mu,k,l}(\cdot)$ and $\beta''_{\mu,k,l}(\cdot)$ have the strictly monotone increasing property. Hence, we have that, for any $\mu^* \in \Omega_{max}$ and $k^* \in \Omega_{min}$,

$$\begin{cases} \eta'_{\mu^*,k^*}(x_{\mu^*}) - \eta'_{\mu^*,k^*}(x_{k^*}) > 0, & \eta'_{\mu^*,k^*}(x_{\mu^*} - x_{k^*}) > 0 \\ & \qquad \text{if } \Omega_{max} \cap \Omega_{min} = \emptyset, \\ \eta'_{\mu^*,k^*}(x_{\mu^*}) - \eta'_{\mu^*,k^*}(x_{k^*}) = 0, & \eta'_{\mu^*,k^*}(x_{\mu^*} - x_{k^*}) = 0 \\ & \qquad \text{if otherwise,} \end{cases}$$

$$\begin{cases} \sum_{l=1}^{n} d_{\mu^* l}(t)[\beta'_{\mu^*,k^*,l}(x_l) - \beta'_{\mu^*,k^*,l}(x_{\mu^*})] \leq 0, \\ \sum_{l=1}^{n} d_{\mu^* l}(t)\beta'_{\mu^*,k^*,l}(x_l - x_{\mu^*})] \leq 0, \end{cases} \qquad (6.14)$$

and

$$\begin{cases} \sum_{l=1}^{n} d_{k^* l}(t)[\beta''_{\mu^*,k^*,l}(x_l) - \beta''_{\mu^*,k^*,l}(x_{k^*})] \geq 0, \\ \sum_{l=1}^{n} d_{k^* l}(t)\beta''_{\mu^*,k^*,l}(x_l - x_{k^*}) \geq 0. \end{cases} \qquad (6.15)$$

Substituting the above inequalities into (6.10) and (6.11) that

$$\frac{d}{dt} L_{\mu^*,k^*}(x_{\mu^*} - x_{k^*}) \leq -\xi'_{\mu^*,k^*}(|x_{\mu^*} - x_{k^*}|) \leq 0,$$

from which $\delta_{max}(t)$ being non-increasing over time can be concluded.

Step 4: Maximum distance $\delta_{max}(t)$ being uniformly and strictly monotone decreasing as long as $D(t)$ is irreducible: based on the derivations in Step 3, we need only show that at least one of the inequalities in (6.14) and (6.15) is strict. To prove our proposition by contradiction, let us assume that both (6.14) and (6.15) be equalities. It follows that, unless $x_{min} = x_{max}$,

(6.15) being equalities $\implies d_{k^* l}(t) = 0$ if $l \in \Omega_{mid} \cup \Omega_{max}$ and $k^* \in \Omega_{min}$,

and that

(6.14) being equalities $\implies d_{\mu^* l}(t) = 0$ if $l \in \Omega_{mid} \cup \Omega_{min}$ and $\mu^* \in \Omega_{max}$.

Recall that, as long as $x_{min} < x_{max}$, index sets Ω_{min}, Ω_{mid} and Ω_{max} are mutually exclusive, and $\Omega_{min} \cup \Omega_{mid} \cup \Omega_{max} = \Omega$. This means that, unless $x_{min} = x_{max}$, there is permutation matrix $P(t)$ under which

$$P(t)D(t)P^T(t) = \begin{bmatrix} E_{11} & 0 & 0 \\ 0 & E_{22} & 0 \\ E_{31} & E_{32} & E_{33} \end{bmatrix} \triangleq E(t), \qquad (6.16)$$

where E_{ii} are square blocks, row indices of E_{11} correspond to those in Ω_{min}, row indices of E_{22} correspond to those in Ω_{max}, and row indices of E_{33} correspond to those in Ω_{mid}. Clearly, the structure of matrix $E(t)$ contradicts the knowledge that $D(t)$ is irreducible. Hence, we know that one of Inequalities 6.14 and 6.15 must be a strict inequality and hence δ_{max} is strictly monotone decreasing. Again, the decrease is uniform with respect to time since, whenever $d_{ij}(t) \neq 0$, $d_{ij}(t)$ is uniformly bounded away from zero.

Step 5: Maximum distance $\delta_{max}(t)$ being uniformly and strictly monotone decreasing if $D(t)$ may not be irreducible but remains to be lower triangularly complete: the proof by contradiction is identical to that of Step 4 because the structure of matrix $E(t)$ in (6.16) cannot be lower triangularly complete.

Step 6: Maximum distance $\delta_{max}(t)$ being strictly decreasing over an infinite sub-sequence of time intervals if $D(t)$ may not be lower triangularly complete over any of the intervals but is sequentially complete: assume that $\delta_{max}(t_0) > 0$. Then the conclusion is established by showing that, given any time instant t_1, there exists a finite duration Δt such that

$$\delta_{max}(t_1 + \Delta t) < \delta_{max}(t_1), \tag{6.17}$$

where $\Delta t > 0$ depends upon changes of $D(t)$ over $[t_1, t_2)$ and the value of $\delta_{max}(t_1)$.

Consider index sets $\Omega_{max}(t_1)$ and $\Omega_{min}(t_1)$. It follows that $\delta_{max}(t_1) = x_{\mu^*}(t_1) - x_{k^*}(t_1)$, where $\mu^* \in \Omega_{max}(t_1)$ and $k^* \in \Omega_{min}(t_1)$. Evolution of $\delta_{max}(t)$ after $t = t_1$ has two possibilities. The first case is that, for every $\mu^* \in \Omega_{max}(t_1)$, there exists $k^* \in \Omega_{min}(t_1)$ such that index pair $\{\mu^*, k^*\}$ belongs to the same lower-triangularly-complete block in the lower triangular canonical form of $D(t_1)$. In this case, it follows from Steps 5 and 3 that $\delta_{max}(t)$ is strictly decreasing at time $t = t_1$ and non-increasing afterwards. Therefore, we know that, for any $\Delta t > 0$, Inequality 6.17 holds.

The second and more general case is that, at time $t = t_1$ as well as in a finite interval afterwards, some of the indices in $\Omega_{max}(t)$ correspond to different diagonal block in the lower triangular canonical form of $D(t)$ than those for all the indices of $\Omega_{min}(t)$. In this case, Steps 4 and 5 are no longer applicable, while Step 3 states that $\delta_{max}(t)$ is non-increasing for all $t \geq t_1$. Nonetheless, the sequence of matrix $D(t)$ over time is uniformly sequentially complete and hence we know that, for any index $i \in \Omega_{mag}$, either $i \in \Omega_{max}(t)$ or $i \in \Omega_{min}(t)$, and set $\Theta'_{i,mag}$ cannot be non-empty except over some finite intervals. It follows from Step 2 that $x_{mag}(t)$ is strictly monotone decreasing over (possibly intermittent) time intervals and hence there exists a finite length Δt such that

$$x_{mag}(t_1 + \Delta t) < 0.5[x_{max}(t_1) - x_{min}(t_1)]. \tag{6.18}$$

Recalling $x_{mag}(t) = \max\{|x_{max}(t)|, |x_{min}(t)|\}$, we know from (6.18) that, for any $\mu \in \Omega_{max}(t_1 + \Delta t)$ and $k \in \Omega_{min}(t_1 + \delta t)$,

$$\begin{aligned}
\delta_{max}(t_1 + \Delta t) &= x_{max}(t_1 + \Delta t) - x_{min}(t_1 + \Delta t) \\
&\leq 2x_{mag}(t_1 + \Delta t) \\
&< x_{max}(t_1) - x_{min}(t_1),
\end{aligned}$$

which establishes Inequality 6.17. In essence, we know that, while $x_{mag}(t)$ (*i.e.*, $\max\{|x_{max}(t)|, |x_{min}(t)|\}$) decreases, the value of $[x_{max}(t) - x_{min}(t)]$

could remain unchanged but only temporarily (and at latest till the instant that $x_{max}(t) = -x_{min}(t)$), and afterwards $\delta_{max}(t)$ must decrease as $x_{mag}(t)$ does. Since t_1 is arbitrary, strictly decreasing of $\delta_{max}(t)$ over an infinite sub-sequence of time intervals is shown.

It is now clear from Steps 3 to 6 that, if $D(t)$ over time is uniformly sequentially complete, $\delta_{max}(t)$ is uniformly asymptotically convergent to zero. Hence, asymptotic cooperative stability is concluded.   □

6.2.2 Generalization

As has been shown in Chapter 5, Item (a) of Theorem 6.5 is a topological requirement on the system network, and it is both necessary and sufficient for uniform asymptotic cooperative stability. On the other hand, Conditions 6.3 and 6.4 can be generalized to the following two conditions on Networked Dynamical System 6.1.

Condition 6.6 *System 6.1 is said to be* amplitude dominant on the diagonal *if, for all i, inequality*
$$x_i F_i^c(x, D(t)) \leq 0$$
holds for all values of $x(t)$ with $|x_i(t)| = \max_{1 \leq j \leq n} |x_j(t)|$.

Condition 6.7 *System 6.1 is said to be* relative amplitude dominant on the diagonal *if, for any index pair $\{\mu, k\}$, inequality*
$$(x_i - x_k)[F_i^c(x, D(t)) - F_k^c(x, D(t))] \leq 0, \tag{6.19}$$
holds for all $x(t)$ with $x_i(t) = \max_{1 \leq j \leq n} x_j(t)$ and $x_k(t) = \min_{1 \leq j \leq n} x_j(t)$. Furthermore, System 6.1 is called relative-amplitude strictly dominant on the diagonal *if Inequality 6.19 holds and it becomes strict whenever $D(t)$ is lower triangularly complete and $\max_i x_i(t) \neq \min_j x_j(t)$.*

The diagonal dominance conditions of 6.6 and 6.7 not only generalize Conditions 6.3 and 6.4 but also are more intuitive to understand and easier to verify. It is worth recalling that the proof of Theorem 6.5 is essentially a combination of amplitude dominance on the diagonal, relative-amplitude dominance on the diagonal, and their relationship to topological condition in terms of non-negative and piecewise-constant matrix $D(t)$. Thus, the following theorem on asymptotic cooperative stability can analogously be concluded, and it is based on Conditions 6.6 and 6.7.

Theorem 6.8. *System 6.1 is both uniformly Lyapunov stable and uniformly asymptotically cooperatively stable if the following conditions are satisfied:*

(a) *Non-negative and piecewise-constant matrix sequence of $D(t)$ over time is uniformly sequentially complete. Furthermore, whenever its element $d_{ij}(t) \neq 0$, it is uniformly bounded from below by a positive constant.*

(b) *System 6.1 is both amplitude dominant on the diagonal and relative-amplitude strictly dominant on the diagonal.*

Without the properties of amplitude/relative-amplitude dominance on the diagonal, Comparison Theorems 6.5 and 6.8 are not expected to hold, neither is the Lyapunov argument in terms of cooperative Lyapunov function components. This is because, as shown in Sections 2.3.4 and 2.4.2, any transient overshoot could induce instability under switching and hence must be explicitly accounted for in a Lyapunov argument but, if a cooperative system has any overshoot, the overshoot cannot be calculated or estimated without the exact knowledge of the switching topologies.

To illustrate further Conditions 6.6 and 6.7 in terms of physical properties of System 6.1, let us first introduce the following two concepts. Then, Lemma 6.11 provides necessary and sufficient conditions on whether a system is minimum-preserving and/or maximum-preserving, and Lemma 6.12 establishes the relationship between amplitude/relative-amplitude dominance on the diagonal and minimum/maximum preservation.

Definition 6.9. *Dynamical System 6.1 is said to be* minimum preserving *if, for any initial condition $x(t_0) \geq c_{min}\mathbf{1}$, its trajectory preserves the minimum as $x(t) \geq c_{min}\mathbf{1}$.*

Definition 6.10. *System 6.1 is said to be* maximum-preserving *if, for any initial condition $x(t_0) \leq c_{max}\mathbf{1}$, its trajectory preserves the maximum as $x(t) \leq c_{max}\mathbf{1}$.*

Lemma 6.11. *System 6.1 is minimum-preserving if and only if, for any $j \in \{1, \cdots, n\}$, inequality*

$$F_j^c(x, D(t)) \geq 0$$

holds for all values of $x(t)$ with $x_j(t) = \min_{1 \leq k \leq n} x_k(t)$. System 6.1 is maximum-preserving if and only if, for any $j \in \{1, \cdots, n\}$, inequality

$$F_j^c(x, D(t)) \leq 0$$

holds for all values of $x(t)$ with $x_j(t) = \max_{1 \leq k \leq n} x_k(t)$.

Proof: For any initial condition $x(t_0) \geq c_{min}\mathbf{1}$ with $x_j(t_0) = c_{min}$, it follows from the property of $F_j^c(\cdot)$ that $\dot{x}_j(t_0) \geq 0$ and hence $x_j(t) \geq c_{min}$. Since j is arbitrary, we know that $x(t) \geq c_{min}\mathbf{1}$. On the other hand, it is apparent that, if $F_j^c(x, D(t)) < 0$ for some j and for some $x(t_0)$ with $x_j(t_0) = c_{min}$, $\dot{x}_j(t_0) < 0$ and hence $x_j(t_0 + \delta t) < c_{min}$ for some sufficiently small $\delta t > 0$. Since t_0 is arbitrary, the minimum-preserving is shown.

Necessity and sufficiency of a maximum-preserving system can be shown in a similar fashion.   □

Lemma 6.12. *If System 6.1 is both minimum-preserving and maximum-preserving, it is both amplitude and relative-amplitude dominant on the diagonal.*

Proof: Let $|x_i(t)| = \max_{1 \leq k \leq n} |x_k(t)| \geq 0$, and there are two cases. The first case is that $x_i = |x_i| \geq 0$ and hence $x_i(t) = \max_{1 \leq k \leq n} x_k(t)$, it follows from Lemma 6.11 that $F_i^c(x, D(t)) \leq 0$. The second case is that $x_i = -|x_i| \leq 0$ and hence $x_i(t) = \min_{1 \leq k \leq n} x_k(t)$, it follows again from Lemma 6.11 that $F_i^c(x, D(t)) \geq 0$. Combining the two case yields Condition 6.6.

Let $x_i(t) = \max_{1 \leq j \leq n} x_j(t)$ and $x_k(t) = \min_{1 \leq j \leq n} x_j(t)$. Then, $(x_i - x_k) \geq 0$. It follows from Lemma 6.11 that Condition 6.7 holds. □

It is worth noting that, if System 6.1 is both amplitude and relative-amplitude dominant on the diagonal, it is maximum-preserving but may not be minimum-preserving. This observation can be verified by considering the special case of $F_j^c(x, D(t)) = -\eta(x_j)$, where $\eta(\cdot)$ is a strictly increasing function with $\eta(0) = 0$. Indeed, with exception of the minimum preservation property, the aforementioned properties are all invariant under the introduction of such term $-\eta(x_j)$ into $F_j^c(x, D(t))$.

Combining Theorem 6.8 and Lemma 6.12 yields the following corollary.

Corollary 6.13. *System 6.1 is both uniformly Lyapunov stable and uniformly asymptotically cooperatively stable if the following conditions are satisfied:*

(a) *Matrix sequence of $D(t)$ over time is non-negative, piecewise-constant and uniformly sequentially complete. Furthermore, whenever its element $d_{ij}(t) \neq 0$, it is uniformly bounded from below by a positive constant.*

(b) *System 6.1 is both minimum-preserving and maximum-preserving and, in (6.19), the inequality of $(x_i - x_k)[F_i^c(x, D(t)) - F_k^c(x, D(t))] < 0$ holds uniformly whenever $\max_i x_i(t) \neq \min_j x_j(t)$ and $D(t)$ is lower triangularly complete.*

An application of Theorem 6.5 or 6.8 typically involves two aspects: directly check Conditions 6.6 and 6.7 (or Conditions 6.3 and 6.4), and verify that matrix $D(t)$ has the same reducibility/irreducibility property as matrix $S(t)$. It follows from Lemma 6.12 that Lemma 6.11, if applicable, can be used to simplify the first aspect and then Corollary 6.13 provides the stability results.

Example 6.14. Consider the following networked non-linear system:

$$\dot{\theta}_\mu = \sum_{j=1}^{q} s_{\mu j}(t)[\gamma_\mu(\theta_j) - \gamma_\mu(\theta_\mu)],$$

where $\mu = 1, \cdots, q$ for some positive integer q, $\gamma_\mu(\cdot)$ are strictly monotone increasing function with $\gamma_\mu(0) = 0$, and $s_{\mu j}(t)$ are the entries of sensing/coomunication matrix $S(t) \in \Re^{q \times q}$. It is straightforward to verify that the system is minimum-preserving, maximum-preserving, and also relative-amplitude strictly dominant on the diagonal. Thus, the networked system is globally uniformly asymptotically cooperatively stable provided that sensing/coomunication matrix $S(t)$ is uniformly sequentially complete.

For the networked system

$$\dot{\theta}_\mu = \sum_{j=1}^{q} s_{\mu j}(t)\gamma_\mu(\theta_j - \theta_\mu), \quad \mu = 1, \cdots, q,$$

global asymptotic cooperative stability can similarly be concluded. The same conclusion can also be drawn for Aggregation Model 1.3 and Synchronization Model 1.7.   ◇

Proof of Theorem 6.5 implies that local Lyapunov stability and local asymptotic cooperative stability can readily be concluded if Conditions 6.6 and 6.7 (or Conditions 6.3 and 6.4) hold only in a neighborhood around the origin.

Example 6.15. Consider the Kuramoto model:

$$\dot{\theta}_\mu = \sum_{j=1}^{q} e_{\mu j}(t)\sin(\theta_j - \theta_\mu),$$

where $\mu = 1, \cdots, q$,

$$e_{\mu j}(t) = \frac{w_{\mu j} s_{\mu j}(t)}{\sum_{i=1}^{q} w_{\mu i} s_{\mu i}(t)},$$

and $w_{\mu i} > 0$ are weighting constants. Since $\sin(\cdot)$ is locally strictly increasing with $\sin 0 = 0$ and since matrix $E(t) = [e_{\mu j}] \in \Re^{q \times q}$ has the same property of irreducibility/reducibility as matrix $S(t)$, the Kuramoto model is locally uniformly asymptotically cooperatively stable if sensing/coomunication matrix $S(t)$ is uniformly sequentially complete.

For the non-linear version of Vicsek model:

$$\dot{\theta}_\mu = -\tan(\theta_\mu) + \frac{1}{\cos(\theta_\mu)} \sum_{j=1}^{q} e_{\mu j}(t)\sin(\theta_j),$$

the same conclusion can be drawn.   ◇

To achieve cooperative stability, an appropriate cooperative control should be designed for every sub-system in the group and, based on the controls designed and on structures of individual systems, matrix $D(t)$ can be found in terms of $S(t)$. In the next section, specific classes of systems and their cooperative control designs are discussed.

6.3 Cooperative Control Design

In this section, the problem of designing cooperative control to achieve asymptotic cooperative stability is discussed for several classes of non-linear systems.

Specifically, cooperative control design is carried out first for relative-degree-one systems, then for systems in the feedback form, and also for the class of affine systems. The design process for non-affine systems will be outlined. Other classes of systems such as non-holonomic systems will be illustrated in Sections 6.5 and 6.6.

6.3.1 Systems of Relative Degree One

Consider the following systems:

$$\dot{y}_i = f_i(y_i) + v_i, \tag{6.20}$$

where $i = 1, \cdots, q$, $y_i \in \Re^m$ is the output of the ith system, and v_i is the control to be designed. System 6.20 is feedback linearizable. Without loss of any generality, assume that control v_i consists of an individual self feedback part and a cooperative control part, that is,

$$v_i(t) = -k_i(y_i)y_i + u_i(y, S(t)) + w_i(t), \tag{6.21}$$

where cooperative control part u_i can simply be set to be (5.11), $S(t)$ is the sensing and communication matrix, $w_i(t)$ presents the noises from measurement and communication channels, and self feedback gain $k_i(\cdot)$ is to be chosen such that inequalities

$$\begin{cases} y_i^T[f_i(y_i) - k(y_i)y_i] \leq 0 \\ (y_i - y_j)^T[f_i(y_i) - k_i(y_i)y_i - f_j(y_j) + k_j(y_j)y_j] \leq 0 \end{cases} \tag{6.22}$$

hold. It is straightforward to verify that, with the inequalities in (6.22) and with $w_i = 0$, Conditions 6.3 and 6.4 are met. Hence, by Theorem 6.5, System 6.20 under Control 6.21 is uniformly asymptotically cooperatively stable if there is no noise and if matrix $S(t)$ is uniformly sequentially complete over time.

There are numerous choices for self feedback gain $k_i(y_i)$ to satisfy the inequalities in (6.22). The simplest choice is $k_i(y_i)y_i = f_i(y_i)$, in which case the overall closed-loop system reduces to a linear cooperative system. If $y_i^T[f_i(y_i) - k_i(y_i)y_i] \leq -\epsilon_1\|y_i\|^{2+\epsilon_2}$ for some constants $\epsilon_l > 0$, the overall closed-loop system is input-to-state stable with respect to noise $w_i(t)$. One such choice is $k_i(y_i)y_i = f_i(y_i) + y_i\|y_i\|^2$, which ensures robustness.

To compare the resulting linear and non-linear cooperative systems, consider the case that $m = 1$, $q = 3$, and $f_i(y_i) = 0$, that the initial condition is $[-0.6 \quad 0.3 \quad 0.9]^T$, that the "noise" if present is a bias as $w(t) = 0.02 \times \mathbf{1}$, and that $S(t)$ is randomly generated as specified in Example 5.9. If $k_i(y_i) = 0$ is chosen, the resulting cooperative system is linear, and its performance is shown in Fig. 6.1. If $k_i(y_i) = y_i^2$ is chosen, the resulting cooperative system is non-linear, and its performance is shown in Fig. 6.2. Clearly, the non-linear feedback term improves convergence rate but more importantly guarantees

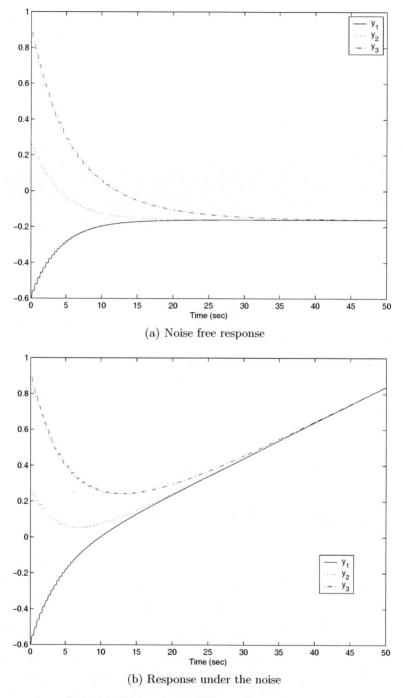

(a) Noise free response

(b) Response under the noise

Fig. 6.1. Performance of linear cooperative system

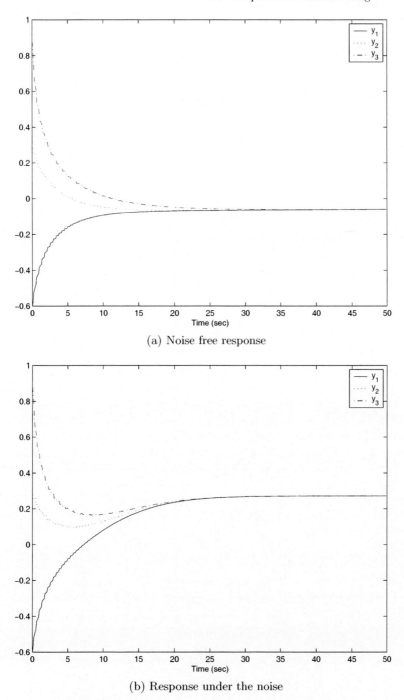

(a) Noise free response

(b) Response under the noise

Fig. 6.2. Performance of non-linear cooperative system

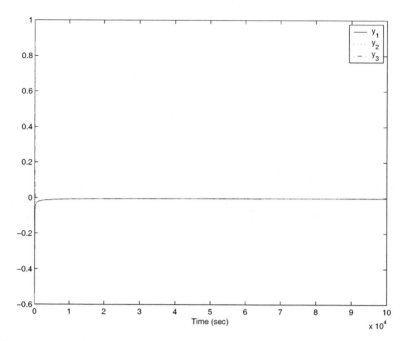

Fig. 6.3. Extended response of the non-linear cooperative system in Fig. 6.2(a)

uniform boundedness and robustness, while Linear Networked System 5.1 inherently has limited robustness in the sense that its state will become unbounded in the presence of persistent uncertainties.

On the other hand, the introduction of self feedback gain $k_i(y_i) = y_\mu^2$ prevents the ith system from having the minimum-preserving property. Consequently, any non-zero consensus reached by the overall system slowly bleeds to zero. As shown in Fig. 6.3, the system is indeed asymptotically stable. Thus, to achieve both asymptotic cooperative stability (but not asymptotic stability) and robustness, the better choice of $k_i(y_i)$ is

$$k_i(y_i) = \begin{cases} \|y_i\|^2 & \text{if } y \notin \Omega_f \\ 0 & \text{else} \end{cases}, \quad (6.23)$$

where $\Omega_f \subset \Re^{mq}$ is a set containing the origin, and it specifies that admissible equilibria of cooperative stability are $c\mathbf{1}$ with $c\mathbf{1} \in \Omega_f$.

6.3.2 Systems in the Feedback Form

Consider first the following group of heterogeneous systems:

$$\begin{cases} y_i = x_{i1}, \quad \dot{x}_{i1} = u_i, \quad i = 1, \cdots, q-1, \\ y_q = x_{q1}, \quad \begin{cases} \dot{x}_{q1} = \zeta_{q1}(\gamma_{q1}(x_{q2}) - \gamma_{q1}(x_{q1})), \\ \dot{x}_{q2} = \zeta_{q2}(\gamma_{q2}(x_{q3}) - \gamma_{q2}(x_{q2})), \\ \dot{x}_{q3} = u_q, \end{cases} \end{cases} \quad (6.24)$$

where $y_j \in \Re$ are the system outputs, $\gamma_{qi}(\cdot)$ and $\zeta_{qi}(\cdot)$ are strictly monotone increasing functions with $\gamma_{qi}(0) = 0$ and $\zeta_{qi}(0) = 0$, and $u_j \in \Re$ are the control inputs to be designed. While the first $(q-1)$ systems are linear, the qth system is cascaded and non-linear, and the cascaded structure and functional dependence of its dynamics can be represented by the non-negative matrix

$$A_q = \begin{bmatrix} 1 & 1 & 0 \\ 0 & 1 & 1 \\ 0 & 0 & 0 \end{bmatrix}.$$

For this group of q systems, non-linear self-state-feedback and neighbor-output-feedback cooperative control can be chosen to be

$$u_i = \begin{cases} \displaystyle\sum_{j=1}^{q} s_{ij}(t)\zeta_{i1}(\gamma_{i1}(y_j) - \gamma_{i1}(y_i)) & \\ & i = 1, \cdots, q-1, \\ \displaystyle\zeta_{q3}(\gamma_{q3}(y_q) - \gamma_{q3}(x_{q3}))\sum_{j=1}^{q} s_{qj}(t) + \sum_{j=1}^{q} s_{qj}(t)\zeta_{q3}(\gamma_{q3}(y_j) - \gamma_{q3}(y_q)) & \\ & i = q, \end{cases}$$

(6.25)

where $s_{ij}(t)$ are the entries of sensing/communication matrix $S(t)$, and $\gamma_{i1}(\cdot)$, $\gamma_{q3}(\cdot)$, $\zeta_{i1}(\cdot)$, and $\zeta_{q3}(\cdot)$ are odd and strictly monotone increasing functions. Substituting the controls into the system yields an overall networked system in the form of (6.1) in which the combined system structure and network matrix is

$$D(t) = \begin{bmatrix} s_{11} & \cdots & s_{1(q-1)}(t) & s_{1q}(t) & 0 & 0 \\ \vdots & & \vdots & \vdots & \vdots & \vdots \\ s_{(q-1)1}(t) & \cdots & s_{(q-1)(q-1)} & s_{(q-1)q}(t) & 0 & 0 \\ 0 & \cdots & 0 & 1 & 1 & 0 \\ 0 & \cdots & 0 & 0 & 1 & 1 \\ s_{q1}(t) & \cdots & s_{q(q-1)}(t) & s_{qq} & 0 & 1 \end{bmatrix}.$$

The above matrix can best be seen from the simplest case of $\zeta_{q3}(\tau) = \tau$ and $\gamma_{ij}(\tau) = \tau$. It follows from (a) of Lemma 5.5 that the above matrix has the same property of irreducibility/reducibility as matrix $S(t)$. For cooperative stability analysis, the following lemma is introduced.

Lemma 6.16. *Consider the following scalar differential equation:*

$$\dot{z}_l = \zeta(\gamma(z_i) - \gamma(z_l))\sum_j \lambda_j + \sum_j \lambda_j \zeta(\gamma(z_j) - \gamma(z_i)),$$

where $z_k \in \Re$, $\zeta(\cdot)$ and $\gamma(\cdot)$ are odd and non-decreasing functions, and $\lambda_j \geq 0$. Then, the differential equation of z_l or simply z_l is both minimum-preserving and maximum-preserving.

Proof: It follows that

$$\dot{z}_l = \sum_j \lambda_j [\zeta(\gamma(z_i) - \gamma(z_l)) + \zeta(\gamma(z_j) - \gamma(z_i))].$$

If $z_l = \max_k z_k$, we have

$$\gamma(z_i) - \gamma(z_l) \le \gamma(z_i) - \gamma(z_j),$$

which implies

$$\zeta(\gamma(z_i) - \gamma(z_l)) + \zeta(\gamma(z_j) - \gamma(z_i)) = \zeta(\gamma(z_i) - \gamma(z_l)) - \zeta(\gamma(z_i) - \gamma(z_j)) \le 0.$$

Similarly, if $z_l = \min_k z_k$, it can be shown that

$$\zeta(\gamma(z_i) - \gamma(z_l)) + \zeta(\gamma(z_j) - \gamma(z_i)) \ge 0.$$

The proof is completed by summarizing the two cases.   □

It is apparent that, for System 6.24 under Control 6.25, differential equations of $\dot{x}_{11}$ up to $\dot{x}_{q2}$ are both minimum-preserving and maximum-preserving. And, by Lemma 6.16, differential equation of $\dot{x}_{q3}$ is both minimum-preserving and maximum-preserving as well. It is straightforward to verify that the closed-loop system is also relative-amplitude strictly dominant on the diagonal. Therefore, by Corollary 6.13, System 6.24 under Cooperative Control 6.25 is uniformly asymptotically cooperatively stable if and only if $S(t)$ is uniformly sequentially complete.

As an example, consider System 6.24 and its Cooperative Control 6.25 with $q = 3$,

$$\zeta_{11}(w) = \zeta_{21}(w) = \zeta_{31}(w) = \zeta_{32}(w) = \zeta_{33}(w) = w,$$

and

$$\gamma_{11}(w) = \gamma_{21}(w) = w^5, \quad \gamma_{31}(w) = w^3, \quad \gamma_{32}(w) = w^{\frac{1}{3}}, \quad \gamma_{33}(w) = w^5.$$

Simulation is carried out with initial condition $[-1.5\ 2\ 0.3\ 0.4\ 1]^T$ and with $S(t)$ being randomly generated as specified in Example 5.9, and the simulation result is shown in Fig. 6.4.

It is straightforward to extend the above design and analysis to those systems in the feedback form. Specifically, consider a group of q systems defined by

$$\begin{cases} \dot{x}_{i1} = \zeta_{i12}(\gamma_{i12}(x_{i2}) - \gamma_{i12}(x_{i1})), \\ \dot{x}_{ij} = \sum_{k=1}^{j-1} a_{i,jk} \zeta_{ik(j+1)}(\gamma_{ik(j+1)}(x_{ik}) - \gamma_{ik(j+1)}(x_{ij})) \\ \qquad\quad + \zeta_{ij(j+1)}(\gamma_{ij(j+1)}(x_{i(j+1)}) - \gamma_{ij(j+1)}(x_{ij})), \\ \dot{x}_{in_i} = u_i, \end{cases} \qquad (6.26)$$

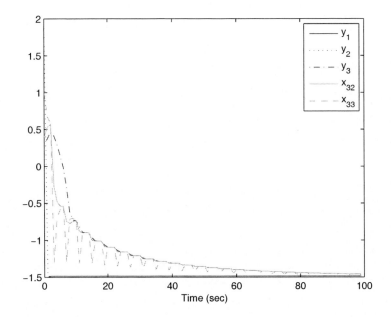

Fig. 6.4. Performance of System 6.24 under Cooperative Control 6.25

where $i = 1, \cdots, q$, $y_i = x_{i1}$ is the output, $a_{i,jk}$ are binary constants, and all the functions have similar properties as those in (6.24). The structure of the non-linear ith system can be represented by the non-negative matrix

$$
A_i = \begin{bmatrix}
1 & 1 & 0 & 0 & 0 \\
a_{i,21} & 1 & 1 & 0 & 0 \\
\vdots & \ddots & & \ddots & \vdots \\
a_{i,(n_i-1)1} & a_{i,(n_i-1)2} & \cdots & 1 & 1 \\
0 & 0 & 0 & \cdots & 0
\end{bmatrix} \in \Re_+^{n_i \times n_i},
$$

whose closed-loop version is always irreducible under any control u_i that contains y_i. Thus, cooperative control can be chosen to be

$$
u_i = \sum_{k=2}^{n_i-1} a_{i,n_ik}\zeta_{ikn_i}(\gamma_{ikn_i}(x_{ik}) - \gamma_{ikn_i}(x_{in_i}))
$$

$$
+ \zeta_{in_i1}(\gamma_{in_i1}(y_i) - \gamma_{in_i1}(x_{in_i})) \sum_{j=1}^{q} s_{ij}(t)
$$

$$
+ \sum_{j=1}^{q} s_{ij}(t)\zeta_{in_i1}(\gamma_{in_i1}(y_j) - \gamma_{in_i1}(y_i)), \tag{6.27}
$$

where $a_{i,n_i k}$ are binary constants that can be arbitrarily chosen. Similar to the previous analysis, one can conclude that System 6.26 under Control 6.27 is uniformly asymptotically cooperatively stable if and only if $S(t)$ is uniformly sequentially complete.

6.3.3 Affine Systems

Among the class of affine systems, the following systems are guaranteed to be minimum-preserving and maximum-preserving: for $i = 1, \cdots, q$ and for $j = 1, \cdots, n_i$,

$$\dot{x}_{ij} = \sum_{k=1}^{n_i} a_{i,jk} \rho_{ijk}(x_i) \zeta_{ik(j+1)} \left(\gamma_{ik(j+1)}(x_{ik}) - \gamma_{ik(j+1)}(x_{ij}) \right) + b_{ij} u_i, \quad (6.28)$$

where $x_i = [x_{i1} \cdots x_{in_i}]^T$, $\rho_{ijk}(x_i) > 0$, $\zeta_{ik(j+1)}(\cdot)$ and $\gamma_{ik(j+1)}(\cdot)$ are strictly monotone increasing functions passing through the origin, $a_{i,jk}$ are binary constants with $a_{i,ji} = 1$, and $b_{ij} = 0$ except for $b_{in_i} = 1$. The internal structure of System 6.28 is captured by the binary matrix

$$A_i = [a_{i,jk}] \in \Re_+^{n_i \times n_i}. \quad (6.29)$$

Now, consider the following neighbor-output-feedback cooperative control:

$$u_i = \sum_{j=1}^{q} s_{ij}(t) \zeta_i'(\gamma_i'(y_j) - \gamma_i'(x_{in_i})), \quad (6.30)$$

where $i = 1, \cdots, q$, and $\zeta_i'(\cdot)$ and $\gamma_i'(\cdot)$ are strictly monotone increasing functions passing through the origin. Substituting Control 6.30 into System 6.28 yields the overall networked system which is in the form of (6.1) and has the combined structural and networking matrix

$$D(t) = \begin{bmatrix} A_1 & \begin{matrix} 0 & 0 \\ s_{12}(t) & 0 \end{matrix} & \begin{matrix} 0 & 0 \\ s_{13}(t) & 0 \end{matrix} & \cdots & \begin{matrix} 0 & 0 \\ s_{1q}(t) & 0 \end{matrix} \\ \begin{matrix} 0 & 0 \\ s_{21}(t) & 0 \end{matrix} & A_2 & \begin{matrix} 0 & 0 \\ s_{23}(t) & 0 \end{matrix} & \cdots & \begin{matrix} 0 & 0 \\ s_{2q}(t) & 0 \end{matrix} \\ \vdots & \ddots & \ddots & \ddots & \vdots \\ \begin{matrix} 0 & 0 \\ s_{q1}(t) & 0 \end{matrix} & \begin{matrix} 0 & 0 \\ s_{q2}(t) & 0 \end{matrix} & \cdots & \begin{matrix} 0 & 0 \\ s_{q(q-1)}(t) & 0 \end{matrix} & A_q \end{bmatrix},$$

where matrices A_i are those given by (6.29). It follows from the preceding analysis and from Corollary 6.13 that the following theorem can be concluded.

Theorem 6.17. *Suppose that matrices A_i in (6.29) are all irreducible. Then, System 6.28 under Control 6.30 is uniformly asymptotically cooperatively stable if and only if $S(t)$ over time is uniformly sequentially complete.*

As an example, consider the following group of non-linear affine systems:

$$\begin{cases} \dot{x}_{11} = x_{12}^7 - x_{11}^7 \\ \dot{x}_{12} = \arctan(x_{13} - x_{12}) \ , \\ \dot{x}_{13} = x_{11}^3 - x_{13}^3 + u_i \end{cases} \quad \begin{cases} \dot{x}_{21} = \arctan(x_{22}^3 - x_{21}^3) \\ \dot{x}_{22} = x_{21}^3 - x_{22}^3 + x_{23} - x_{22} \ , \\ \dot{x}_{23} = x_{21}^7 - x_{23}^7 + u_2 \end{cases}$$

and

$$\begin{cases} \dot{x}_{31} = x_{32}^3 - x_{31}^3 + \sinh(x_{33} - x_{31}) \\ \dot{x}_{32} = x_{31} - x_{32} + \sinh(x_{33}^3 - x_{32}^3) \ . \\ \dot{x}_{33} = x_{31}^3 - x_{33}^3 + x_{32} - x_{33} + u_3 \end{cases}$$

Apparently, the above systems are in the form of (6.28), their internal dynamical interconnections are captured by matrices

$$A_1 = \begin{bmatrix} 1 & 1 & 0 \\ 0 & 1 & 1 \\ 1 & 0 & 1 \end{bmatrix}, \quad A_2 = \begin{bmatrix} 1 & 1 & 0 \\ 1 & 1 & 1 \\ 1 & 0 & 1 \end{bmatrix}, \quad A_3 = \begin{bmatrix} 1 & 1 & 1 \\ 1 & 1 & 1 \\ 1 & 1 & 1 \end{bmatrix},$$

and these structural matrices are all irreducible. Thus, Control 6.30 with $\zeta_i'(w) = w$ and $\gamma_i'(w) = w^5$ can be applied. Its simulation results with initial condition $[-1.5 \ 2 \ 0.1 \ 2 \ 0.4 \ 1 \ 1 \ -0.2 \ 2]^T$ are shown in Fig. 6.5, and all the state variables converge to 1.8513.

6.3.4 Non-affine Systems

To illustrate the process of designing cooperative control for non-affine systems, consider heterogeneous non-linear systems:

$$\dot{x}_i = F_i(x_i, v_i) \quad y_i = C_i x_i, \quad i = 1, \cdots, q, \tag{6.31}$$

where $x_i \in \Re^{n_i}$ is the state, $y_i \in \Re$ is the output, $v_i \in \Re$ is the control input, and $C_i = \begin{bmatrix} 1 & 0 & \cdots & 0 \end{bmatrix}$. Requiring only real-time knowledge of $S_i(t)$ (the ith row of sensing/communication matrix $S(t)$), control v_i at the ith system can be chosen to be of form

$$v_i = -K_i(x_i)x_i + u_i(x_i, y, S_i(t)), \tag{6.32}$$

where $K_i(x_i)$ is the self-feedback control gain matrix, and cooperative control part $u_i(\cdot)$ has a similar expression as (6.30). As such, u_i has the properties that $u_i(1, 1, S_i(t)) = 0$ and that, if $s_{ij}(t) = 0$ for all $j \neq i$, $u_i(x_i, y, S_i(t)) = 0$. Thus, for the overall system to exhibit cooperative behavior, it is necessary that the isolated dynamics of the ith system, $\dot{x}_i = F_i(x_i, -K_i(x_i)x_i)$, be asymptotically cooperatively stable by themselves. Accordingly, Control 6.32 should generally be synthesized in two steps:

Step 1: Find $K_i(x_i)$ such that isolated dynamics of $\dot{x}_i = F_i(x_i, -K_i(x_i)x_i)$ be uniformly asymptotically cooperatively stable.
Step 2: Determine $u_i(\cdot)$ in a way parallel to that in (6.27) or (6.30).

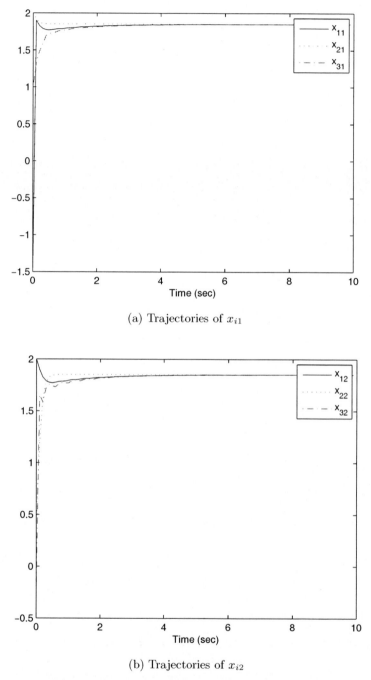

(a) Trajectories of x_{i1}

(b) Trajectories of x_{i2}

Fig. 6.5. Performance under Cooperative Control 6.30

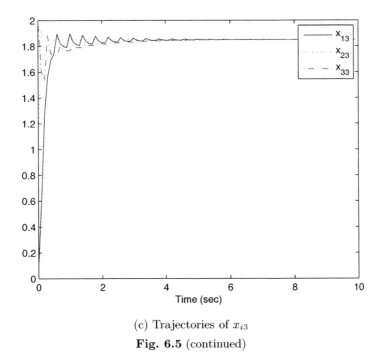

(c) Trajectories of x_{i3}

Fig. 6.5 (continued)

In both steps, Theorem 6.8 or Corollary 6.13 can be applied.

Definition 6.18. *Dynamical system* $\dot{z} = F(z, w)$ *is said to be* positive *if, for any non-negative initial condition* $x(t_0) \geq 0$ *and for any non-negative input* $w(t) \geq 0$, *its trajectory is also non-negative, i.e.,* $x(t) \geq 0$.

It is apparent that any minimum-preserving closed-loop system is always positive and that Lemma 6.11 can be applied to conclude whether a non-linear system is positive. An asymptotically cooperatively stable system may likely be positive but is not always; for instance, System 6.20 under Control 6.21 but with its self feedback term $-k_i(y_i)y_i$ redesigned (in a way parallel to (6.23)) to make admissible equilibrium set Ω_f contain only $c\mathbf{1}$ where $c < 0$. If Corollary 6.13 is successfully applied to System 6.31 under Control 6.32, the system is both asymptotically cooperatively stable and positive.

6.3.5 Output Cooperation

In the preceding subsections, cooperative control is designed to make all the state variables exhibit the same cooperative behavior denoted by $x \to c\mathbf{1}$. As such, system dynamics are required to meet stringent conditions. In many applications, what need to be cooperative are system outputs rather than the vectors of their whole state. In these cases, cooperative control can be designed by combining an output consensus algorithm and a standard tracking control. Such a combination is conceptually simpler, easier to design and implement, and requires less restrictive conditions on system dynamics.

To be specific, reconsider the collection of the systems in (6.31), and the output cooperation problem is to ensure $y_i(t) \to c$ for all i. Assume that, given any set of initial conditions of system outputs $\{y_i(t_0), \ i = 1, \cdots, q\}$, the desired output cooperative behavior be generated by the following discrete average algorithm:

$$y_i^d(t_{k+1}) = \frac{1}{s_{i1}(t) + \cdots + s_{iq}(t)} \sum_{j=1}^{q} s_{ij}(t_{k+1}) y_j^d(t_k), \quad y_i^d(t_0) = y_i(t_0), \quad (6.33)$$

where $\{t_k : k \in \aleph\}$ is the sequence of time instants at which there are changes in sensing/communication matrix $S(t)$. It follows from the discussion in Section 5.2.4 that, as long as matrix $S(t)$ over time is uniformly sequentially complete, average consensus of the desired outputs is ensured in the sense that

$$y_i^d(t_{k+1}) \to [y_1(t_0) + \cdots + y_q(t_0)]/q.$$

Upon prescribing Discrete Algorithm 6.33, the output cooperation problem of Systems 6.31 reduces to the problem of designing a tracking control $u_i = u_i(x_i, y_i^d)$ under which output tracking $y_i \to y_i^d$ is achieved for any piecewise-constant desired trajectory y_i^d. Such a tracking control can be designed using one of the standard techniques such as input-output feedback linearization and the backstepping design, and these techniques have been summarized in Chapter 2.

6.4 Discrete Systems and Algorithms

In general, a discrete non-linear networked system can be expressed as, given $x(k) = [x_1(k) \ \cdots \ x_n(k)]^T$,

$$x_i(k+1) = F_i^c(x(k), D(k)), \tag{6.34}$$

where $x(k) \in \Re^n$ is the state, $D(k) = [d_{ij}(k)] \in \Re_+^{n \times n}$ is a non-negative and diagonally positive matrix, and $F_i^c(\cdot)$ is the closed-loop dynamics of the ith state variable. Then, Discrete System 6.34 can be analyzed using components of cooperative control Lyapunov function in a way parallel to the analysis of Continuous-time System 6.1 in Section 6.2. Specifically, diagonal dominance conditions as well as minimum-preserving and maximum-preserving properties can analogously be defined, and Theorem 6.21 can be concluded in the same way as Theorem 6.8. It is straightforward to apply Theorem 6.21 to discrete non-linear systems such as Flocking Model 1.4. Also, Theorem 6.21 should be used to design discrete cooperative controls for various classes of networked systems (as done in Section 6.3).

Condition 6.19 *System 6.34 is called* amplitude dominant on the diagonal *or, equivalently,* maximum preserving *if, for all i, inequality*

$$F_i^c(x(k), D(k)) - x_i(k) \leq 0$$

holds for all values of $x(k)$ *with* $|x_i(k)| = \max_{1 \leq j \leq n} |x_j(k)|$.

Condition 6.20 *System 6.34 is called* minimum preserving *if, for all i, inequality*

$$F_i^c(x(k), D(k)) - x_i(k) \geq 0$$

holds for all values of $x(k)$ *with* $|x_i(k)| = \min_{1 \leq j \leq n} |x_j(k)|$. *Furthermore, System 6.34 is called* relative-amplitude strictly dominant on the diagonal *if inequality*

$$F_i^c(x(k), D(k)) - F_j^c(x(k), D(k)) - [x_i(k) - x_j(k)] \leq 0 \tag{6.35}$$

holds for all $x(k)$ *with* $x_i(k) = \max_{1 \leq l \leq n} x_l(k)$ *and* $x_j(k) = \min_{1 \leq l \leq n} x_l(k)$ *and if it becomes strict whenever* $\max_l x_l(k) \neq \min_l x_l(k)$ *and* $D(k)$ *is lower triangularly complete.*

Theorem 6.21. *System 6.34 is both uniformly Lyapunov stable and uniformly asymptotically cooperatively stable if the following conditions are satisfied:*

(a) Non-negative and diagonally positive matrix sequence of $D(k)$ *over time is uniformly sequentially complete. Furthermore, whenever its element* $d_{ij}(k) \neq 0$, *it is uniformly bounded from below by a positive constant.*

(b) System 6.34 is both amplitude dominant on the diagonal and relative-amplitude strictly dominant on the diagonal.

Non-linear cooperative systems arise naturally if one of non-linear consensus algorithms is applied. To focus upon non-linear consensus algorithms, let us assume for simplicity that the collection of agents have scalar and linear dynamics as

$$x_i(k+1) = u_i(k), \quad i = 1, \cdots, q. \tag{6.36}$$

Consider the so-called *modified circumcenter algorithm*: given any value of sensing/communication matrix $S(t) = [s_{ij}(t)]$ and for some scalar weight $0 \leq \beta \leq 1$,

$$x_i(k+1) = u_i(k) \triangleq \beta \max_{j: s_{ij}(k) \neq 0} x_j(k) + (1-\beta) \min_{j: s_{ij}(k) \neq 0} x_j(k), \tag{6.37}$$

which specifies the location of a shifted circumcenter. If $\beta = 0.5$, Algorithm 6.37 reduces to the standard circumcenter algorithm, that is,

$$x_i(k+1) = \frac{1}{2} \left[\max_{j: s_{ij}(k) \neq 0} x_j(k) + \min_{j: s_{ij}(k) \neq 0} x_j(k) \right].$$

If $\beta = 1$, it reduces to the *maximum algorithm*:

$$x_i(k+1) = \max_{j:s_{ij}(k)\neq 0} x_j(k);$$

and, if $\beta = 0$, it becomes the *minimum algorithm*:

$$x_i(k+1) = \min_{j:s_{ij}(k)\neq 0} x_j(k).$$

The modified circumcenter algorithm has several useful properties as summarized in the following lemma.

Lemma 6.22. *Consider Modified Circumcenter Algorithm 6.37. Then, its convergence properties are given as follows:*

(a) If $\beta \in (0,1)$, System 6.37 is uniformly asymptotically cooperatively stable if and only if its associated sensing/communication matrix is uniformly sequentially complete.

(b) If $\beta \in (0,1)$, System 6.37 is uniformly asymptotically cooperatively stable and its convergence to a consensus is in a finite time if its associated sensing/communication matrix $S(k)$ is symmetrical and the cumulative sensing/communication matrices over time intervals are irreducible.

(c) If $\beta = 0$ or $\beta = 1$, System 6.37 is uniformly asymptotically cooperatively stable and it converges to a consensus in a finite time if its associated cumulative sensing/communication matrices over time intervals are irreducible.

(d) If $\beta = 0$ or $\beta = 1$, System 6.37 may not converge to a consensus if its associated sensing/communication matrix is only uniformly sequentially complete or if the cumulative sensing/communication matrices over time intervals are only lower triangularly complete (over time).

Proof: Statement (a) is established by invoking Theorem 6.21 and noting that the dynamics of (6.37) are both minimum-preserving and maximum-preserving and that the strict inequality of (6.35) is satisfied unless

$$\max_{j:s_{ij}(k)\neq 0} x_j(k) = \min_{j:s_{ij}(k)\neq 0} x_j(k).$$

Statement (b) is due to the fact that, if $S(k) \in \Re^{q \times q}$ is symmetric, all the state variables associated with the same irreducible block converge to their modified circumcenter in no more than $(q-1)$ steps.

Statement (c) is obvious because, with $S(k)$ being irreducible and of order q, the maximum or the minimum propagates to all the state variables within $(q-1)$ steps.

To show statement (d), consider the case that $\beta = 1$ and that $S(t)$ is lower triangularly complete as

$$S(t) = \begin{bmatrix} S_{11}(t) & 0 & 0 & \\ S_{21}(t) & S_{22}(t) & 0 & 0 \\ \vdots & & \ddots & 0 \\ S_{p1} & \cdots & S_{p(p-1)} & S_{pp}(t) \end{bmatrix},$$

where $S_{ii}(t) \in \Re^{r_i \times r_i}$ are irreducible, $S_{ij}(t) \neq 0$ for some $j < i$, $x_i(t_0) <$ $\max_i x_i(t_0)$ for all $i \leq r_1$. It is apparent that the first group of agents never receives the maximum information from other agents and hence the maximum consensus cannot be reached. Indeed, given the above structural of $S(t)$, the system can reach its maximum consensus only if $x_i(t_0) = \max_i x_i(t_0)$ for some $i \leq r_1$. The same conclusion can be drawn for the minimum consensus algorithm.   □

In Section 5.4, the circumcenter algorithm is used to select the weighting coefficients in a linear cooperative control and to maintain network connectivity. In the next section, the maximum or minimum algorithm is used to design cooperative controls for non-holonomic systems.

6.5 Driftless Non-holonomic Systems

Consider a group of heterogeneous vehicles which move in the 2-D space and whose models are already transformed into the chained form

$$\dot{z}_{i1} = u_{i1}, \quad \dot{z}_{i2} = u_{i1}z_{i3}, \quad \cdots \quad \dot{z}_{i(n_i-1)} = u_{i1}z_{in_i}, \quad \dot{z}_{n_i} = u_{i2}, \qquad (6.38)$$

where $i \in \{1, \cdots, q\}$ is the index of the vehicles, $z_i = [z_{i1}, \cdots, z_{in_i}]^T \in \Re^{n_i}$ is the state of the ith vehicle, $u_i = [u_{i1}, u_{i2}]^T \in \Re^2$ is its control input, $y_i = [z_{i1}, z_{i2}]^T \in \Re^2$ is its output, and (z_{i1}, z_{i2}) are the x-axis and y-axis coordinates of the ith vehicle in a fixed world coordinate system, respectively. Vehicle motion in the 3-D space can be handled in a similar fashion.

The cooperative control design problem discussed in this section is to achieve consensus for vehicles' outputs. Note that, as shown in Section 5.3, combining consensus and geometrical displacements yields a solution to such real-world problems as formation control, etc. Let $S(t) = [s_{ij}(t)]$ and $\{t_k : k \in \aleph\}$ denote the matrix of the sensing/communication network among the vehicles and the time sequence of its topology changes, respectively. Then, there are two distinct cases for the consensus problem of non-holonomic systems.

6.5.1 Output Rendezvous

For any rendezvous control, $u_{i1}(t)$ must be vanishing over time. As shown in Chapter 3, a control design for the ith vehicle alone needs to be done carefully according to how $u_{i1}(t)$ converges zero. Here, we need not only synthesize an appropriately convergent control u_i but also ensure position consensus of y_i. To this end, let us define the following auxiliary signal vector based on the maximum algorithm: for all $t \in [t_k, t_{k+1})$ and for all $k \in \aleph$,

$$y_i^d(t) = y_i^d(k) = \begin{bmatrix} y_{i1}^d(k) \\ y_{i2}^d(k) \end{bmatrix}, \qquad (6.39)$$

where $\zeta > 0$ and $\epsilon > 0$ are design parameters, $\beta(k)$ is a scalar positive sequence defined by

$$\beta(k) \triangleq \epsilon e^{-\zeta t_k}, \quad k \in \aleph$$

and the desired rendezvous location for the ith vehicle and in (6.39) is updated by

$$y_i^d(-1) \triangleq \begin{bmatrix} z_{i1}(t_0) \\ z_{i2}(t_0) \end{bmatrix},$$
$$y_i^d(k) \triangleq \begin{bmatrix} \max\limits_{j:s_{ij}(k)\neq 0} y_{j1}^d(k-1) + \beta(k) \\ \max\limits_{j:s_{ij}(k)\neq 0} y_{j2}^d(k-1) \end{bmatrix} \quad k \in \aleph. \qquad (6.40)$$

It is obvious that, if $t_{k+1} - t_k \geq c$ for some constant $c > 0$, the sum of sequence $\beta(k)$ is convergent as

$$\sum_{k=0}^{\infty} \beta(k) = \sum_{k=0}^{\infty} e^{-\zeta t_k} \leq e^{-\zeta t_0} \sum_{k=0}^{\infty} e^{-k\zeta c} = \frac{e^{-\zeta t_0}}{1 - e^{-\zeta c}}.$$

To simplify the rendezvous design for non-holonomic systems, $y_j^d(k)$ rather than $y_j(t)$ is the information being shared through the network and used in the cooperative control design. Discrete Algorithm 6.40 of generating the desired rendezvous location $y_j^d(k)$ is a combination of the maximum consensus algorithm and bias $\beta(k)$ and hence, according to Lemma 6.22, it has following properties:

(a) Update $y_i^d(k)$ of desired location is piecewise-constant, non-decreasing and uniformly bounded. Also, its first element $y_{i1}^d(t_k)$ (i.e., $z_{i1}^d(t_k)$) is strictly increasing with respect to k.

(b) Excluding $\beta(k)$, the maximum algorithm of $y_i(t_0)$ converges to the maximum consensus in a finite time if cumulative sensing/communication matrix of $S(t)$ over time intervals is irreducible. The consensus of Biased Maximum Algorithm 6.40 is the maximum of vehicles' initial conditions plus the limit of $\sum_{k=0}^{\infty} \beta(k)$.

Accordingly, the consensus error system corresponding to System 6.38 is given by: for $t \in [t_k, t_{k+1})$,

$$\begin{cases} \dot{e}_{i1} = u_{i1}, \\ \dot{e}_{i2} = u_{i1} z_{i3}, \end{cases} \quad \cdots, \quad \dot{z}_{i(n_i-1)} = u_{i1} z_{in_i}, \quad \dot{z}_{in_i} = u_{i2}, \qquad (6.41)$$

where

$$e_{i1}(t) = z_{i1}(t) - y_{i1}^d(k), \quad e_{i2}(t) = z_{i2}(t) - y_{i2}^d(k).$$

Choose the following control for the first sub-system in (6.41):

$$u_{i1}(t) = -\zeta[z_{i1}(t) - y_{i1}^d(k)], \qquad (6.42)$$

under which the solution during the interval $t \in [t_k, t_{k+1})$ is

$$e_{i1}(t) = e^{-\zeta(t-t_k)}[z_{i1}(t_k) - y_{i1}^d(k)]. \qquad (6.43)$$

Recalling from (6.40) that $[z_{i1}(t_0) - y_{i1}^d(0)] \le -\beta(0)$, we know from (6.43) that $e_{i1}(t) < 0$ for all $t \in [t_0, t_1)$. Hence, in the interval of $[t_0, t_1)$, $u_{i1}(t) > 0$, $z_{i1}(t)$ is monotone increasing from $z_{i1}(t_0)$, and

$$z_{i1}(t_1) - y_{i1}^d(0) \le 0.$$

It follows from the above inequality and again from (6.40) that

$$z_{i1}(t_1) - y_{i1}^d(1) \le y_{i1}^d(0) - y_{i1}^d(1) \le -\beta(1).$$

Now, repeating the same argument inductively with respect to k, we can conclude the following facts using Solution 6.43 under Control 6.42:

(a) $u_{i1}(t) > 0$ for all finite instants of time, and it is monotone decreasing for $t \in [t_k, t_{k+1})$.

(b) For $t \in [t_k, t_{k+1})$, $z_{i1}(t)$ is monotone increasing from $z_{i1}(t_k)$, and the following two inequalities hold:

$$z_{i1}(t_k) - y_{i1}^d(k) \le -\beta(k), \quad \text{and} \quad z_{i1}(t) \le y_{i1}^d(k).$$

(c) It follows that, for all $t \in [t_k, t_{k+1})$ and no matter what (finite or infinite) value t_{k+1} assumes,

$$e^{\zeta t}u_{i1}(t) = -e^{\zeta t_k}\zeta[z_{i1}(t_k) - y_{i1}^d(k)] \ge \zeta e^{\zeta t_k}\beta(k) = \zeta\epsilon > 0.$$

(d) Output error component $e_{i1}(t)$ converges to zero exponentially.

Utilizing Property (c) above, one can design feedback control component $u_{2i}(t)$ as that defined by (3.84) (together with (3.82), (3.83) and (3.85)) to make output error component $e_{i2}(t)$ converge to zero. In summary, we have the following result.

Lemma 6.23. *Consider the group of vehicles in (6.38). Then, output rendezvous of $(y_i - y_j) \to 0$ is ensured under Biased Maximum Algorithm 6.40, Cooperative Control 6.42, and Control 3.84 if the cumulative versions of sensing and communication matrix $S(t)$ over composite time intervals are irreducible.*

To illustrate the performance under Biased Output Maximum Algorithm 6.40, consider three vehicles described by (6.38) of order $n_i = 3$, and choose

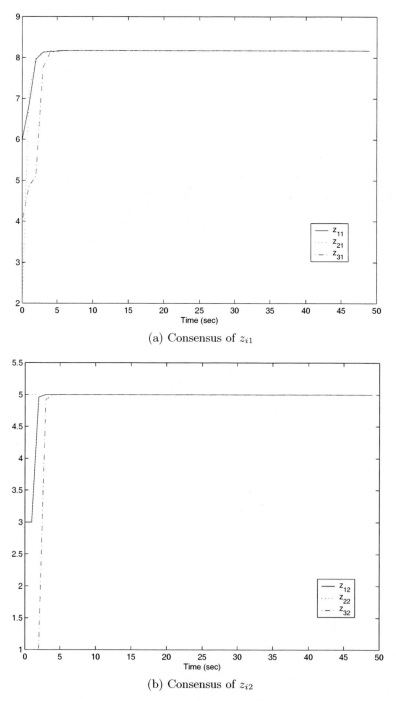

(a) Consensus of z_{i1}

(b) Consensus of z_{i2}

Fig. 6.6. Output consensus of three vehicles

$\zeta = 2$ and $\epsilon = 1$. Sensing/communication matrix $S(t)$ is randomly generated as specified in Example 5.9. Simulation is done for the initial output conditions of $y_1(t_0) = [6 \ 3]^T$, $y_2(t_0) = [2 \ 5]^T$, and $y_3(t_0) = [4 \ 1]^T$. As shown in Fig. 6.6, output variables y_{i2} follow $y_{i2}^d(k)$ and converge to the maximum of initial values of $z_{i2}(t_0)$, while output variables y_{i1} follow $y_{i1}^d(k)$ and converge to the maximum of initial values of $z_{i1}(t_0)$ plus a positive bias. Feedback control inputs u_{ij} corresponding to tracking $y_{ij}^d(k)$ are shown in Fig. 6.7.

In fact, any of the linear or non-linear consensus algorithms can be used to generate $y_{i2}^d(k)$. It is straightforward to develop a biased minimum algorithm to replace Biased Maximum Algorithm 6.40. The minimum or maximum algorithm is preferred for generating $y_{i1}^d(k)$ since these two algorithms render the nice property that $u_{i1}(t)$ has a fixed sign. An appropriate bias should always be added in the algorithm of generating $y_{i1}^d(k)$ since, if not, consensus for y_{i2} is not achievable when $y_{i1}(t_0)$ are all identical (in which case $u_{i1}(t)$ is zero under any of unbiased standard consensus algorithms).

6.5.2 Vector Consensus During Constant Line Motion

Consider the vehicles described by (6.38) and assume their non-holonomic virtual leader is described by

$$\dot{z}_{01} = u_{10}, \quad \dot{z}_{02} = u_{01}z_{03}, \quad \dot{z}_{03} = u_{02}. \tag{6.44}$$

The virtual leader is to command the group of vehicles through intermittent communication to some of the vehicles, and it is assumed to have a constant line motion as

$$u_{10}(t) = v_0, \quad u_{02}(t) = 0, \quad z_{03}(t_0) = 0, \tag{6.45}$$

where $v_0 \neq 0$ is a constant. The control objective is to ensure the vector consensus of $(z_{ij} - z_{kj}) \to 0$ for all available j (since models of different vehicles may be of different order) and for all $i, k = 0, 1, \cdots, q$ and $k \neq i$. Due to Chained Structure 6.38, the standard single consensus of $(z_{ij} - z_{il}) \to 0$ for $j \neq l$ does not make any sense physically and hence is not considered.

As analyzed in Chapter 3, each system in the chained form of (6.44) and (6.38) can be divided into two sub-systems. To facilitate the development of vector consensus control, a special form is developed for the second sub-system. For the virtual leader, $\mathrm{sign}(u_{01}(t))$ is constant, and the following state transformation can be defined:

$$w_{02}(t) = \mathrm{sign}(u_{01}(t))z_{02}(t), \quad w_{03}(t) = z_{03}(t) + w_{02}(t),$$

which is also one-to-one. Hence, the second sub-system of Virtual Leader 6.44 is mapped into

$$\dot{w}_{02} = |u_{01}|[w_{03}(t) - w_{02}(t)], \quad \dot{w}_{03}(t) = -w_{03}(t) + w_{02}(t), \tag{6.46}$$

provided that the second command component $u_{02}(t)$ is chosen as

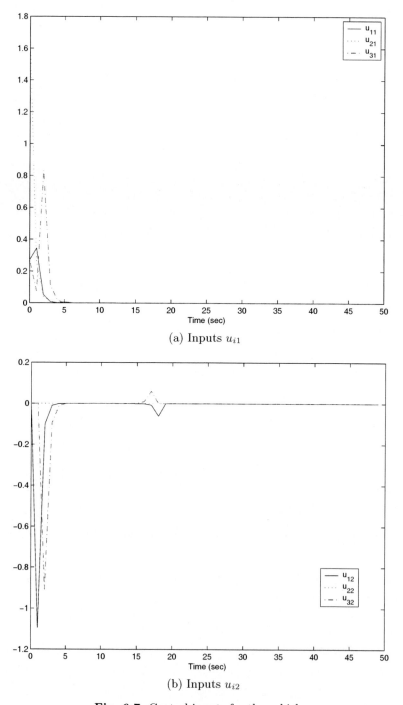

(a) Inputs u_{i1}

(b) Inputs u_{i2}

Fig. 6.7. Control inputs for the vehicles

$$u_{02}(t) = -z_{03}(t) - \text{sign}(u_{01}(t))z_{02}(t) - |u_{01}(t)|z_{03}(t) + w_{02}(t). \qquad (6.47)$$

It is straightforward to verify that, given (6.45), $u_{02}(t) = 0$ for all t.

The first sub-system of the ith vehicle in (6.38) is simply

$$\dot{z}_{i1} = u_{i1}. \qquad (6.48)$$

It follows from the discussion in Section 5.8 that the corresponding cooperative control component to ensure vector consensus of x-axis position z_{i1} and the corresponding velocity v_i should be chosen as

$$
\begin{aligned}
u_{i1} &\triangleq \frac{1}{\displaystyle\sum_{l=0}^{q} s_{il}(t)} \sum_{j=0}^{q} s_{ij}(t)[z_{j1} - z_{i1}] + v_i, \\
\dot{v}_i &= \frac{1}{\displaystyle\sum_{l=0}^{q} s_{il(t)}} \sum_{j=0}^{q} s_{ij}(t)(v_j - v_i),
\end{aligned}
\qquad (6.49)
$$

where $s_{ij}(t)$ is the element of binary augmented sensing/communication matrix $\overline{S}(t) \in \Re^{(q+1)\times(q+1)}$.

Control $u_{i1}(t)$ defined in (6.49) is piecewise-continuous with respect to the changes of sensing/communication topology and, within each of the corresponding time interval $[t_k, t_{k+1})$, $\text{sign}(u_{01}(t))$ is also piecewise-constant. For the second sub-system in (6.38), one can find (as for the virtual leader) a piecewise-differentiable state transformation from variables of $(z_{i2}, \cdots, z_{in_i})$ to those of $(w_{i2}, \cdots, w_{in_i})$ and a control transformation from u_{i2} to u'_{i2} such that

$$
\begin{cases}
\dot{w}_{i2} = |u_{i1}|(w_{i3} - w_{i2}), \\
\vdots \\
\dot{w}_{i(n_i-1)} = |u_{i1}|[w_{in_i} - w_{i(n_i-1)}], \\
\dot{w}_{in_i} = -w_{in_i} + u'_{i2}.
\end{cases}
\qquad (6.50)
$$

For instance, if $n_i = 3$, the state and control mappings are

$$w_{i2} = \text{sign}(u_{i1}(t))z_{i2}, \quad w_{i3} = z_{i3} + w_{i2}, \quad u_{i2} = -w_{i3} - |u_{i1}(t)|z_{i3} + u'_{i2}.$$

Similarly, if $n_i = 4$, the transformations are

$$w_{i2} = \text{sign}(u_{i1}(t))z_{i2}, \quad w_{i3} = z_{i3} + w_{i2}, \quad w_{i4} = \text{sign}(u_{i1}(t))z_{i4} + z_{i3} + w_{i3},$$

and

$$u_{i2} = \text{sign}(u_{i1}(t))[-w_{i3} - \text{sign}(u_{i1}(t))z_{i4} - u_{i1}(t)(w_{i4} - w_{i3}) + u'_{i2}],$$

which are one-to-one whenever $u_{i1}(t) \neq 0$. Then, the cooperative control component u_{i2} can be designed in terms of u'_{i2} as

$$u'_{i2} = \frac{1}{\displaystyle\sum_{l=0}^{q} s_{il(t)}} \sum_{j=0}^{q} s_{ij}(t) w_{j2}. \tag{6.51}$$

Having known vector consensus of z_{i1} and v_i and applying Theorem 6.8 or Corollary 6.13 to Sub-systems 6.46 and 6.50, we can conclude the following result on vector consensus of non-holonomic systems.

Lemma 6.24. *Consider the group of vehicles in (6.38) and suppose that their Virtual Leader 6.44 undertakes a line motion defined by (6.45). If the augmented sensing/communication matrix $\overline{S}(t)$ is uniformly sequentially complete, state consensus of $(z_{ij} - z_{kj}) \to 0$ is ensured under Cooperative Controls 6.49 and 6.51.*

To illustrate vector consensus, consider again three chained systems in the form of (6.38) and with $n_i = 3$. Virtual Leader 6.44 is to undertake a constant line motion specified by (6.45) and with $v_0 = 1$. Simulation is done with augmented sensor/communication matrix

$$\overline{S} = \begin{bmatrix} 1 & 0 & 0 & 0 \\ 1 & 1 & 1 & 0 \\ 0 & 0 & 1 & 1 \\ 0 & 1 & 0 & 1 \end{bmatrix}$$

and initial conditions of $z_0(t_0) = [0 \ -1 \ 0]$, $z_1(t_0) = [6 \ 3 \ 0]^T$, $z_2(t_0) = [2 \ 5 \ 0]^T$ and $z_3(t_0) = [4 \ 1 \ 0]^T$. Vector consensus is shown in Fig. 6.8, and the corresponding cooperative controls are shown in Fig. 6.9.

It is not difficult to show that, if $u_{10}(t)$ and $u_{20}(t)$ are polynomial functions of time, integral cooperative control laws can be designed along the line of (6.49) to achieve vector consensus for chained systems. In general, motion inputs $u_{10}(t)$ and $u_{20}(t)$ of the virtual leader may be generated by certain exogenous dynamics, and the above cooperative control design can be extended to yield vector consensus for chained systems.

6.6 Robust Cooperative Behaviors

A cooperative control design is successful if the resulting cooperative system has the following features:

(a) It is robust to the changes of the sensing and communication network, and it ensures asymptotic cooperative stability whenever achievable.
(b) It is pliable to physical constraints (such as non-holonomic constraints).
(c) It complies with the changes in the environment.

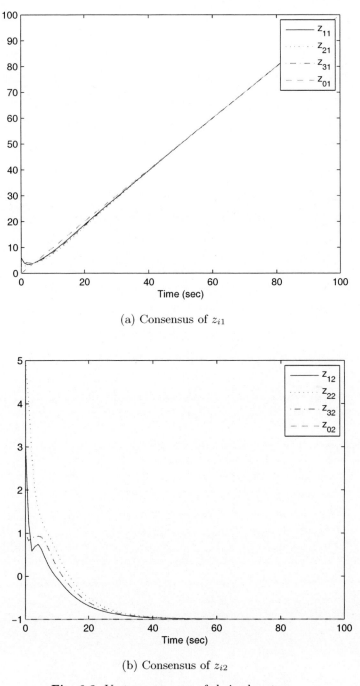

(a) Consensus of z_{i1}

(b) Consensus of z_{i2}

Fig. 6.8. Vector consensus of chained systems

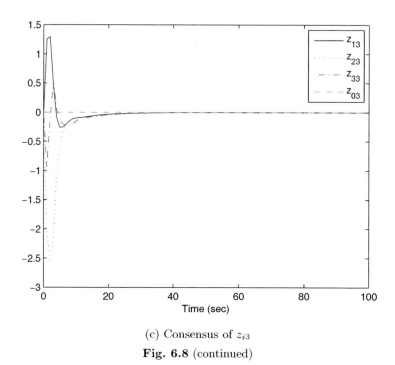

(c) Consensus of z_{i3}

Fig. 6.8 (continued)

Item (a) is achieved by applying the analysis and design tools of the matrix-theoretical approach in Chapter 4 and the topology-based Lyapunov argument in Section 6.2. In Section 6.5, cooperative control of non-holonomic systems is addressed. In what follows, the cooperative control design is shown to be robust to latency in the sensing/communication network and, by incorporating the reactive control design in Section 3.4.2, it can be also be made robust to environmental changes for autonomous vehicles.

6.6.1 Delayed Sensing and Communication

Consider the following group of heterogeneous non-linear systems:

$$\dot{z}_j(t) = f_j(z_j(t), v_j(t)) \quad y_j(t) = h_j(z_j(t)), \tag{6.52}$$

where $j = 1, \cdots, q$, $z_j(t) \in \Re^{n_j}$ is the state, $y_j \in \Re$ is the output, and $v_j \in \Re$ is the control input. Without loss of any generality, control input v_j can be assumed to be of form

$$v_j(t) = -g_j(z_j(t)) + u_j(z_j(t), s_{j1}(t)y_1(t), \cdots, s_{j(j-1)}(t)y_{j-1}(t),$$
$$s_{j(j+1)}(t)y_{j+1}(t), \cdots, s_{jq}(t)y_q(t)), \tag{6.53}$$

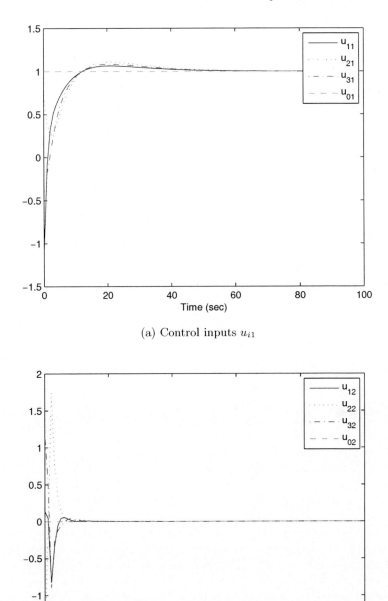

(a) Control inputs u_{i1}

(b) Control inputs u_{i2}

Fig. 6.9. Cooperative Control 6.49 and 6.51

where $g_j(z_j(t))$ is the self-feedback control part, $s_{jl}(t)$ is the (j,l)th element of sensing/communication matrix $S(t)$, and $u_j(\cdot)$ is the cooperative control part. Suppose that, under Control 6.53, System 6.52 is both amplitude dominant on the diagonal and relative-amplitude strictly dominant on the diagonal as defined by Conditions 6.6 and 6.7. Thus, by Theorem 6.8, the closed-loop system is uniformly asymptotically cooperatively stable if the network is uniformly sequentially complete.

Now, consider the case that there are sensing and communication delays in the network. In this case, Control 6.53 has to be modified to be

$$v_j(t) = -g_j(z_j(t)) + u_j(z_j(t), s_{j1}(t)y_1(t - \tau_{j1}), \cdots, s_{j(j-1)}(t)y_{j-1}(t - \tau_{j(j-1)}),$$
$$s_{j(j+1)}(t)y_{j+1}(t - \tau_{j(j+1)}), \cdots, s_{jq}(t)y_q(t - \tau_{jq})), \tag{6.54}$$

in which time delays $\tau_{ji} \in [0, \overline{\tau}]$ are explicitly accounted for. Without loss of any generality, $\overline{\tau}$ is assumed to be finite (otherwise, those sensing/communication channels may never be connected). If Control 6.54 ensures asymptotic cooperative stability for all $\tau_{ji} \in [0, \overline{\tau}]$, it is robust to network latency.

To conclude delay-independent cooperative stability, let us express the closed-loop dynamics of System 6.52 under Control 6.54 as

$$\dot{x}_i(t) = F_i^c(x(t), s_{j1}(t)y_1(t - \tau_{j1}), \cdots, s_{j(j-1)}(t)y_{j-1}(t - \tau_{j(j-1)}),$$
$$s_{j(j+1)}(t)y_{j+1}(t - \tau_{j(j+1)}), \cdots, s_{jq}(t)y_q(t - \tau_{jq})),$$

where $x = [z_1^T \cdots z_q^T]^T$, and x_i is an entry in sub-vector z_j of state x. Note that, in $F_i^c(\cdot)$, $x_i(t)$ appears without any delay, that is, $\tau_{jj} = 0$ for all j. It follows from Condition 6.6 that, since $F_i^c(\cdot)$ is amplitude dominant with respect to $|x_i(t)| = \max_l |x_l(t)|$ for the case that $\tau_{jl} = 0$ for all l, $F_i^c(\cdot)$ must be amplitude dominant with respect to $|x_i(t)| = \max_l \max_{\tau_{jl} \in [0,\overline{\tau}]} |x_l(t - \tau_{jl})|$ for the case that τ_{jl} may be non-zero. This implies that, in the presence of latency, Condition 6.6 remains except that the norm of $\max_l |x_l(t)|$ is replaced by functional norm of $\max_l \max_{\tau_{jl} \in [0,\overline{\tau}]} |x_l(t - \tau_{jl})|$. Similarly, Condition 6.7 can be extended. Upon extending Conditions 6.6 and 6.7, Theorem 6.8 can be proven as before to conclude delay-independent cooperative stability. Extensions of the maximum-preserving property, the minimum-preserving property and Corollary 6.13 are straightforward as well. In short, one can show using the concepts of amplitude dominance on the diagonal and relative-amplitude strict dominance on the diagonal that stability properties of System 6.52 under Control 6.53 are retained for System 6.52 under Control 6.54.

As an example, the following three systems

$$\begin{cases} y_1 = x_{11}, & \dot{x}_{11} = u_1, \\ y_2 = x_{21}, & \dot{x}_{21} = u_2, \\ y_3 = x_{31}, & \dot{x}_{31} = x_{32}^3 - x_{31}^3, \quad \dot{x}_{32} = x_{33}^{1/3} - x_{32}^{1/3}, \quad \dot{x}_{33} = y_3^5 - x_{33}^5 + u_3, \end{cases}$$

are considered in Section 6.3.2, and its neighbor-output-feedback cooperative control is selected to be

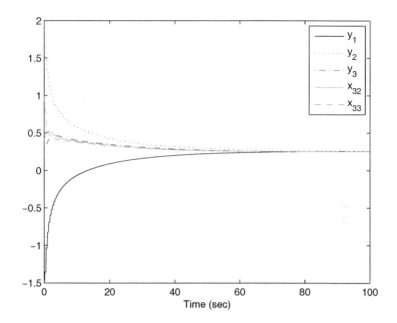

Fig. 6.10. Consensus under latency

$$u_i(t) = \frac{1}{\sum_{l \neq i} s_{il}(t)} \sum_{j=1,\, j \neq i}^{3} s_{ij}(t)[y_j^3(t) - y_i^3(t)], \quad i = 1, 2, 3.$$

In the presence of network delays, the cooperative control should be

$$u_i(t) = \frac{1}{\sum_{l \neq i} s_{il}(t)} \sum_{j=1,\, j \neq i}^{3} s_{ij}(t)[y_j^3(t - \tau_{ij}) - y_i^3(t)], \quad i = 1, 2, 3.$$

It is obvious that the overall system with delays is maximum-preserving, minimum-preserving, and relative-amplitude strictly dominant on the diagonal. Hence, asymptotic cooperative stability is delay-independent. To demonstrate this result, simulation is done with $\tau_{ij} = 0.5$ and under the same setting as that in Section 6.3.2 and, as shown in Fig. 6.10, consensus of all the state variables are achieved.

It is worth recalling that, using the concept of quasi diagonal dominance, delay-independent asymptotic stability is concluded for linear time-invariant systems [132] and for time-invariant interconnected systems with non-linear interconnections [166]. With the aforementioned functional norm, Theorem 6.8 can be used to conclude delay-independent asymptotic cooperative stability for non-linear and time-varying systems. On the other hand, if the self-feedback term in Control 6.54 is subject to time delay as $g_j(z_j(t - \tau_{jj}))$, the closed-loop dynamics no longer have delay-independent stability. In this case, cooperative stability of linear cooperative systems has been analyzed using the Nyquist stability criterion [62, 221], and may be achieved using predictive control [203].

6.6.2 Vehicle Cooperation in a Dynamic Environment

Consider a group of n_q vehicles operating in the following setting:

(a) There are many static and/or moving obstacles, and each of the vehicles can only detect those obstacles in its vicinity;

(b) The vehicles should follow a continual motion of their virtual leader according to a prescribed formation, while the augmented sensing and communication matrix $\overline{S}(t)$ capturing the information flow within the whole group is intermittent and time-varying;

(c) Control of each vehicle should ensure cooperation among the vehicles, avoid any collision with obstacles or other vehicles, and require only locally available information.

Given such a dynamically changing environment, the best option is to design a cooperative and reactive control. Since the vehicles are to move continually, it follows from the discussions in Section 3.1.3 that motion equations of the vehicles can be transformed into

$$\ddot{q}_i = u_i, \tag{6.55}$$

where $q_i \in \Re^l$ (with either $l = 2$ or $l = 3$) is the configuration coordinate of the ith vehicle, $i = 1, \cdots, n_q$, and u_i is the control to be designed. It follows from the discussions in Section 5.3.4 that the cooperative formation control is given by $u_i = u_i^c$, where

$$u_i^c = \frac{1}{s_{i0}(t) + s_{i1}(t) + \cdots + s_{in_q}(t)} \sum_{l=0}^{n_q} s_{il}(t) \left(q_l - q_i + \sum_{j=1}^{m} \alpha'_{ilj} \hat{e}'_{ilj}(t) \right), \tag{6.56}$$

e'_{ilj} are the estimates of orthonormal basis vectors of the moving frame $\mathcal{F}_l(t)$ at the lth vehicle, α'_{ilj} are the coordinates of the desired location for the ith vehicle with respect to the desired formation in frame $\mathcal{F}_i^d(t)$. Note that, by incorporating the virtual leader as a virtual vehicle, Cooperative Control 6.56 is not a tracking control and does not require the knowledge of $q^d(t)$ (or $q_i^d(t)$) all the time or for all the vehicles, nor is $\dot{q}^d$ or $\ddot{q}^d(t)$ needed. Accordingly, by setting $q_i^d = 0$ and $\xi_i(\cdot) = 0$, Reactive Control 3.118 reduces to

$$u_i = -\frac{\partial P_a(q_i)}{\partial q_i} + u_i^r, \tag{6.57}$$

where u_i^r is the repulsive control component defined by

$$u_i^r = -\sum_{j\in N_i} \frac{\partial P_{r_j}(q_i - q_{o_j})}{\partial q_i} - k\sum_{j\in N_i} \frac{\partial^2 P_{r_j}(q_i - q_{o_j})}{\partial(q_i - q_{o_j})^2}(\dot{q}_i - \dot{q}_{o_j})\|\dot{q}_{o_j}\|^2$$

$$-2k\sum_{j\in N_i} \frac{\partial P_{r_j}(q_i - q_{o_j})}{\partial(q_i - q_{o_j})}\dot{q}_{o_j}^T\ddot{q}_{o_j} + \varphi_i(q_i^*, \dot{q}_i, \dot{q}_i - \dot{q}_{o_j}), \tag{6.58}$$

N_i is the index set of all the entities (obstacles and other vehicles) in the vicinity of the ith vehicle, P_{r_j} is the repulsive potential field function for the jth entity listed in N_i, and $\varphi_i(\cdot)$ is a uniformly bounded term to overcome any saddle equilibrium point. Apparently, Cooperative Control 6.56 can be combined into Reactive Control 6.57 by piecewise defining the following co-operative attractive potential field function:

$$P_a(q_i) = \frac{1}{2\displaystyle\sum_{j=0}^{n_q} s_{ij}(t)}\sum_{l=0}^{n_q} s_{il}(t)\left(q_l - q_i + \sum_{j=1}^{m}\alpha_{ilj}'\hat{e}_{ilj}'(t)\right)^2, \tag{6.59}$$

where $s_{ij}(t)$ are piecewise-constants, and so may be $\hat{e}_{ilj}'(t)$. Therefore, cooperative and reactive control should be chosen to be

$$u_i = u_i^c + u_i^r, \tag{6.60}$$

where u_i^c is the cooperative control component defined by (6.56), and u_i^r is the repulsive control component defined by (6.58).

The autonomous vehicle system of (6.55) under Control 6.60 is obviously non-linear. It follows from the discussions in Section 5.6 and the preceding sections in this chapter that there is a cooperative control Lyapunov function $V_c(q)$ and that a sub-set of the Lyapunov function components correspond to Cooperative Potential Field Function 6.59 and hence to cooperative control component u_i^c. In a way parallel to the proof of Theorem 3.21 and to the preceding analysis, one can show that, away from the obstacles, cooperative behavior can be ensured if $\overline{S}(t)$ is uniformly sequentially complete and that, during the transient, there is no collision among the vehicles or between a vehicle and an obstacle. That is, the robust cooperative behavior is achieved. Should the obstacles form a persistent trap, the cooperative behavior of following the virtual leader may not be achievable.

To demonstrate the performance of Cooperative Reactive Control 6.60, consider a team of three physical vehicles plus a virtual vehicle as their virtual leader. The virtual leader and obstacles (one static and the other moving) in the environment are chosen to be the same as those in Section 3.4.2. Accordingly, all the reactive control terms are calculated using the same repulsive potential function in (3.122) with $\lambda_{r_j} = 5$ for all the obstacles and vehicles. For avoiding collision with one of the obstacles, $\rho_j = \epsilon_j = 0.3$. For avoiding collision among vehicles, $\rho_j = \epsilon_j = 0.2$. The vehicle team has initial positions of $q_1(t_0) = [3\ \ 3]^T$, $q_2(t_0) = [3.5\ \ 2]^T$ and $q_3(t_0) = [2\ \ 1]^T$ and is to

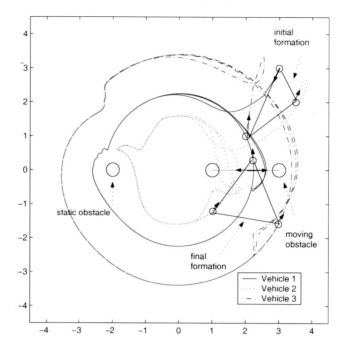

Fig. 6.11. Performance under Cooperative Reactive Control 6.60

follow the virtual leader in a formation while avoiding any collision. The sensing/communication matrix and the vehicle formation are chosen to be same as those in the simulation of Cooperative Control 5.29 in Section 5.3.4. Thus, formation coordinates α'_{ilj} and basis vectors $\hat{e}'_{ijl}$ are the same except that, since trajectory of the virtual leader is different,

$$
\begin{cases}
\alpha'_{101} = 1, \quad \alpha'_{102} = 0, \\
\hat{e}'_{101} = \begin{bmatrix} -\sin(\frac{\pi}{20}t) \\ \cos(\frac{\pi}{20}t) \end{bmatrix}, \\
\hat{e}'_{102} = \begin{bmatrix} -\cos(\frac{\pi}{20}t) \\ -\sin(\frac{\pi}{20}t) \end{bmatrix}.
\end{cases}
$$

Under this setting, Cooperative Reactive Control 6.60 is simulated, the vehicle system response is provided in Fig. 6.11, and the corresponding vehicle controls are shown in Fig. 6.12. To see better vehicles' responses, several snapshots over consecutive time intervals are provided in Figs. 6.13 - 6.15.

6.7 Notes and Summary

The standard comparison theorem is typically used to conclude qualitative properties from a scalar differential inequality by solving its equality version.

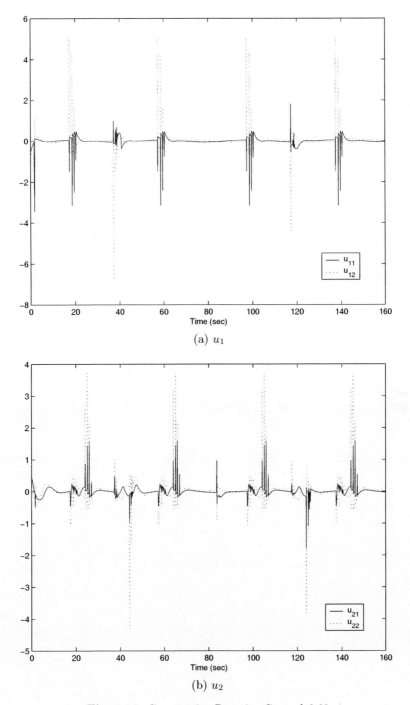

(a) u_1

(b) u_2

Fig. 6.12. Cooperative Reactive Control 6.60

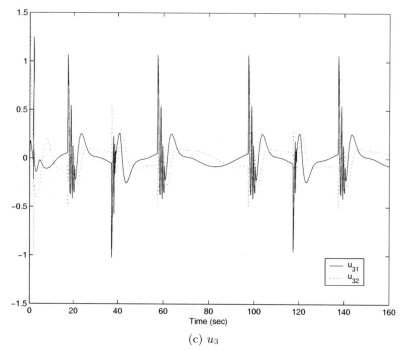

(c) u_3

Fig. 6.12 (continued)

This comparative argument is always an integrated part of Lyapunov direct method and has been widely used in stability analysis and control design [108]. A comparison theorem in terms of vector differential inequalities is also available (as theorem 1.5.1 on page 22 of [118]), and it requires the so-called quasi-monotone property (which does not hold for cooperative systems). Along this line, Theorems 6.5 and 6.8 enrich Lyapunov stability theory for the purpose of cooperative stability analysis of non-linear networked systems and their cooperative control designs. They can be referred to as comparison theorems for cooperative systems [193] and are used to design non-linear cooperative controls [194, 197].

Cooperative stability of discrete non-linear systems can be analyzed [161] using a combination of graph theory, convexity, and non-smooth analysis with non-differentiable Lyapunov function $V(x) = \max_i x_i - \min_j x_j$. The other approach is to conduct cooperative stability analysis in terms of the known components of cooperative control Lyapunov function, and proof and applications of Theorems 6.5 and 6.8 involve only a smooth and topology-based Lyapunov argument and do not require any convexity condition on system dynamics. Despite the unpredictable changes in $S(t)$ and hence in $D(t)$, the corresponding conditions in Theorems 6.5 and 6.8 can be easily checked and also used to proceed with cooperative control design for classes of continuous and discrete non-linear systems. Indeed, Theorems 6.5 and 6.8 also provide a smooth anal-

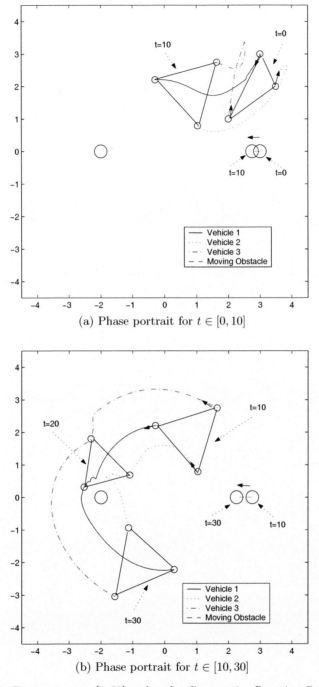

(a) Phase portrait for $t \in [0, 10]$

(b) Phase portrait for $t \in [10, 30]$

Fig. 6.13. Responses over $[0, 30]$ and under Cooperative Reactive Control 6.60

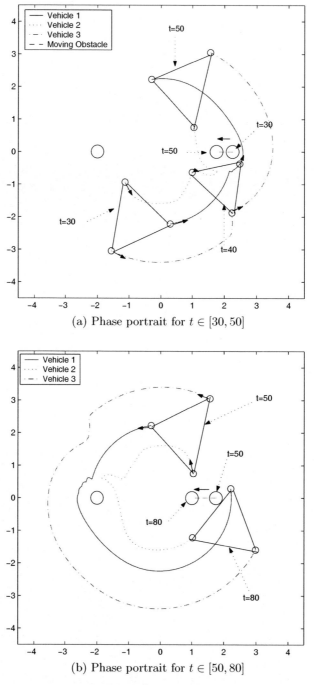

(a) Phase portrait for $t \in [30, 50]$

(b) Phase portrait for $t \in [50, 80]$

Fig. 6.14. Responses over $[30, 80]$ and under Cooperative Reactive Control 6.60

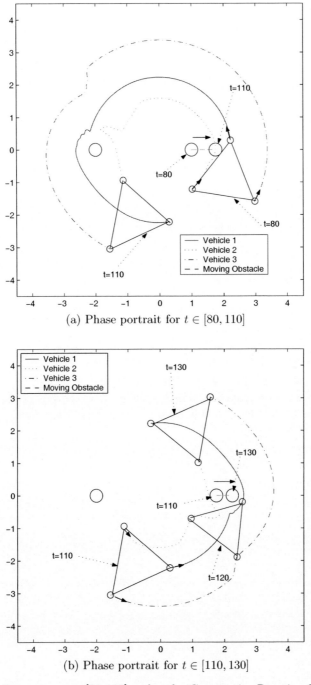

(a) Phase portrait for $t \in [80, 110]$

(b) Phase portrait for $t \in [110, 130]$

Fig. 6.15. Responses over $[80, 160]$ and under Cooperative Reactive Control 6.60

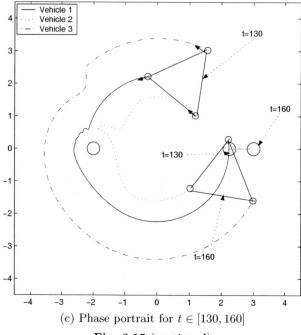

(c) Phase portrait for $t \in [130, 160]$

Fig. 6.15 (continued)

ysis and explicit conditions on whether non-differentiable Lyapunov function $V(x) = \max_i x_i - \min_j x_j$ (or equivalently, every of the known components of a cooperative control Lyapunov function) keeps decreasing along the trajectory of a networked system.

References

1. P. Alexandroff and H. Hopf. *Topologie.* J. Springer, Berlin, 1935
2. K. T. Alligood, T. D. Sauer, and J. A. Yorke. *Chaos: An Introduction to Dynamical Systems.* Springer-Verlag, New York, NY, 1997
3. B. D. O. Anderson and J. B. Moore. *Optimal Control Linear Quadratic Methods.* Prentice-Hall, Inc., Engelwood Cliffs, NJ, 1989
4. H. Ando, Y. Oasa, I. Suzuki, and M. Yamashita. Distributed memoryless point convergence algorithm for mobile robots with limited visibility. *IEEE Transactions on Robotics and Automation,* 51:818–828, 1999
5. T. Arai, E. Pagello, and L. E. Parker. Editorial: Advances in multi-robot systems. *IEEE Transactions on Robotics and Automation,* 18:655–661, 2002
6. T. Arai, E. Pagello, and L.E. Parker. Editorial: Advances in multi-robot systems. *IEEE Transactions on Robotics and Automation,* 18:655–661, 2002
7. R. Arkin. *Behavior-Based Robotics.* MIT Press, Cambridge, MA, 1998
8. Z. Artstein. Stabilization with relaxed control. *Nonlinear Analysis, TMA,* 7:1163–1173, 1983
9. A. Astolfi. Discontinuous control of nonholonomic systems. *Systems & control letters,* 27:37–45, 1996
10. M. Athens and P. L. Falb. *Optimal Control.* McGraw-Hill, New York, NY, 1966
11. J. B. Aubin and A. Cellina. *Differential Inclusions.* Springer-Verlag, Berlin, 1984
12. A. Bacciotti. *Local Stabilizability of Nonlinear Control Systems.* World Scientific, Singapore, 1992
13. T. Balch and R. C. Arkin. Behavior-based formation control for multirobot teams. *IEEE Transactions on Robotics and Automation,* 14:926–939, 1998
14. R.B. Bapat and T.E.S. Raghavan. *Nonnegative Matrices and Applications.* Cambridge University Press, Cambridge, 1997
15. J. Barraquand and J.-C. Latombe. Nonholonomic multibody mobile robots: controllability and motion planning in the presence of obstacles. In *Proceedings of the IEEE International Conference on Robotics and Automation,* pages 2328–2335, Sacramento, CA, April 1991
16. R. E. Bellman and S. E. Dreyfus. *Applied Dynamic Programming.* Princeton University Press, Princeton, NJ, 1962

17. G. Beni. The concept of cellular robot. In *Proceedings of 3rd IEEE Symposium on Intelligent Control*, pages 57–61, Arlington, Virginia, 1988

18. A. Berman and R. J. Plemmons. *Nonnegative Matrices in the Mathematical Sciences*. Academic Press, Inc., New York, 1979

19. N. Biggs. *Algebraic Graph Theory*. Cambridge University Press, Cambridge, UK, 1974

20. G. Birkhoff and R. S. Varga. Reactor criticality and nonnegative matrices. *Journal of the Society for Industrial and Applied Mathematics*, 6 (4):354–377, 1958

21. A. Bloch and S. Drakunov. Stabilization and tracking in the nonholonomic integrator via sliding modes. *Systems & Control letters*, 29:91–99, 1996

22. A. Bloch, M. Reyhanoglu, and N. H. McClamroch. Control and stabilization of nonholonomic dynamic systems. *IEEE Transactions on Automatic Control*, 37:1746–1757, 1992

23. A. M. Bloch. *Nonholonomic Mechanics and Control*. Springer, New York, 2003

24. V. D. Blondel, J. M. Hendrickx, A. Olshevsky, and J. N. Tsitsiklis. Convergence in multiagent coordination, consensus, and flocking. In *Proceedings of the 44th IEEE Conference on Decision and Control*, pages 2996 – 3000, Seville, Spain, Dec 2005

25. J. D. Boissonnat, A. Cerezo, and J. Leblond. Shortest paths of bounded curvature in the plane. In *Proceedings of IEEE International Conference on Robotics and Automation*, pages 2315–2320, Nice,France, May 1992

26. J. Borenstein and Y. Koren. The vector field histogram — fast obstacle avoidance for mobile robots. *IEEE Transactions Robotics and Automation*, 7:278–288, 1991

27. S. Boyd, L. E. Ghaoui, E. Feron, and V. Balakrishnan. *Linear Matrix Inequalities in System and Control Theory*. SIAM, 1994

28. M. S. Branicky. Multiple lyapunov functions and other analysis tools for switched and hybrid systems. *IEEE Transactions on Automatic Control*, 43:475 – 482, 1998

29. C. M. Breder. Equations descriptive of fish schools and other animal aggregations. *Ecology*, 3:361–370, 1954

30. R. W. Brockett. Asymptotic stability and feedback stabilization. In *Differential Geometric Control Theory*, volume R. W. Brockett, R. S. Millman, and H. J. Sussmann, Eds, pages 181–191, 1983

31. R.A. Brooks. A robust layered control system for a mobile robot. *IEEE Journal of Robotics and Automation*, 2:14–23, 1986

32. L. E. J. Brouwer. Ueber abbildungen von mannigfaltigkeiten. *The Mathematische Annalen*, 71:97115, 1912

33. A. E. Bryson and Y. C. Ho. *Applied Optimal Control*. Hemisphere Publishing Corp., Bristol, PA, 1975

34. L. Bushnell, D. Tilbury, and S. S. Sastry. Steering three-input chained form nonholonomic systems using sinusoids: The firetruck example. In *Proceedings of European Controls Conference*, pages 1432–1437, 1993

35. H. Caswell. *Matrix Population Models*. Sinauer Associates, Sunderland, MA, 2001

36. W. W. Charles. A technique for autonomous underwater vehicle route planning. *IEEE Transactions of Oceanic Engineering*, 3:199–204, 1990

37. B. Charlet, J. Levine, and R. Marino. On dynamic feedback linearization. *System & Control Letters*, 13:143151, 1989

38. B. Charlet, J. Levine, and R. Marino. Sufficient conditions for dynamic state feedback linearization. *SIAM Journal on Control and Optimization*, 29:3857, 1991

39. C. T. Chen. *Linear System Theory and Design*. Holt, Rinehart and Winston, Inc., New York, 1984

40. W. L. Chow. Uber systemen von linearren partiallen differentialgleichungen erster ordnung. *The Mathematische Annalen*, 117:98–105, 1939

41. L. O. Chua and D. N. Green. A qualitative analysis of the behavior of dynamic nonlinear networks: stability of autonomous networks. *IEEE Transactions on Circuits and Systems*, 23:355–379, 1976

42. J. Chuyuan, Z. Qu, E. Pollak, and M. Falash. A new multi-objective control design for autonomous vehicles. In *Proceedings of the 8th International Conference on Cooperative Control and Optimization*, University of Florida, Gainesville, Florida, January 30 - February 1 2008

43. J. Cortes, S. Martinez, and F. Bullo. Robust rendezvous for mobile autonomous agents via proximity graphs in arbitrary dimensions. *IEEE Transactions on Automatic Control*, 51:1289–1298, 2006

44. J. Cortes, S. Martinez, T. Karatas, and F. Bullo. Coverage control for mobile sensing networks. *IEEE Transactions on Robotics and Automation*, 20:243–255, 2004

45. I. D. Couzin, J. Krause, N. R. Franks, and S. A. Levin. Effective leadership and decision-making in animal groups on the move. *Nature*, 433:513–516, 2005

46. F. Cucker and S. Smale. Emergent behavior in flocks. *IEEE Transactions on Automatic Control*, 52:852–862, 2007

47. A. Czirok, A. L. Barabasi, and T. Vicsek. Collective motion of self-propelled particles: kinetic phase transition in one dimension. *Physical Review Letters*, 82:209–212, 1999

48. I. Daubechies and J. C. Lagaias. Sets of matrices all infinite products of which converge. *Linear Algebra and Applications*, 161:227–263, 1992

49. W. P. Dayawansa and C. F. Martin. A converse lyapunov theorem for a class of dynamical systems which undergo switching. *IEEE Transactions on Automatic Control*, 44:751–760, 1999

50. J. P. Desai, J. Ostrowski, and V. Kumar. Controlling formations of multiple mobile robots. In *Proceedings of IEEE Conference on Robotics and Automation*, pages 2864–2869, Leuven, Belgium, May 1998

51. R. Diestel. *Graph Theory*. Springer-Verlag, New York, 2000

52. D. V. Dimarogonal and K. J. Kyriakopoulos. On the rendezvous problem for multiple nonholonomic agents. *IEEE Transactions on Automatic Control*, 52:916–922, 2007

53. A. W. Divelbiss and J. T. Wen. A path space approach to nonholonomic motion planning in the presence of obstacles. *IEEE Transactions on Robotics and Automation*, 13:443–451, 1997

54. R. L. Dobrushin. Central limit theorem for nonstationary markov chains, parts i and ii. *Theory of Probability and Its Applications*, 1, no.4:65–80 and 329–383, 1956

55. B. Donald, P. Xavier, J. Canny, and J. Reif. Kinodynamic motion planning. *Journal of the ACM*, 40:1048–1066, 1993

56. L. E. Dubins. On curves of minimal length with a constraint on average curvature, and with prescribed initial and terminal positions and tangents. *American Journal of Mathematics*, 79:497–516, 1957

57. O. Egeland, E. Berglund, and O. J. Sordalen. Exponential stabilization of a nonholonomic underwater vehicle with constant desired configuration. In *Proceedings of IEEE Conference on Robotics and Automation*, pages 20–25, S. Diego, CA, May 1994

58. M. Erdmann and T. Lozano-Perez. On multiple moving objects. In *Proceedings of IEEE International Conference on Robotics and Automation*, pages 1419–1424, San Francisco, CA, Apr. 1986

59. H. Eren, C. F. Chun, and J. Evans. Implementation of the spline method for mobile robot path control. In *Proceedings of the 16th IEEE Instrumentation and Measurement Technology Conference*, volume 2, pages 739–744, May 1999

60. L. Farina and S. Rinaldi. *Positive Linear Systems: Theory and Applications.* John Wiley & Sons, Inc., New York, 2000

61. L. Faubourg and J. B. Pomet. Control lyapunov functions for homogeneous jurdjevic-quinn systems. *ESAIM: Control Optimization and Calculus of Variations*, 5:293–311, 2000

62. J. A. Fax and R. M. Murray. Information flow and cooperative control of vehicle formations. *IEEE Transactions on Automatic Control*, 49:1465–1476, 2004

63. W. Feller. *An Introduction to Probability Theory and Its Applications.* Wiley, New York, NY, 1957

64. C. Fernandes, L. Gurvits, and Z. Li. Near-optimal nonholonomic motion planning for a system of coupled rigid bodies. *IEEE Transactions on Automatic Control*, 39:450–463, 1994

65. A. F. Filippov. *Differential Equations with Discontinuous Righthand Sides.* Kluwer Academic Publishers, Dordrecht, Netherlands, 1988

66. P. Fiorini and Z. Shiller. Motion planning in dynamic environments using velocity obstacles. *International Journal of Robotics Research*, 17:760–772, 1998

67. M. Fliess, J. Levine, P. Martin, and P. Rouchon. Flatness and defect of nonlinear systems: Introductory theory and examples. *International Journal of Control*, 61:1327–1361, 1995

68. T. I. Fossen. *Guidance and Control of Ocean Vehicles.* Wiley, New York, 1994

69. D. Fox, W. Burgard, H. Kruppa, and S. Thrun. A probabilistic approach to collaborative multi-robot localization. *Autonomous Robots*, 8:325–344, 2000

70. M. Frechet. Theorie des evennments en chaine dans le cas d'um nombre fini d'etats possible. *Recherches theoriques Modernes sur le cacul des probabilites*, 2, 1938

71. J. Fredslund and M. J. Mataric. A general algorithm for robot formations using local sensing and minimal communication. *IEEE Transactions on Robotics and Automation*, 18:837–846, 2002

72. R. A. Freeman and P. V. Kokotovic. *Robust nonlinear control design: state-space and Lyapunov techniques.* Birkhauser, Boston, MA, 1996

73. G. Frobenius. Über Matrizen aus nicht-negativen Elementen. *S.-B. Preuss. Akad. Wiss. (Berlin)*, pages 456–477, 1912

74. F. R. Gantmacher. *The Theory of Matrices, vol.II.* Chelsea, New York, NY, 1959

75. V. Gazi and K. M. Passino. Stability of a one-dimensional discrete-time asynchronous swarm. *IEEE Transactions on Systems, Man, and Cybernetics: Part B*, 35:834–841, 2005

76. V. Gazi and K.M. Passino. Stability analysis of swarms. *IEEE Transactions on Automatic Control*, 48:692–697, 2003

77. J.-M. Godhavn and O. Egeland. A lyapunov approach to exponential stabilization of nonholonomic systems in power form. *IEEE Transactions on Automatic Control*, 42:1028–1032, 1997

78. C. Godsil and G. Royle. *Algebraic Graph Theory*. Springer-Verlag, New York, 2001

79. D. T. Greenwood. *Classical Mechanics*. Prentice Hall, Englewood Cliffs, NJ, 1977

80. D. T. Greenwood. *Principles of Dynamics*. Prentice Hall, Engelwood Cliffs, NJ, 1988

81. L. Grune and O. Junge. A set oriented approach to global optimal control. *System & Control Letters*, 54:169–180, 2005

82. Y. Guo and Z. Qu. Control of friction dynamics of an one-dimensional particle array. *Automatica*, 44:2560–2569, 2008

83. Y. Guo, Z. Qu, and Z. Zhang. Lyapunov stability and precise control of the frictional dynamics of an one-dimensional particle array. *Physics Review Letter B*, 73, no.9, paper # 094118, March, 2006

84. L. Gurvits. Stability of discrete linear inclusion. *Linear Algebra and its Applications*, 231:47–85, 1995

85. J. Hajnal. The ergodic properties of non-homogeneous finite markov chains. *Proc. Cambridge Philos. Soc.*, 52:67–77, 1956

86. J. Hajnal. Weak ergodicity in non-homogeneous markov chains. *Proc. Cambridge Philos. Soc.*, 54:233–246, 1958

87. J. K. Hale. *Ordinary Differential Equations*. Wiley Interscience, New York, NY, 1969

88. H. J. Herrmann. Spontaneous density fluctuations in granular flow and traffic. In *Nonlinear Physics of Complex Systems: Current Status and Future Trends*, volume 476, pages 23–34, Lecture Notes in Physics, Berlin: Springer, 1997

89. H. R. Hertz. *Gesammelte Werke, Band III. Die Prinzipien der Mechanik in neuem Zusammenhange dargestellt*. Barth, Leipzig, 19894, English translation Macmillan, London, 1899, reprint, Dover, NY, 1956

90. R. A. Horn and C. R. Johnson. *Matrix Analysis*. Cambridge University Press, Cambridge, 1985

91. L. Hou, A. N. Michel, and H. Ye. Stability analysis of switched systems. In *Proceedings of the 35th IEEE Conference on Decision and Control*, pages 1208–1212, Kobe, Japan, December 1996

92. M. Howard, Z. Qu, K. Conrad, and J. Kaloust. A multi-objective decision making and learning model for increased autonomy of unmanned vehicles. In *Proceedings of AUVSI's Unmanned Systems North America*, San Diego, California, June 2008

93. D. Hsu, R. Kindel, J-C. Latombe, and S. Rock. Randomized kinodynamic motion planning with moving obstacles. *International Journal of Robotics Research*, 21:233–255, 2002

94. L. R. Hunt, R. Su, and G. Meyer. Global transformations of nonlinear systems. *IEEE Transactions on Automatic Control*, 28:24–31, 1983

95. Y. K. Hwang and N. Ahuja. A potential field approach to path planning. *IEEE Transactions on Robotics and Automation*, 8:23–32, 1992

96. A. Isidori. *Nonlinear Control Systems*. 3rd ed., Springer-Verlag, Berlin, 1995

97. D. H. Jacobson. *Extensions of Linear Quadratic Control, Optimization and Matrix Theory.* Academic Press, New York, NY, 1977

98. A. Jadbabaie, J. Lin, and A.S. Morse. Coordination of groups of mobile autonomous agents using nearest neighbor rules. *IEEE Transactions on Automatic Control*, 48:988–1001, 2003

99. B. Jakubczyk and W. Respondek. On linearization of control systems. *Bull. Acad. Pol. Sci. Ser. Sci. Math.*, 28:517–522, 1980

100. Z. P. Jiang. Robust exponential regulation of nonholonomic systems with uncertainties. *Automatica*, 36:189–209, 2000

101. D. W. Jordan and P. Smith. *Nonlinear Ordinary Differential Equations.* 2nd edition, Oxford University Press, New York, 1989

102. T. Kailath. *Linear Systems.* Prentice Hall, Englewood Cliffs, NJ, 1980

103. R. E. Kalman. Contributions to the theory of optimal control. *Boletin de las sociedad Matematica Mexicana*, 5:102–119, 1960

104. R. E. Kalman and J. E. Bertram. Control system analysis and design via the "second method" of lyapunov i. continuous-time systems. *Transactions of the ASME, Journal of Basic Engineering*, 82D:371–393, 1960

105. W. Kang, N. Xi, and A. Sparks. Theory and applications of formation control in a perceptive referenced frame. In *Proceedings of IEEE Conference on Decision and Control*, pages 352–357, Sydney, Australia, Dec. 2000

106. K. Kant and S. W. Zucker. Planning collision free trajectories in time-varying environments: a two level hierarchy. In *Proceedings of IEEE International Conference on Robotics and Automation*, pages 1644–1649, Raleigh, NC, 1988

107. M. H. Kaplan. *Modern Spacecraft Dynamics and Control.* John Wiley, New York, 1976

108. H. K. Khalil. *Nonlinear Systems.* Prentice Hall, 3rd ed., Upper Saddle River, NJ, 2003

109. O. Khatib. Real-time obstacle avoidance for manipulator and mobile robots. *International Journal of Robotics Research*, 5:90–98, 1986

110. M. Kisielewicz. *Differential Inclusions and Optimal Control.* Kluwer Academic Publishers, Dordrecht, Netherlands, 1991

111. I. Kolmanovsky and N. H. McClamroch. Hybrid feedback laws for a class of cascade nonlinear control systems. *IEEE Transactions on Automatic Control*, 41:1271–1282, 1996

112. I. Kolmanovsky, M. Reyhanoglu, and N. H. McClamroch. Switched mode feedback control laws for nonholonomic systems in extended power form. *Systems & Control letters*, 27:29–36, 1996

113. G. Kreisselmeier and T. Birkholzer. Numerical nonlinear regulator design. *IEEE Transactions on Automatic Control*, 39:33–46, 1994

114. A. J. Krener. A generalization of chow's theorem and the bang-bang theorem to nonlinear control systems. *SIAM Journal on Control*, 12:43–52, 1974

115. M. J. B. Krieger, J.-B. Billeter, and L. Keller. Ant-like task allocation and recruitment in cooperative robots. *Nature*, 406:992–995, 2000

116. M. Krstic, I. Kanellakopoulos, and P. V. Kokotovic. *Nonlinear and Adaptive Control Design.* Wiley, New York, 1995

117. M. C. Laiou and A. Astolfi. Exponential stabilization of high-order chained system. *Systems & Control Letters*, 37:309–322, 1999

118. V. Lakshmikantham and S. Leela. *Differential and Integral Inequalities: Theory and Application.* Academic Press, Vol.I, New York, 1969

119. V. Lakshmikantham, S. Leela, and M. N. Oguztoreli. Quasi-solutions, vector lyapunov functions, and monotone method. *IEEE Transactions on Automatic Control*, 26:1149–1153, 1981

120. A. N. Langville and C. D. Meyer. A survey of eigenvector methods for web information retrieval. *SIAM Review*, 47:135–161, 2005

121. A. Lasota and M. C. Mackey. *Chaos, Fractals, and Noise: Stochastic Aspects of Dynamics*. John Wiley & Sons, New York, NY, 1989

122. J. Latombe. *Robot Motion Planning*. Kluwer Academic Publisher, 1998, Boston

123. J.-P. Laumond. *Robot Motion Planning and Control*. Springer-Verlag, 1998, London

124. J.-P. Laumond, P. E. Jacobs, M. Taix, and R. M. Murray. A motion planner for nonholonomic mobile robots. *IEEE Transactions on Robotics and Automation*, 10:577–593, 1994

125. S. Lavalle and J. Kuffner. Randomized kinodynamic planning. *International Journal of Robotics Research*, 20:378–400, 2001

126. J. L. Lazaro and A. Gardel. Adaptive workspace modeling, using regression methods, and path planning to the alternative guide of mobile robots in environments with obstacles. In *Proceedings of the 7th IEEE International Conference on Emerging Technologies and Factory Automation*, volume 1, pages 529–534, October 1999

127. D. D. Lee and H. S. Seung. Learning the parts of objects by non-negative matrix factorization. *Nature*, 401 (6755):788791, 1999

128. N. E. Leonard and E. Fiorelli. Virtual leaders, artificial potentials and coordinated control of groups. In *Proceedings of IEEE Conference on Decision and Control*, pages 2968–2973, Orlando, FL, Dec. 2001

129. W. W. Leontief. *The structure of the American economy 1919-1929*. Harvard University Press, Cambridge, MA, 1941

130. F.L. Lewis, C.T. Abdallah, and D.M. Dawson. *Control of Robot Manipulators*. Macmillan, New York, 1993

131. M. A. Lewis and K. H. Tan. High precision formation control of mobile robots using virtual structures. *Autonomous Robots*, 4:387–403, 1997

132. R. M. Lewis and B. D. O. Anderson. Necessary and sufficient conditions for delay-independent stability of linear autonomous systems. *IEEE Transactions on Automatic Control*, 25:735–739, 1980

133. D. Liberzon. *Switching in Systems and Control*. Birkhauser, Boston, MA, 2003

134. J. Lin, A. Morse, and B. Anderson. The multi-agent rendezvous problem, part 2: The asynchronous case. *SIAM Journal on Control and Optimization*, 46:2120–2147, 2007

135. Z. Lin, B. Francis, and M. Maggiore. Necessary and sufficient graphical conditions for formation control of unicycles. *IEEE Transactions on Automatic Control*, 50:121–127, 2005

136. Z. Lin, B. Francis, and M. Maggiore. State agreement for continuous-time coupled nonlinear systems. *SIAM Journal on Control and Optimization*, 46:288–307, 2007

137. Y. Liu and K. M. Passino. Stability analysis of one-dimensional asynchronous swarms. *IEEE Transactions on Automatic Control*, 48:1848–1854, 2003

138. Y. Liu, K. M. Passino, and M. M. Polycarpou. Stability analysis of m-dimensional asynchronous swarms with a fixed communication topology. *IEEE Transactions on Automatic Control*, 48:76–95, 2003

139. J. Luo and P. Tsiotras. Exponentially convergent control laws for nonholonomic systems in power form. *Systems & control letters*, 35:87–95, 1998

140. J. Luo and P. Tsiotras. Control design for chained-form systems with bounded inputs. *Systems & control letters*, 39:123–131, 2000

141. A. M. Lyapunov. *Stability of Motion.* Academic Press, New York, NY, 1966

142. J. L. Mancilla-Aguilar and R. A. Garcia. A converse lyapunov theorem for nonlinear switched systems. *Systems & Control Letters*, 41:67–71, 2000

143. N. March and M. Alamir. Discontinuous exponential stabilization of chained form systems. *Automatica*, 39:343–348, 2003

144. R. Marino. On the largest feedback linearizable subsystem. *System & Control Letters*, 6:245–351, 1986

145. J. E. Marsden and T. S. Ratiu. *Introduction to Mechanics and Symmetry.* Springer, New York, 1999

146. S. Martinez. Practical rendezvous through modified circumcenter algorithms. In *Proceedings of IEEE Conference on Decision and Control*, New Orleans, LA, 2007

147. S. Martinez, F. Bullo, J. Cortes, and E. Frazzoli. On synchronous robotic networks — part ii: Time complexity of rendezvous and deployment algorithms. *IEEE Transactions on Automatic Control*, 52:2214–2226, 2007

148. S. Mastellone, D. M. Stipanovic, C. R. Graunke, K. A. Intlekofer, and M. W. Spong. Formation control and collision avoidance for multi-agent nonholonomic systems: Theory and experiments. *International Journal of Robotics Research*, 27:107–126, 2008

149. M. Mataric. Minimizing complexity in controlling a mobile robot population. In *Proceedings of IEEE International Conference on Robotics and Automation*, pages 830–835, Nice, France, 1992

150. M. J. Mataric. Issues and approaches in the design of collective autonomous agents. *Robotics and Autonomous Systems*, 16:321–331, 1995

151. M. J. Mataric. Reinforcement learning in the multi-robot domain. *Autonomous Robots*, 4:73–83, 1997

152. F. Mazenc and M. Malisoff. Further constructions of control lyapunov functions and stabilizing feedbacks for systems satisfying jurdjevic-quinn conditions. *IEEE Transactions on Automatic Control*, 51:360–365, 2006

153. R. T. M'Closkey and R. M. Murray. Exponential stabilization of driftless nonlinear control systems using homogeneous feedback. *IEEE Transactions on Automatic Control*, 42:614–628, 1997

154. R. T. MCloskey and R. M. Murray. Exponential stabilization of driftless nonlinear control systems using homogeneous feedback. *IEEE Transactions on Automatic Control*, 42:614–628, 1997

155. A. M. Meilakhs. Design of stable control systems subject to parameter perturbation. *Automat. Remote Control*, 39:1409–1418, 1978

156. A. N. Michel and R. K. Miller. *Qualitative Analysis of Large Scale Dynamical Systems.* Academic Press, New York, NY, 1977

157. H. Minc. *Nonnegative Matrices.* John Wiley & Sons, New York, 1988

158. A. P. Molchanov and Y. S. Pyatnitskiy. Criteria of absolute stability of differential and difference inclusions encountered in control theory. *Systems & Control Letters*, 13:59–64, 1989

159. S. Monaco and D. Normand-Cyrot. An introduction to motion planning under multirate digital control. In *Proceedings of IEEE International Conference on Decision and Control*, pages 1780–1785, Tucson, AZ, 1992

160. P. K. Moore. An adaptive finite element method for parabolic differential systems: some algorithmic considerations in solving in three space dimensions. *SIAM Journal on Scientific Computing*, 21:1567–1586, 2000

161. L. Moreau. Stability of multiagent systems with time-dependent communication links. *IEEE Transactions on Automatic Control*, 50:169–182, 2005

162. P. Morin and C. Samson. Control of nonlinear chained systems: from the routh-hurwitz stability criterion to time-varying exponential stabilizers. *IEEE Transactions on Automatic Control*, 45:141–146, 2000

163. P. Morin and C. Samson. Practical stabilization of driftless systems on lie groups: the transverse function approach. *IEEE Transactions on Automatic Control*, 48:1496–1508, 2003

164. N. Moshtagh and A. Jadbabaie. Distributed geodesic control laws for flocking of nonholonomic agents. *IEEE Transactions on Automatic Control*, 52:681–686, 2007

165. E. Moulay and W. Perruquetti. Stabilization of nonaffine systems: A constructive method for polynomial systems. *IEEE Transactions on Automatic Control*, 50:520–526, 2005

166. P. J. Moylan. A connective stability result for interconnected passive systems. *IEEE Transactions on Automatic Control*, 25:812–813, 1980

167. R. R. Murphy. *Introduction to AI Robotics*. MIT Press, Cambridge, MA, 2000

168. R. M. Murray, Z. Li, and S. S. Sastry. *A Mathematical Introduction to Robotic Manipulation*. CRC Press, Inc., 1994, Boca Raton

169. R. M. Murray and S. S. Sastry. Nonholonomic motion planning: Steering using sinusoids. *IEEE Transactions on Automatic Control*, 38:700–716, 1993

170. R. M. Murray and S. S. Sastry. Nonholonomic motion planning: steering using sinusoids. *IEEE Transactions on Automatic Control*, 38:700–716, 1993

171. Y. Nakamura and S. Savant. Nonlinear tracking control of autonomous underwater vehicles. In *Proceedings of IEEE Conference on Robotics and Automation*, pages A4–A9, Nice, France, May 1992

172. K. S. Narendra and A. M. Annaswamy. *Stable Adaptive Systems*. Prentice-Hall, Englewood Cliffs, NJ, 1989

173. K. S. Narendra and J. Balakrishnan. A common lyapunov function for stable lti systems with commuting a-matrices. *IEEE Transactions on Automatic Control*, 39:2469–2471, 1994

174. R. C. Nelson. *Flight Stability and Automatical Control*. McGraw-Hill, New York, NY, 1989

175. H. Nijmeijer and A. J. van der Schaft. *Nonlinear Dynamical Control Systems*. Springer-Verlag, New York, 1990

176. K. Ogata. *Modern Control Engineering*. Prentice Hall, Englewood Cliffs, NJ, 1990

177. P. Ogren, E. Fiorelli, and N. E. Leonard. Cooperative control of a mobile sensor networks: Adaptive gradient climbing in a distributed environment. *IEEE Transactions on Automatic Control*, 49:1292–1302, 2004

178. A. Okubo. Dynamical aspects of animal grouping: swarms, schools, flocks and herds. *Advances in Biophysics*, 22:1–94, 1986

179. R. Olfati-Saber and R.M. Murray. Distributed cooperative control of multiple vehicle formations using structural potential functions. In *Proceedings of the 15th Triennial World Congress*, Barcelona, Spain, 2002

180. R. Olfati-Saber and R.M. Murray. Flocking with obstacle avoidance: cooperation with limited communication in mobile network. In *Proceedings of IEEE Conference on Decision and Control*, Maui, Hawaii, USA, 2003

181. G. Oriolo and Y. Nakamura. Free-joint manipulators: motion control under second-order nonholonomic constraints. In *Proceedings of IEEE/RSJ International Workshop on Intelligent Robots and Systems (IROS'91)*, pages 1248–1253, Osaka: Japan, 1991

182. A. Pant, P. Seiler, and K. Hedrick. Mesh stability of look-ahead interconnected systems. *IEEE Transactions on Automatic Control*, 47:403–407, 2002

183. L. E. Parker. Alliance: An architecture for fault-tolerant multi-robot cooperation. *IEEE Transactions on Robotics and Automation*, 14:220–240, 1998

184. L. E. Parker. Current state of the art in distributed autonomous mobile robotics. In *Distributed Autonomous Robotic Systems 4*, pages 3–12, L.E. Parker, G. Bekey and J. Barhen (Eds.), New York: Springer-Verlag, 2000

185. P. Peleties and R. A. DeCarlo. Asymptotic stability of m-switching systems using lyapunov-like functions. In *Proceedings of American Control Conference*, pages 1679–1684, Boston, MA, June 1991

186. O. Perron. Zur theorie der matrizen. *The Mathematische Annalen*, 64:248–263, 1907

187. I.P. Petrov. *Variational Methods in Optimum Control*. Academic Press, New York, NY, 1968

188. A. Piazzi, C. G. Lo Bianco, M. Bertozzi, A. Fascioli, and A. Broggi. Quintic g2-splines for the iterative steering of vision-based autonomous vehicles. *IEEE Transactions on Intelligent Transportation Systems*, 3:27–36, 2002

189. J.-B. Pomet. Explicit design of time-varying stabilizing control laws for a class of controllable systems without drift. *Systems & control letters*, 18:147–158, 1992

190. S. Prajna, P. A. Parrilo, and A. Rantzer. Nonlinear control synthesists by convex optimization. *IEEE Transactions on Automatic Control*, 49:310–314, 2004

191. C. Prieur and A. Astolfi. Robust stabilization of chained systems via hybrid control. *IEEE Transactions on Automatic Control*, 48:1768–1772, 2003

192. Z. Qu. *Robust Control of Nonlinear Uncertain Systems*. Wiley Interscience, New York, 1998

193. Z. Qu. A comparison theorem for cooperative control of nonlinear systems. In *Proceedings of American Control Conference*, Seattle, Washington, June 2008

194. Z. Qu, J. Chunyu, and J. Wang. Nonlinear cooperative control for consensus of nonlinear and heterogeneous systems. In *Proceedings of the 46th IEEE Conference on Decision and Control*, pages 2301–2308, New Orleans, Louisiana, December 2007

195. Z. Qu and D. M. Dawson. *Robust Tracking Control of Robot Manipulators*. IEEE Press, 1996, New York

196. Z. Qu, C. M. Ihlefeld, J. Wang, and R. A. Hull. A control-design-based solution to robotic ecology: Autonomy of achieving cooperative behavior from a high-level astronaut command. In *Proceedings of Robosphere 2004: A Workshop on Self-Sustaining Robotic Systems*, NASA Ames Research Center, Nov. 2004

197. Z. Qu, J. Wang, and J. Chunyu. Lyapunov design of cooperative control and its application to the consensus problem. In *Proceedings of IEEE Multi-conference on Systems and Control*, Singapore, October 2007

198. Z. Qu, J. Wang, and R. A. Hull. Cooperative control of dynamical systems with application to mobile robot formation. In *Proceedings of the 10th IFAC/IFORS/IMACS/IFIP Symposium on Large Scale Systems: Theory and Applications*, Japan, July 2004

199. Z. Qu, J. Wang, and R. A. Hull. Leaderless cooperative formation control of autonomous mobile robots under limited communication range constraints. In *Proceedings of the 5th International Conference on Cooperative Control and Optimization*, Gainesville, FL, Jan 2005

200. Z. Qu, J. Wang, and R. A. Hull. Multi-objective cooperative control of dynamical systems. In *Proceedings of the 3rd International Multi-Robot Systems Workshop*, Naval Research Laboratory, Washington DC, March 2005

201. Z. Qu, J. Wang, and R. A. Hull. Products of row stochastic matrices and their applications to cooperative control for autonomous mobile robots. In *Proceedings of American Control Conference*, Portland, Oregon, June 2005

202. Z. Qu, J. Wang, and R. A. Hull. Cooperative control of dynamical systems with application to autonomous vehicles. *IEEE Transactions on Automatic Control*, 53:894–911, 2008

203. Z. Qu, J. Wang, R. A. Hull, and J. Martin. Cooperative control design and stability analysis for multi-agent systems with communication delays. In *Proceedings of IEEE International Conference on Robotics and Automation*, pages 970–975, Orlando, FL, May 2006

204. Z. Qu, J. Wang, R. A. Hull, and J. Martin. Continuous control designs for stabilizing chained systems: A global state scaling transformation and a time scaling method. *Optimal Control, Applications and Methods*, 30:to appear, 2009

205. Z. Qu, J. Wang, and X. Li. Quadratic lyapunov functions for cooperative control of networked systems. In *Proceedings of IEEE International Conference on Control and Automation*, pages 416–421, Guangzhou, China, May 2007

206. Z. Qu, J. Wang, and C. E. Plaisted. A new analytical solution to mobile robot trajectory generation in the presence of moving obstacles. In *Proceedings of Florida Conference on Recent Advances in Robotics*, FAU, FL, May 2003

207. Z. Qu, J. Wang, and C. E. Plaisted. A new analytical solution to mobile robot trajectory generation in the presence of moving obstacles. *IEEE Transactions on Robotics*, 20:978–993, 2004

208. Z. Qu, J. Wang, C. E. Plaisted, and R. A. Hull. A global-stabilizing near-optimal control design for real-time trajectory tracking and regulation of nonholonomic chained systems. *IEEE Transactions on Automatic Control*, 51:1440–1456, September 2006

209. Z. Qu, J. Wang, and E. Pollak. Cooperative control of dynamical systems and its robustness analysis. In *Proceedings of the 45th IEEE Conference on Decision and Control*, pages 3614–3621, San Diego, CA, December 2006

210. A. Rantzer. A dual to lyapunov's stability theorem. *System & Control Letters*, 42:161–168, 2001

211. J. A. Reeds and R. A. Shepp. Optimal paths for a car that goes both forward and backwards. *Pacific Journal of Mathematics*, 145:367–393, 1990

212. J. H. Reif and H. Wang. Social potential fields: a distributed behavioral control for autonomous robots. *Robotics and Autonomous Systems*, 27:171–194, 1999

213. M. W. Reinsch. A simple expression for the terms in the baker-campbell-hausdorff series. *Journal of Mathematical Physics*, 41:2434–2442, 2000

214. W. Ren and R. W. Beard. Consensus seeking in multiagent systems under dynamically changing interaction topologies. *IEEE Transactions on Automatic Control*, 50:655–661, 2005

215. M. Reyhanoglu. Exponential stabilization of an underactuated autonomous surface vessel. *Automatica*, 33:2249–2254, 1997

216. M. Reyhanoglu, A. V. Der Schaft, N. H. McClamroch, and I. Kolmanovsky. Dynamics and control of a class of underactuated mechanical systems. *IEEE Transactions on Automatic Control*, 44:1663–1671, 1999

217. C. W. Reynolds. Flocks, herds, and schools: a distributed behavioral model. *Computer Graphics (ACM SIGGRAPH 87 Conference Proceedings)*, 21:25–34, 1987

218. E. Rimon and D. E. Koditschek. Exact robot navigation using artificial potential functions. *IEEE Transactions Robotics and Automation*, 8:501–518, 1992

219. R. T. Rockafellar. *Convex Analysis*. Princeton University Press, Princeton, NJ, 1970

220. W. Rudin. *Functional Analysis*. McGraw-Hill, New York, NY, 1973

221. R. O. Saber and R. M. Murray. Consensus problems in networks of agents with switching topology and time-delays. *IEEE Transactions on Automatic Control*, 49:1520–1533, 2004

222. C. Samson. Control of chained systems: Application to path following and time-varying point-stabilization of mobile robots. *IEEE Transactions on Automatic Control*, 40:64–77, 1995

223. S. Sastry. *Nonlinear Systems: Analysis, Stability and Control*. Springer-Verlag, New York, 1999

224. C. P. Schenk, P. Schutz, M. Bode, and H. G. Purwins. Interaction of self organized quasi particles in a two dimensional reactive-diffusion system: The formation of molecules. *Physical Review E.*, 58:6480–6486, 1998

225. E. Seneta. On the historical development of the theory of finite inhomogeneous markov chains. *Proc. Cambridge Philos. Soc.*, 74:507–513, 1973

226. E. Seneta. *Non-negative Matrices and Markov Chain*. Springer, New York, NY, 1981

227. R. Sepulchre, A. Paley, and N. E. Leonard. Stabilization of planar collective motion: all-to-all communication. *IEEE Transactions on Automatic Control*, 52:916–922, 2007

228. D. D. Siljak. Connective stability of competitive equilibrium. *Automatica*, 11:389–400, 1975

229. D. D. Siljak. *Decentralized Control of Complex Systems*. Academic press, 1991

230. G. M. Siouris. *Missile Guidance and Control Systems*. Springer, New York, 2004

231. R. Skjetne, S. Moi, and T.I. Fossen. Nonlinear formation control of marine craft. In *Proceedings of IEEE Conference on Decision and Control*, pages 1699–1704, Las Vegas, Nevada, Dec. 2002

232. J. J. Slotine and W. Li. *Applied Nonlinear Control*. Prentice-Hall, Englewood Cliffs, NJ, 1991

233. W. M. Sluis. A necessary condition for dynamic feedback linearization. *System & Control Letters*, 21:277283, 1993

234. E. D. Sontag. A universal construction of artstein's theorem on nonlinear stabilization. *System & Control Letters*, 13:117–123, 1989

235. E. D. Sontag. *Mathematical Control Theory*. Springer-Verlag, New York, NY, 1991

236. E. D. Sontag. Control lyapunov function. In *Open Problems in Mathematical Systems and Control Theory*, pages 211–216, V. D. Blondel, E. D. Sontag, M. Vidyasagar, and J. C. Willems (Eds.), London: Springer-Verlag, 1999

237. O. J. Sordalen. Conversion of the kinematics of a car with n trailers into a chained form. In *Proceedings of IEEE Conference on Robotics and Automation*, pages 382–387, Atlanta, Georgia, May 1993

238. O. J. Sordalen. Conversion of the kinematics of a car with n trailers into a chained form. In *Proceedings of the IEEE International Conference on Robotics and Automation*, pages 382–387, 1993

239. O. J. Sordalen and O. Egeland. Exponential stabilization of nonholonomic chained systems. *IEEE Transactions on Automatic Control*, 40:35–49, 1995

240. O. Soysal and E. Sahin. Probabilistic aggregation strategies in swarm robotic systems. In *Proceedings of the IEEE Swarm Intelligence Symposium*, Pasadena, CA., 2005

241. M. W. Spong. Energy based control of a class of underactuated mechanical systems. In *Proceedings of IFAC World Congress*, pages 431–435, San Francisco, CA, 1996

242. M. W. Spong and M. Vidyasagar. *Robot Dynamics and Control*. John Wiley & Sons, New York, NY, 1989

243. A. Stentz. Optimal and efficient path planning for partially-known environments. In *Proceedings of IEEE International Conference on Robotics and Automation*, volume 4, pages 3310–3317, San Diego, CA, May 1994

244. A. Stentz. The focussed d* algorithm for real-time replanning. In *Proceedings of the International Joint Conference on Artificial Intelligence*, Pittsburgh, Pennsylvania, August 1995

245. B. L. Stevens and F. L. Lewis. *Aircraft Control and Simulation*. Wiley, Hoboken, N.J., 2003

246. D. M. Stipanovic, P. F. Hokayem, M. W. Spong, and D. D. Siljak. Cooperative avoidance control for multi-agent systems. *Journal of Dynamic Systems, Measurement, and Control*, 129:699–707, 2007

247. S. H. Strogatz. From kuramoto to crawford: Exploring the onset of synchronization in populations of coupled oscillators. *Physics D*, 143:1–20, 2000

248. K. Sugihara and I. Suzuki. Distributed motion coordination of multiple mobile robots. In *Proceedings of the 5th IEEE International Symposium on Intelligent Control*, pages 138–143, Philadelphia, PA, 1990

249. S. Sundar and Z. Shiller. Optimal obstacle avoidance based on the hamilton-jacobi-bellman equation. *IEEE Transactions on Robotics and Automation*, 13:305–310, 1997

250. H. J. Sussmann and W. Liu. Limits of highly oscillatory controls and the approximation of general paths by admissible trajectories. In *Tech. Rep. SYSCON-91-02*, Rutgers Ctr. Syst. and Contr., Piscataway, NJ, Feb. 1991

251. J. Sussmann and G. Tang. Shortest paths for the reeds-shepp car: A worked out example of the use of geometric techniques in nonlinear optimal control. In *SYCON, Rutgers Ctr. Syst. Contr. Tech. Rep.*, New Brunswick, NJ, Oct. 1991

252. I. Suzuki and M. Yamashita. Distributed anonymous mobile robots: formation of geometric patterns. *SIAM Journal on Computing*, 28:1347–1363, 1999

253. D. Swaroop and J. Hedrick. String stability of interconnected systems. *IEEE Transactions on Automatic Control*, 41:349–357, 1996

254. A. Takayama. *Mathematical Economics*. Cambridge University Press, New York, NY, 1985

255. H. G. Tanner, A. Jadbabaie, and G. J. Pappas. Stable flocking of mobile agents, part i: fixed topology. In *Proceedings of IEEE Conference on Decision and Control*, pages 2010–2015, Maui, Hawaii, 2003

256. H. G. Tanner, A. Jadbabaie, and G. J. Pappas. Stable flocking of mobile agents, part ii: dynamic topology. In *Proceedings of IEEE Conference on Decision and Control*, pages 2016–2021, Maui, Hawaii, 2003

257. H. G. Tanner, G. J. Pappas, and V. Kumar. Leader-to-formation stability. *IEEE Transactions on Robotics and Automation*, 20:433–455, 2004

258. A. R. Teel, R. M. Murray, and G. Walsh. Nonholonomic control systems: from steering to stabilization with sinusoids. In *Proceedings of the 31st IEEE Conference on Decision and Control*, pages 1603–1609, Tucson, Arizona, Dec. 1992

259. D. Tilbury, R. M. Murray, and S. S. Sastry. Trajectory generation for the n-trailer problem using goursat normal form. *IEEE Transactions on Automatic Control*, 40:802–819, 1995

260. J. Toner and Y. Tu. Flocks, herds, and schools: A quantitative theory of flocking. *Physical Review E.*, 58:4828–4858, 1998

261. J. Tsinias. Existence of control lyapunov functions and applications to state feedback stabilizability of nonlinear systems. *SIAM Journal on Control and Optimization*, 29:457–473, 1991

262. R. S. Varga. *Matrix Iterative Analysis*. Prentice Hall, New Jersey, 1962

263. T. Vicsek, A. Czirok, E. B. Jacob, I. Cohen, and O. Shochet. Novel type of phase transition in a system of self-driven particles. *Physical Review Letters*, 75:1226–1229, 1995

264. M. Vidyasagar. *Nonlinear Systems Analysis*. Prentice Hall, Englewood Cliffs, NJ, 1978

265. Nguyen X. Vinh. *Flight Mechanics of High Performance Aircraft*. Cambridge University Press, Great Britain, 1993

266. G. C. Walsh and L. G. Bushnell. Stabilization of multiple input chained form control systems. *Systems & Control Letters*, pages 227–234, 1995

267. J. Wang, Z. Qu, R. A. Hull, and J. Martin. Cascaded feedback linearization and its application to stabilization of nonholonomic systems. *System & Control Letters*, 56:285–295, 2007

268. K. Wang and A. N. Michel. Qualitative analysis of dynamical systems determined by differential inequalities with applications to robust stability. *IEEE Transactions on Automatic Control*, 41:377–386, 1994

269. P. K. C. Wang. Navigation strategies for multiple autonomous mobile robots. In *Proceedings of IEEE/RSJ International Conference on Intelligent Robots and Systems (IROS)*, pages 486–493, Tsukuba, Japan, 1989

270. K. Warburton and J. Lazarus. Tendency-distance models of social cohesion in animal groups. *Journal of Theoretical Biology*, 150:473–488, 1991

271. K. Y. Wichlund, O. J. Sordalen, and O. Egeland. Control of vehicles with second-order nonholonomic constraints: underactuated vehicles. In *Proceedings of European Control Conference*, pages 3086–3091, Rome, Italy, 1995

272. B. Wie. *Space Vehicle Dynamics and Control*. AIAA, Reston, Virginia, 1998

273. H. Wielandt. Unzerlegbare, nicht negative matrizen. *Math. Z.*, 52:642–648, 1950

274. J. Wolfowitz. Products of indecomposable, aperiodic, stochastic matrices. *Proc. Amer. Mathematical Soc.*, 14:733–737, 1963

275. S. Wolfram. *Cellular Automata and Complexity.* Addison-Wesley, Reading, MA, 1994

276. C. W. Wu. Synchronization in arrays of coupled nonlinear systems with delay and nonreciprocal time-varying coupling. *IEEE Transactions on Circuits and Systems II: Express Brief*, 52:282–286, 2005

277. C. W. Wu and L. O. Chua. Synchronization in an array of linearly coupled dynamical systems. *IEEE Transactions on Circuits and Systems I: Fundamental theory and applications*, 42:430–447, 1995

278. H. Yamaguchi. A distributed motion coordination strategy for multiple nonholonomic mobile robots in cooperative hunting operations. *Robotics and Autonomous Systems*, 43:257–282, 2003

279. J. Yang, A. Daoui, Z. Qu, J. Wang, and R. A. Hull. An optimal and real-time solution to parameterized mobile robot trajectories in the presence of moving obstacles. In *Proceedings of IEEE International Conference on Robotics and Automation*, pages 4423–4428, Barcelona,Spain, April 18-22 2005

280. J. Yang, Z. Qu, and R. Hull. Trajectory planning for ugvs in an environment with 'hard' and 'soft' obstacles. In *Proceedings of AIAA GNC Conference*, Keystone, CO., August 2006

Index